Sustainable Development and Biodiversity

Volume 9

Series editor

Kishan Gopal Ramawat
Botany Department, M.L. Sukhadia University, Udaipur, Rajasthan, India

This book series provides complete, comprehensive and broad subject based reviews about existing biodiversity of different habitats and conservation strategies in the framework of different technologies, ecosystem diversity, and genetic diversity. The ways by which these resources are used with sustainable management and replenishment are also dealt with. The topics of interest include but are not restricted only to sustainable development of various ecosystems and conservation of hotspots, traditional methods and role of local people, threatened and endangered species, global climate change and effect on biodiversity, invasive species, impact of various activities on biodiversity, biodiversity conservation in sustaining livelihoods and reducing poverty, and technologies available and required. The books in this series will be useful to botanists, environmentalists, marine biologists, policy makers, conservationists, and NGOs working for environment protection.

More information about this series at http://www.springer.com/series/11920

Dilip Nandwani
Editor

Organic Farming for Sustainable Agriculture

Springer

Editor
Dilip Nandwani
Department of Agricultural and
Environmental Sciences
College of Agriculture, Human and Natural
Sciences
Tennessee State University
Nashville, TN
USA

ISSN 2352-474X ISSN 2352-4758 (electronic)
Sustainable Development and Biodiversity
ISBN 978-3-319-26801-9 ISBN 978-3-319-26803-3 (eBook)
DOI 10.1007/978-3-319-26803-3

Library of Congress Control Number: 2015957221

Printed on acid-free paper

This Springer imprint is published by SpringerNature
The registered company is Springer International Publishing AG Switzerland

Foreword

Organic agriculture has emerged from the public desire to obtain safe and healthy food and to realize the long-term sustainability of food production. Although organic agriculture is growing in popularity, it faces substantial challenges to become a major source of food and fiber. This book, "Organic Farming for Sustainable Agriculture," presents a collection of chapters that addresses successes and challenges to organic agriculture.

A general consensus exists that current farming practices are disrupting the natural cycling of nutrients and soil, are not sustainable, and must be transformed. Excessive soil and nutrients are lost from farmland. Under conventional agriculture, the soil is often inadequately protected against wind, rain, and irrigation-induced erosion. Nutrients applied in excess of plant needs are lost in runoff, volatilization, and leaching. The loss of soil carries topsoil with its many nutrients and organic matter toward streams, rivers, lakes, and reservoirs. Organic agriculture seeks to simulate the natural cycling of nutrients, so that off-site losses are minimal. The effectiveness of nutrient cycling may be enhanced through a myriad of practices, many of which are discussed here. The use of mulch, cover crops, minimum tillage systems, and no tillage systems enhance soil cover.

There is substantial public concern that many of the products used in conventional agriculture may provide risks to human and environmental health. The difficulty is to discover which products and methods used in conventional agriculture are adequately safe in the long run. Scientific concerns exist that the widespread use of antibiotics for animal production will substantially shorten the useful life of these antibiotics to protect human health. The public has many disagreements on the chemicals and practices that ought to be used in food production.

Within organic agriculture, the intent is that organic food and fiber production practices will be sustainable, fundamentally sound in the long term. Organic agriculture places value not only on the production of the present but also on future production and the future capability of the earth to sustain mankind and diverse ecosystems. Organic agriculture embraces the ethical standard that current human needs can be met while sustaining the productivity of the land and conducting all activities so as to support a wide variety of other terrestrial and aquatic species.

Modern human activities may contribute significantly to greenhouse gases, with the possibility of aggravating climate change. Organic husbandry and production of crops seek to effectively use energy inputs and bring energy use within a range of long-term sustainability.

With the growth of human population, increasing pressures are placed upon the limited resources of the planet. To answer these needs, the use of more and more resource inputs at greater and greater levels of intensity is tempting. As the world's population is expected to reach 9 or 10 billion inhabitants, substantially more food and fiber will be required. Many of the people in the world are escaping poverty. In their attempt to feed and clothe themselves more adequately, they are calling on more of the world's resources. As higher incomes allow consumers to satisfy more of their food requirement with meat and milk products, more resources are needed. Meat and milk products inherently require more energy and nutritional resources to produce in a farm setting. Consequently, the pressures on the Earth's environments in the twenty-first century come not only from a growing human population, but also from the changing consumption patterns of that population.

One challenge for organic farming is that many of the current system designs are low input and modest output. Transforming highly productive conventional farming into moderately productive organic farming brings about a potential conflict between the enormity of human needs and the ability of organic agriculture to supply the current demands of food and fiber for the Earth's population. Simply transforming conventional farming into organic farming with lower input and lower output systems creates a contradiction; less efficient organic agriculture could create pressure to incorporate additional marginal acreage into production, placing additional pressure on endangered species and fragile environments. Consequently, organic agriculture needs substantial new developments so that it can effectively and efficiently supply a large proportion of human needs. The creation of organic farming systems that have high productivity or higher productivity than the systems that they replace will require creativity and innovation.

The natural world contains much to discover. The science of the effective use of microorganisms to control pathogenic bacteria, fungi, and viruses in animals and plants is in its infancy. Naturally occurring bacteria, fungi, and viruses may hold promise. Examples of this work are presented in this book. Many plants contain secondary metabolites and these secondary metabolites could provide a vast array of natural chemical substances with possible effectiveness against pathogens or possible opportunities to improve productivity.

Dr. Dilip Nandwani has a unique background and experience that makes him particularly well suited to assemble chapters on the topic of Organic Farming for Sustainable Agriculture. Dr. Nandwani grew up in India, where he earned his bachelors, masters, and doctoral degrees. His work experience includes the public and private sectors with pertinent work in India, Hawaii, Micronesia, the Mariana Islands, the US Virgin Islands, Bangladesh, and Tennessee.

Dr. Nandwani's own research and extension have addressed many key issues and problems that are important components of organic agriculture and sustainable production systems. His personal research has examined botanical pesticides, plant tissue culture and plant propagation in general, alternative medicines for livestock, crop improvement, plant disease resistance, biodiesel production, hydroponic production, and hybrid seed production. His research has been converted into practical results and pertinent extension publications. The social dimensions of Dr. Nandwani's work include the interests of small farmers and women farmers, including their food security, nutrition, and income. His broad experience provides wisdom and insight to assemble and edit this book.

Dr. Clinton C. Shock
Malheur Experiment Station, Oregon State University

Preface

Organic agriculture has emerged as a sustainable farming system which has great impact on socioeconomic status of small farmers in rural areas, particularly of developing countries. This system of farming enhances soil and ecosystem health by avoiding the use of chemical fertilizers and pesticides and recycles farm waste, making the system self-sustainable. With increasing health awareness, current increasing interest in organic agriculture is favored for health as well as environmental and food quality concerns. Currently, the land under organic farming is ~40 m ha with a market value of ~US$65 billion. Australia, China, and India are increasingly growing farming under organic. I am pleased to present a timely compilation on "Organic Farming for Sustainable Agriculture" under the series on "Sustainable Development and Biodiversity" after a very successful first book on *Sustainable Horticultural Systems: Issues, Technology and Innovation* (2014).

"Organic Farming for Sustainable Agriculture" book is contributed by authors from the entire gamut of agricultural disciplines who are distributed throughout the globe, particularly developing countries, which is the region impacted the most by climate change and excessive use of chemical farming. Organic farming practices are resilient and becoming increasingly important due to pressing needs to protect the air, soils, and water; to improve socioeconomic conditions of farmers, farm workers, and rural communities; and to provide healthy, safe, and nutritious horticultural products to a rapidly increasing world population.

This book gathers review articles that analyze current organic agriculture practices, principles, knowledge, and proposed solutions. This book is the most up-to-date and comprehensive review of our knowledge on the use of innovative technologies and issues in organic farming systems with case studies from various regions of the world. It contains fifteen reviews on the production, practices, urban agriculture and integrated pest management, breeding for organic farming, safety issues, organic meat, organic certification, and health and nutrition.

The book is designed to cater to the needs of undergraduates and postgraduates studying organic agriculture, horticulture, sustainable crop production, crop protection, agricultural sciences, integrated pest management, and plant sciences.

Research scientists in such fields as horticulture, vegetables, agriculture, and crop protection will also find this book as a useful compilation of review articles. Libraries in all universities and research establishments where agricultural and horticultural sciences are studied and taught should have multiple copies of this valuable book on their shelves.

August 2015

Dr. Dilip Nandwani
College of Agriculture, Tennessee State University
Nashville, TN, USA

Acknowledgments

Editor gratefully acknowledges valuable suggestions; encouragement and support received from K.G. Ramawat (Series editor), Clinton Shock, and George E. Boyhan; contributions from the authors; Chandra Reddy, College of Agriculture, Tennessee State University; and my organic agriculture research team. Sincere appreciation is to Springer team, Takeesha Moerland-Torpey, Valeria Rinaudo, Izabela Witkowska who worked with me diligently throughout the period. Special thanks to my family members, wife Varsha, children (Gayatri and Rahul), and parents for their love and patience that enabled me to complete this project successfully.

Nashville, USA
2015

Dr. Dilip Nandwani
Editor

Contents

Editor and Contributors

About the Editor

Dr. Dilip Nandwani Associate Professor with MS and Ph.D. from the University of Jodhpur (now Jai Narayan Vyas University), Jodhpur, India. He joined the Tennessee State University as Associate Professor of Organic Agriculture in 2014. Prior to joining Tennessee State University, he served University of the Virgin Islands as Research Associate Professor and Program Leader of Horticulture and spent ten years in Micronesia. He has been working with Land Grant institutions for two decades in agricultural research, extension, and teaching. He is widely published in the area of production agriculture, micropropagation, and cultivar evaluations of fruits, vegetables, and root crops. His research and extension work has resulted in excess of 103 articles (peer-reviewed, extension booklets, and conference proceedings). He is the editor of two books and Certified Professional Horticulturist (CPH) from the American Society for Horticultural Science (ASHS). He worked with United Nations Environment Programme (UNEP-GEF) as Regional Advisor. He teaches principles of organic agriculture and principles of crop science courses. Dr. Nandwani has received several awards, obtained over $2.3M grants from regional, national, and international organizations, consulted, and travelled globally.

Contributors

C. Aguilar Campus de Ciencias Biológicas Y Agropecuarias, Universidad Autónoma de Yucatán, Mérida, Yucatán, Mexico

A.A. Ansari Department of Biology, University of Guyana, Georgetown, Guyana

A. Ayala Campus de Ciencias Biológicas Y Agropecuarias, Universidad Autónoma de Yucatán, Mérida, Yucatán, Mexico

Rasheed B. Balogun Department of Horticulture and Landscape Technology, Federal College of Agriculture(FCA), Ishiagu, Nigeria

Saikat Kumar Basu Department of Biological Sciences, University of Lethbridge, Lethbridge, Canada

Devendra Bhandari Department of Agriculture and Environmental Science, College of Agriculture, Human and Natural Sciences, Tennessee State University, Tennessee, USA

Dila Ram Bhandari Department of Agriculture, Ministry of Agricultural Development, Lalitpur, Nepal

Sarabjit Bhatti Agricultural Biotechnology, Department of Agriculture and Environmental Sciences, College of Agriculture, Human and Natural Sciences, Tennessee State University, Nashville, TN, USA

Matthew W. Blair Department of Agriculture and Environmental Science, College of Agriculture, Human and Natural Sciences, Tennessee State University, Tennessee, USA

George E. Boyhan Department of Horticulture, University of Georgia, Athens, GA, USA

Shabeg S. Briar Central Agricultural Research Center, Montana State University, Moccasin, MT, USA

Fur-Chi Chen Agricultural Biotechnology, Department of Agriculture and Environmental Sciences, College of Agriculture, Human and Natural Sciences, Tennessee State University, Nashville, TN, USA

Khem Raj Dahal Institute of Agriculture and Animal Science, Tribhuvan University, Rampur, Chitwan, India

Kathleen Delate Department of Horticulture, Iowa State University, Ames, IA, USA

J.A. Erales Campus de Ciencias Biológicas Y Agropecuarias, Universidad Autónoma de Yucatán, Mérida, Yucatán, Mexico

W.B. Evans Mississippi Agricultural and Forestry Experiment Station, Truck Crops Branch, Crystal Springs, MS, USA

K. Fiedler Fulbright Foundation and International Institute of Education, New York, NY, USA

Junjie Hao Qingdao Academy of Agricultural Sciences, Qingdao, Shandong, China

Prabodh Illukpitiya Department of Agricultural and Environmental Sciences, Nashville, TN, USA

M.A. Islam Department of Horticulture, Faculty of Agriculture, Bangladesh Agricultural University (BAU), Mymensingh, Bangladesh

S. Jaikishun Department of Biology, University of Guyana, Georgetown, Guyana

Risqat B. Kalejaiye-Matti Department of Animal Production Technology, Federal College of Agriculture (FCA), Ishiagu, Nigeria

Vanaja Kankarta Department of Agricultural and Environmental Sciences, College of Agriculture, Human and Natural Sciences, Tennessee State University, Nashville, TN, USA

Pramatma Khanal Department of Agricultural and Environmental Sciences, Nashville, TN, USA

Sarada Krishnan Department of Horticulture and Center for Global Initiatives, Denver Botanic Gardens, Denver, CO, USA

J.C. Ku Campus de Ciencias Biológicas Y Agropecuarias, Universidad Autónoma de Yucatán, Mérida, Yucatán, Mexico

S.K. Kuri Faculty of Agriculture, Department of Agricultural Extension Education, Bangladesh Agricultural University (BAU), Mymensingh, Bangladesh

Ajay Nair Department of Horticulture, Iowa State University, Ames, IA, USA

Dilip Nandwani Department of Agricultural and Environmental Sciences, College of Agriculture, Human and Natural Sciences, Tennessee State University, Nashville, TN, USA

Sochinwechi Nwosisi Department of Agricultural and Environmental Sciences, College of Agriculture, Human and Natural Sciences, Tennessee State University, Nashville, TN, USA

Justin U. Ogbu Department of Horticulture and Landscape Technology, Federal College of Agriculture (FCA), Ishiagu, Nigeria

Ankit Patras Department of Agricultural and Environmental Sciences, College of Agriculture, Human and Natural Sciences, Nashville, TN, USA

L. Ramírez Campus de Ciencias Biológicas Y Agropecuarias, Universidad Autónoma de Yucatán, Mérida, Yucatán, Mexico

Gadi V.P. Reddy Western Triangle Agricultural Research Center, Montana State University, Conrad, MT, USA

Basu Dev Regmi Green Grand Australia, Osborne Park, WA, Australia

L. Sarabia Campus de Ciencias Biológicas Y Agropecuarias, Universidad Autónoma de Yucatán, Mérida, Yucatán, Mexico

Krishna Prasad Sharma Department of Agriculture, Ministry of Agricultural Development, Lalitpur, Nepal

George Smith Department of Agricultural and Environmental Sciences, College of Agriculture, Human and Natural Sciences, Tennessee State University, Nashville, TN, USA

F.J. Solorio Campus de Ciencias Biológicas Y Agropecuarias, Universidad Autónoma de Yucatán, Mérida, Yucatán, Mexico

Suzanne P. Stone Department of Horticulture, University of Georgia, Athens, GA, USA

Ephraim C. Umeokechukwu Department of Horticulture and Landscape Technology, Federal College of Agriculture(FCA), Ishiagu, Nigeria

Li Wang Department of Agricultural and Environmental Sciences, College of Agriculture, Human and Natural Sciences, Nashville, TN, USA

Shu Wei State Key Laboratory of Tea Plant Biology and Utilization, College of Tea and Food Science and Technology, Anhui Agricultural University, Hefei, Anhui, China

David Wichman Central Agricultural Research Center, Montana State University, Moccasin, MT, USA

J. Wright Centre for Agroecology, Water and Resilience, Coventry University, Coventry, UK

Xingbo Wu Department of Agriculture and Environmental Science, College of Agriculture, Human and Natural Sciences, Tennessee State University, Tennessee, USA

Ying Wu Department of Agricultural and Environmental Sciences, College of Agriculture, Human and Natural Sciences, Nashville, TN, USA

Peiman Zandi Department of Agronomy, Takestan Branch, Islamic Azad University, Takestan, Iran

Xiaoyan Zhang Qingdao Academy of Agricultural Sciences, Qingdao, Shandong, China

Suping Zhou Agricultural Biotechnology, Department of Agriculture and Environmental Sciences, College of Agriculture, Human and Natural Sciences, Tennessee State University, Nashville, TN, USA

Chapter 1
Global Trends in Organic Agriculture

Dilip Nandwani and Sochinwechi Nwosisi

Abstract Organic agriculture (OA) advocates against the application of chemical and genetically modified (GM) materials on farms except those approved by the United States Department of Agriculture (USDA) National Organic Standards Board (NOSB) which consists of a voluntary team of 15 advisors selected by the secretary of the United States of Agriculture. Globally, OA has grown approximately by 20 % yearly as consumers and growers make healthier food choices and show more concern about the impacts of our actions on the environment. OA attempts to increase the level of food security and create a more sustainable environment for future generations. Nevertheless, OA is not without its challenges. In this chapter, we take a look at the history, the present, and the future of OA. Our focus is on outlining briefly the status of OA on various continents, their certification processes, global challenges, benefits and impacts, the way forward, the position of the government, policies and the institutions, and consumer behavior toward OA.

Keywords Sustainable · Organic certification · Organic management practices environment · Global impacts

1.1 Introduction

As the demand for food production increases in the world community and the limits of the Earth's resources become more apparent, sustainable agriculture is gaining increased attention. Sustainable agriculture integrates the disciplines of food security, nutrient cycling, water quality and supply, soil health, energy efficiency, pest control, breeding, animal and plant physiology, and ecology. Sustainable

D. Nandwani (✉) · S. Nwosisi
Department of Agricultural and Environmental Sciences, College of Agriculture, Human and Natural Sciences, Tennessee State University, Nashville, TN 37209, USA
e-mail: dnandwan@tnstate.edu

D. Nandwani (ed.), *Organic Farming for Sustainable Agriculture*,
Sustainable Development and Biodiversity 9,
DOI 10.1007/978-3-319-26803-3_1

agriculture employs ecological principles and use of environmentally friendly and renewable strategies in agriculture to protect biodiversity. New eco-friendly solutions have been developed based on findings from various fields of science including botany, agronomy, ecology, and food and entomological sciences.

Organic agriculture (OA) is a management system that avoids the use of synthetic pesticides, inorganic fertilizers, and genetically modified organisms (GMOs) and that seeks to reduce pollution (air, soil, and water) and optimize the health and productivity of interdependent communities of plants, animals, and humans. OA has emerged as an option to the problems of chemical usage by conventional agriculture. To meet these objectives, organic farmers need to implement a series of practices that optimize nutrient and energy flows and minimize risk. Organic practices include crop rotations, enhanced crop diversity, different combinations of livestock and crop production, symbiotic nitrogen fixation with legumes, efficient utilization of organic manure and other crop waste streams, and biological pest control (Müller-Lindenlauf 2009). Prior to the arrival of synthesized fertilizers, biocides, medicines, farm mechanization, and fossil fuels, organic agriculture was the sole option (Kristiansen and Merfield 2006). Farmers had no alternative but to work within natural constraints.

Conventional agriculture has some potentially dire consequences which may include environmental degradation (Sununtapongsak 2006), economic problems, and increased health risks (Jitsanguan 2001; Sharma 2006; Tancho 2006; Uphoff 2002). Several problems have at times been related to conventional agricultural practices, such as decreased prices of agricultural products and increased costs of inputs which have led many farmers who have adopted such practices into bankruptcy (Pattanapant 2009). Agriculture has posed many threats to the environment; such effects include air pollution from greenhouse gases (GHGs); land degradation as a result of clearing, cultivation of sloping land, and salinity; water pollution from fertilizers, pesticides, overuse, and wetland draining; and the loss of biological and ecological diversity (Norse and Tschirley 2003). Sharma (2001) stated that organic farming is the most widely recognized alternative farming system for sustainable production without many detrimental effects in the environment and ecology. For OA to experience growth, it relies on producers' and consumers' awareness, availability of good infrastructures, and the willingness of the consumers to purchase the organic products (Sharma 2001). Broadly speaking, OA is widely supported and encouraged by nongovernmental organizations (NGOs), government agencies, associations of farmers, and consumers who are concerned about the detrimental effects of chemicals on human health. Commercially available organic products are sold both in domestic and in international markets.

OA can be viewed in different ways, depending on the person's perspective. Firstly, it can be viewed as an alternative in opposition to the mainstream conventional farming. Secondly, it can be viewed as a self-organizing system based on common organic values, and thirdly, OA can be viewed as a market opportunity. These three perspectives are developed on the basis of collected experiences with organic research, practice, and discourse. The four well-known "No's" in OA are as follows: no use of synthetic fertilizers, no pesticides, no food additives, and (the more recent) no use of GMOs (Alroe and Noe 2008). There are detrimental relationships between the exposure to pesticides and human health. Pesticide exposures have been associated with acute health problems such as "nausea, skin and eye problems, dizziness, vomiting, headaches, and abdominal pain" (Ecobichon 1996).

Organic foods have several major benefits which include low pesticide residue, good taste, and increased nutritious values. A study conducted by an EU group of researchers on the benefits of organic food suggests some organic foods such as fruits, vegetables, and milk containing more nutrients than inorganic ones (Paddock 2007). Reports indicate a high concentration of antioxidants, minerals, and other healthy chemicals in organic fruits (Green 2004), vegetables, and dairy (Ungoed-Thomas 2007). Many consumers have preferences of organic fruits not only because they are more nutritious, but also because they taste much better. Some researchers attribute the differences of taste to the better soil quality in organic farming techniques compared to conventional farming (University of Maryland 2009; Green 2004). The benefits of organic farming to the wider environment have been reviewed, and the major advantages are biodiversity, high soil quality, and lower energy use. On average, organic farms provide more natural habitat for wildlife (Shepherd et al. 2009). The soil and water on organic farms generally contain low pesticide residues, and the absence of hazardous chemicals from pesticides avoids killing harmless insects and plants (University of Maryland 2009). A system with the coexistence of diverse species tends to yield better quality crops.

Apart from the preservation of biodiversity, organic farming also improves the quality of soil. Soil organic matter is essential for soil to produce high-quality crops (University of Maryland 2009). Another important benefit of organic farming to the environment is its lower emission of carbon dioxide. In a study by Pattanapant (2009) on the opportunities and constraints of OA in Chiang Mai Province with regard to organic production processes, the organic farmers noted that many problems could occur, such as growth of weeds, especially in the rainy season. Thus, they spent more labor and time on weed management, resulting in higher labor costs. They also stated that off-season vegetables could not be grown due to the unfavorable environmental conditions. Such concerns contributed to the limited production of organic vegetables, which could not meet the market demand.

Certification bodies (CB) play a vital role in the organic supply chain by guaranteeing product integrity and a flow of information from the producers to consumers. Certification can be defined as the procedure by which a third party, i.e., certification body, gives written assurance that a clearly identified process has been systematically assessed in a way that provides adequate confidence that specified the conformity of the products to specified standards (Haas et al. 2010). Those standards may involve products, processes, systems, or persons. It is a general belief that organic principles based on natural methods and means are environmentally friendly and thus superior to systems based on artificial methods

(Bergström et al. 2008). This overview summarizes OA, its practices, status, certification procedures, and agencies in various countries. The topics of food safety, health, nutrition, environmental quality, system sustainability, and energy consumption were also discussed.

1.2 Principles of Organic Agriculture

OA is an alternative production system that avoids the use of synthetic pesticides and fertilizers, relies on biological pest control, and relies on crop rotation, green manure, compost, and other recycled wastes to maintain soil fertility (adapted from Goh 2011). OA is a holistic production management system which promotes and enhances agroecosystem health, including biodiversity, biological cycles, and soil biological activity (Haas et al. 2010). It emphasizes the use of management practices in preference to the use of off-farm synthetic inputs, taking into account that regional conditions require locally adapted systems (Haas et al. 2010). This is accomplished by using, where possible, agronomic, biological, and mechanical methods, as opposed to using synthetic materials, to fulfill any specific function within the system (FAO 1999). The fundamental aim and objective of OA is to enhance the effectiveness of health and productivity of interdependent communities of soil life, plants, animals, and people.

OA is based on certain principles as stated by Lynch and Truro (2009):

- Protect the environment, minimize soil degradation and erosion, decrease pollution, and optimize biological activity and health,
- Maintain soil fertility by optimizing conditions for biological activity within the soil,
- Maintain biological diversity within the system,
- Recycle materials and resources to the greatest extent possible within the enterprise, and
- Rely on renewable resources in locally produced organic food systems.

Gomiero et al. (2011) espoused four basic principles to guide OA. The first point is based on health. OA should sustain and enhance the health of soil, plant, animal, human, and planet. The second point is ecology. OA should be based on living ecological systems and cycles, increasing soil organic matter, working with them, and helping to sustain them. The third point is fairness. OA should build on relationships that ensure fairness with regard to the common environment and life opportunities. Lastly, he expressed the fourth point as care. OA should be managed in a precautionary and responsible manner to protect the health and well-being of current and future generations and the environment. These aforementioned principles encompass fundamental goals and caveats pertinent for producing high-quality food, fiber, and other goods in an environmentally sustainable way (Kristiansen et al. 2006).

1.3 History

The term "OA" was introduced by a British agriculturalist (Lord Walter Northbourne) in 1940 in his book "Look to the Land" (Paull 2010a). Northbourne stated the idea that the agricultural space is a competitive space between organic farming and chemical farming. Three decades after Northbourne's concept of OA, all streams of agriculture that eschewed synthetic fertilizers and pesticides—including biodynamic, organic, biological, and ecological—were united in France under the auspices of the newly formed International Federation of Organic Agriculture Movement (IFOAM) (Paull 2010b) (Tables 1.1 and 1.2). This development laid the groundwork for sharing, extending, and harmonizing local innovations, including discoveries of agricultural practice, standards and certifications, labeling, training, and advocacy into the international arena.

Rudolf Steiner in his agricultural course in 1924 at Kobierzyce, Poland, expressed disquiet at the new directions in commercial agriculture and laid the foundations for the development of an alternative agriculture and a differentiated food stream (Paull 2011c). Steiner's course of 1924 eventually led to the publication of the widely read book "Bio-Dynamic Farming and Gardening" which appeared simultaneously in English, German, Dutch, French, and Italian editions (Paull 2011b). The experimental circle of anthroposophical farmers immediately tested Steiner's indications in daily farming practice. Three years later, an organization was established to market biodynamic products forming the association Demeter (Gomiero et al. 2011). In 1928, the first standards for Demeter quality control were formulated. Biodynamic agriculture, as this method was named, was

Table 1.1 Organic agriculture movements (Jones 2012; Kirchmann et al. 2008)

Period	Type of movement	Year	Milestones achieved
1900s to 1960	Reform movement	1924 1946	Introduction of biodynamic farming soil association was founded, spiritual food production, healthy food production
1960 to 1990	Environmental movement	1962 1968 1972 1980s	"Silent Spring" by Carson was published, bio-organic farming was introduced International Federation of Organic Agriculture Movement (IFOAM) was founded definition as "eco-agriculture" against pesticides and proenvironment holistic food production standardization, lobbying for worldwide adaptation, marketing environmental superiority
1990	Political movement through the present	1990	Government support promotion, subsidies, funding of research, currently being presented as a solution to the environmental problems caused by conventional agriculture

Table 1.2 Pioneers in organic agriculture (Jones 2012; Kirchmann et al. 2008)

Year	Pioneer	Contribution
1924	Rudolf Steiner	He gave a series of lectures on the spiritual foundations for the renewal of agriculture with instructions on how to produce organic food
1927/28	Community of Natural Farming and Settlement	Focus on natural method of fruit and vegetable production without the application of artificial fertilizers and pesticides
1940s	Sir Albert Howard and Lady Eve Balfour	Founded the British Soil Association in 1946, claimed that healthy soils are the basis for human health. Lady Balfour published a book entitled "The Living Soil" in which she pointed out the nutritional superiority of organically grown food
1950s	Hans-Peter Rusch and Maria Müller	Application of natural principles in agriculture, based on the fact that nature is our superior
1962	Rachel Carson	Her book "Silent Spring" pointed out the detrimental effects of pesticide use
1972	International Federation of Organic Agriculture Movement (IFOAM) was founded	Promoted its worldwide adoption, set standards, drawn up certification procedures, etc.
1972	Meadows et al. (1972)	Focused on population growth, resource depletion, and the impact of modern agriculture
2002	George Oshawa	Asian forms of organic agriculture according to Zen macrobiotic farming
1978, 1989, 1991	Japanese Masanobu Fukuoka	Asian forms of organic agriculture according to Buddhism

grounded in the practical aspects of adding manure into the soil, which is the bedrock of organic farming, but it also dealt with lunar and astrological scheduling, communication with "nature spirits," and the use of special preparations, which are described as alchemical means (Koepf 1976, 2006; Conford 2001). These latter considerations are not easily verified in quantified scientific evidence.

In 1929, Pfeiffer and several other leading European biodynamics experts from Switzerland and Holland led the Betteshanger Summer School on biodynamic farming at the Kent farm of the English agriculturalist Northbourne (Paull 2011d). While Rudolf Steiner was establishing the foundation for the growth of the biodynamic movement, Sir Albert Howard (1873–1947), a British agronomist based in India, was trying to develop a scientific based system for preserving soil and crop health (Gomiero et al. 2011). Upon his return to the UK, he worked to promote his soil and crop health approach (Howard 1943; Conford 2001). He was convinced that most agricultural problems were due to soil mismanagement and that reliance on chemical fertilization could not solve problems such as loss of soil fertility and pest management. He maintained that the new agrochemical approach was misguided and that it was a product of reductionism by "laboratory hermits" who paid no attention to how nature worked (Gomiero et al. 2011). In his milestone book, An Agricultural Testament (1943), Howard described a concept that was to become central to organic farming: "The Law of Return" (a concept expressed also by Steiner). The Law of Return states the importance of recycling all organic waste materials, including sewage sludge, back to farmland to maintain soil fertility and the land humus content (Howard 1943; Conford 2001).

Northbourne elaborated on the notion of a farm as an "organic whole," where farming has to be performed as a biologically complete process. The term "organic" then, in its original sense, describes a holistic approach to farming: fostering diversity, maintaining optimal plant and animal health, and recycling nutrients through complementary biological interactions (Conford 2001). In 1943, Lady Eve Balfour (1899–1990) published the book "The Living Soil" in which she described the direct relationship between farming practice and plant, animal, human, and environmental health (Gomiero et al. 2011). The book exerted a significant influence on public opinion, leading in 1946 to the foundation of "The Soil Association" in the UK by a group of farmers, scientists, and nutritionists. In the subsequent years, the organization also developed organic standards and its own certification body. Eve Balfour, who was one of IFOAM's founders, claimed that: "The criteria for a sustainable agriculture can be summed up in one word—permanence, which means adopting techniques that maintain soil fertility indefinitely, that utilize only renewable resources; to avoid those that contaminate the environment; and that foster biological activity throughout the cycles of all the involved food chains" (Balfur 1977).

In 1940, in an article published in Fact Digest, Jerome I. Rodale introduced the term "OA" and techniques such as crop rotation and mulching that have, since then, become accepted organic practices around the world. Rodale (1945) expanded Albert Howard's ideas in his book Pay Dirt, adding a number of other good farming practices. The existence and availability of organic foods for consumers is a result of a prolong series of events generally thought to have started between the 1940s and 1950s (Pearson et al. 2011). These included "ecological agriculture" in Switzerland, which is associated with the writings of the Muellers, Rudolf Steiner's teachings on "biodynamic agriculture" in Germany and Austria, Jerome Rodale's writings on soil and health in the USA, and Albert Howard and Eve Balfor's work in the UK on what came to be known as "organic farming" (Pearson et al. 2011).

The development of the organic movement continued during the 1960s and 1970s when there was increasing consumer activism associated with concern about anthropomorphic changes to the natural environment (Pearson et al. 2011). However, it was not until the 1990s that organic received formal recognition as a food production system in many countries. It was at this point that it started to move from the fringes into a significant activity in the mainstream food industry. In addition to becoming a possible food production system for the masses, academic research on the organic production also started to gain attention (Pearson et al. 2011).

There is now a significant body of international research which includes comparisons of many facets of organic farming, including crop production, benefits to biodiversity and soil health (Fuller et al. 2005; Mader et al. 2002), and demographics and motives of organic consumers (Hughner et al. 2007a, b). Rodale published the first edition of his periodical Organic Farming and Gardening in 1942. The Australian Organic Farming and Gardening Society were founded in Sydney in 1944 and published the periodical Organic Farming Digest (Paull 2008). A major milestone in OA occurred in 1972 with the founding of theIFOAM at Versailles, France, to unite and foster the organic cause (Paull 2010b). The vision of IFOAM (2011) was the worldwide adoption of OA.

1.4 Organic Farming Practices

Arrays of definitions have been attempted to describe organic farming. One of the most popular definitions is the USDA National Organic Standards Board's (1995) definition, "An ecological production management system that promotes and enhances biodiversity, biological cycles, and soil biological activity" (Gold 2007). Organic farming is based on reduced off-farm inputs and on management practices that restore, maintain, or enhance ecological harmony (Haas et al. 2010).

Most recent practices of organic farming in the USA can be traced back to the farming movements practiced between the 1920s and 1950s across Europe (Pamela 2012). Increase in the use of synthetic fertilizers and pesticides has resulted in the evolution of these movements.

The proponents of humus farming believed that the highest quality food and the sustainability of agriculture were achieved by "feeding the soil," which invariably results in soil fertility. Their goal was to increase the humus in order to fully decompose organic matter that had reached a stable state in the soil. Humus farming requires little or no fertilizers or pesticides, because the soil is healthy.

As public concern in the USA over pesticide use increased in the 1960s and 1970s, attention was drawn to organic farming systems. The growth of the organic industry during this period led to the establishment of standards and third-party certifications (Pamela 2012).

The five major characteristics of the OA system are (1) respect for the environment and animals, (2) promotion of sustainable cropping methods, (3) use of non-chemical fertilizers and pest/disease/weed control means, (4) production of high-quality foodstuffs, and (5) no use of genetically modified (GM) crops (Lairon 2010). OA can provide food through a unique combination of environmentally sound practices with low external inputs (Zundel et al. 2007). The organic agricultural community developed a broad range of practices for enhancing productivity without relying on external agricultural inputs (Müller-Lindenlauf 2009). Organic farming systems can be characterized by (i) efficient soil fertility management, (ii) crop, livestock, farm, and landscape management, and (iii) efficient use of nutrients (Schmid et al. 2009). Since its inception, OA has been based on the principle of sustainability (Pamela 2012). Each farm is personal to individual farmers because most vital management decisions are based on the singular conditions found in their own operations.

Each season, farmers encounter a one-of-a-kind set of natural resource and environmental conditions to which they must adapt and respond. In an organic production system, the farmer will consider how soil, water, plants, animals, insects, bacteria, fungi, and all other parts of the system will interact to cause or prevent problems. Not only will a successful organic farmer be interested in building healthy soil, but he will also focus on preventing problems, rather than reacting to them. For example, organic farmers prevent insect pest invasion by providing habitat for beneficial insects that could check the populations of harmful ones (Pamela 2012). There are generally three types of organic farming: (1) substitution-based operations (replace synthetic insecticides with organically certified and approved materials such as botanical insecticides); (2) holistic systems such as the incorporation of a wide range of soil management and cropping practices targeted at preventing insect pest outbreaks; and (3) subsistence cropping, which relies on cultural pest control methods, in part because growers have no access to synthetic inputs (Letourneau and Bothwell 2008). In organic farming, soil nutrients can be enhanced by crop rotation, intercropping, polyculture, and mulching. Pest control is achieved by using appropriate cropping techniques, biological control, and natural pesticides (Gomiero et al. 2011). A major concern in organic farming is weed control. Some partial weed control solutions are timing of planting, mechanic cultivation, mulching, transplanting, and flaming (Howard 1943; Altieri 1987; Lampkin 2002; Lotter 2003; Altieri and Nichols 2004; Koepf 2006; Kristiansen et al. 2006; Gliessman 2007).

In the USA, Congress passed the Organic Foods Production Act (OFPA) in 1990. The OFPA required the US Department of Agriculture (USDA) to develop national standards for organically produced agricultural products, to assure consumers that organic products meet consistent standards. The OFPA in conjunction with the National Organic Program (NOP) regulations requires that products be labeled as originating from farms certified by a state or private entity that has been accredited by the USDA (Gold and Gates 2007).

While there is emerging indications that organic foods are a healthier option, organic certification and production standards worldwide are predicated on promoting the holistic benefits of the farming system itself, rather than exclusively focusing on the benefits to the end users of organic products (Winter and Davis 2006; Lynch and Truro 2009).

1.4.1 Insect Pest Management

The insect control strategy in organic farming is targeted at preventing and reducing the aggregation of insect populations (Texas A&M AgriLife Extension 2015). The risks of pest outbreaks are often averted by practices developed by organic farmers over time. Practices include but are not limited to soil enrichment with compost, conservation tillage, crop rotation, and intercropping (Niggli 2009). A pest control strategy could include a number of cultural practices in combination with the use of a limited number of organically derived pesticides. Among the effective control measures available to organic producers are trap crops, field scouting and insect trapping, and application of biological control methods (such as the introduction of beneficial insects and use of natural enemies to reduce pest populations), soaps, oils, and diatomaceous earth (Dainello 1998).

1.4.2 Disease Management

Fungicides are capable of successfully controlling most diseases if they are properly and effectively applied. Since organic farming does not encourage the application of fungicides, they depend heavily on preventing diseases from occurring rather than controlling them after they occur (Dainello 1998). The most successful disease management strategies consist of three major components: genetic resistance through breeding, avoidance techniques, and approved fungicidal products (Texas A&M AgriLife Extension 2015). A few of the more commonly used fungicidal compounds in OA are elemental sulfur, copper compounds, Bordeaux mixture, and fungicidal soaps.

1.5 Impacts of Organic Agriculture

1.5.1 Food Quality, Nutrition, and Safety

Food quality may be defined as "everything a consumer would find desirable in a food product" (Grunert 2005). This definition implies that provision of quality from a grower entails meeting the needs, wants, and expectations of customers (Haas et al. 2010). Therefore, the concept of quality is dependent on customer preferences, which may be highly subjective (Haas et al. 2010). This explains why there is a correlation between marketing and quality. With reference to recent findings from different sources, product-related criteria for organic food can be categorized as follows: price, brand/label, safety, nutrition, enjoyment/pleasure, organic integrity (Kahl et al. 2012), as well as cosmetic appearance.

In the past few years, organic food has been one of the fastest growing food sectors, an alternative option to conventional food (Koch 1998). Both empirical data and consumer preferences reveal that the benefits of organic food include less hazardous pesticide, more nutrition, and a farming technique that is more environmentally friendly than non-organic food (Koch 1998; Hughner et al. 2007a, b; Pearson et al. 2007).

Although consumers consistently refer to health as a major reason for patronizing organic food, this perception is not consistently supported by scientific research (Benbrook et al. 2008; Burton 2006). Other less commonly mentioned reasons why people buy organic food include animal welfare (Chang and Zepeda 2005; Lea and Worsley 2005) and fashion (Lockie et al. 2002; Hughner et al. 2007a, b). Concern about animal welfare is more important for particular organic products and countries where intensive animal farming systems are commonly used (Pearson et al. 2011). This includes poultry and eggs, pork products, and, to a lesser extent, beef and dairy products (Pearson et al. 2011).

The challenge of understanding the complexity of organic food benefits to consumers is available from numerous research activities (Pearson et al. 2011). There is a general consensus in the literature on the reasons why people buy organic food (Pearson et al. 2011). These have remained stable over time, although there are some slight differences between countries and for particular products. The main reasons, in order of priority, are personal health, product quality, and concern about degradation of the natural environment (Pearson et al. 2011). The Food Standards Agency in the UK has recently reported on a meta-analysis of the scientific evidence that examines the potential human health benefits of consuming organic food (FSA 2009). Most of the reviewed papers report significant differences between organically and conventionally produced vegetables and fruits regarding dry matter, total sugars, vitamin C, and polyphenolic substances (Kahl et al. 2012). There is a preponderance of evidence to support the fact that organic farming is less damaging to the natural environment (Fuller et al. 2005; Mader et al. 2002). Indeed, the environmental benefits of organic production methods are the rules given by the government in the UK for providing additional financial support to organic farmers (Defra 2004).

1.5.2 Who Buys Organic Food?

In some empirical studies conducted in Europe, income has been found to be a salient factor in explaining organic food purchases (Torjusen et al. 2001; Millock et al. 2004; Kuhar and Juvancic 2005; Tsakiridou et al. 2006). European studies indicate that consumers with higher income are more likely to purchase organic food products (Gracia and Magistris 2007). However, studies carried out in the USA show that income has not been statistically significant in determining US organic food purchases (Loureiro et al. 2001; Durham and Andrade 2005; Onyango et al. 2006; Zepeda and Lin 2007). Sociodemographic characteristics are significant factors in explaining US consumer decisions to buy organic foods (Thompson 1998; Thompson and Kidwell 1998; Blend and Ravenswaay 1999; Wessells et al. 1999; Loureiro et al. 2001; Onyango et al. 2006; Zepeda and Lin 2007), while in Europe only age, education, and household size were significant factors (Millock et al. 2004; Lockie et al. 2004; and Tsakiridou et al. 2006). Recent findings stated that older, more educated consumers and those living in larger households are more likely to purchase organic food products (Gracia and Magistris 2007). The knowledge of organic product is an important factor because it is the only means whereby consumers have to differentiate it from conventional ones and to form positive attitudes and quality perceptions toward organic products (Gracia and Magistris 2007). The level of organic product knowledge will depend on sociodemographic characteristics, lifestyles, and information on organic products available on the market. Most consumers purchase organic products, but only infrequently, and hence, they switch between organic and conventional products on a regular basis (Pearson et al. 2011).

Moreover, organic consumers are hampered by the lack of consistent, objective (scientific) evidence of the benefits (taste, health, and environment) which leaves most consumers confused and some to develop skepticism (Fearne 2008).

1.5.3 Organic Agriculture and the Environment

Attention has recently been drawn to the issues regarding food production as a result of increasing environmental awareness and has strengthened the environmental lobby, especially those groups that favor OA (Anderson and Nielsen 2000). Sloan (2002) and Cahill et al. (2010) suggested that media coverage of pesticides, genetic engineering, and environmental degradation might have played a role in encouraging consumers to purchase organic products.

1.5.4 Climate Change and Global Warming Impacts

Climate change is a major challenge. Some geographical regions (Caribbean region) and countries (Cuba, Haiti, and Jamaica,) are already experiencing the effects of climate change, which is making it difficult to maintain stable food production. Pest and diseases are affecting the crops, and for many problems, solutions have not yet been found (Garibay and Ugas 2009). For example, in cacao, the Monilia pod rot (*Moniliophthora roreri*) is a serious fungal disease. Damage caused by the disease varies from less than 25 % in some regions to a total loss of production in other regions. Climate change mitigation is urgent, and adaptation to climate change is crucial, particularly in agriculture, where food security is at stake (Muller et al. 2012). In both developed and developing countries, OA has considerable potential in mitigating climate change due largely to its ability to reduce GHG emissions and to enhance carbon sequestration in soils (Goh 2011). The potential of OA to mitigate climate change is mostly claimed on the basis of assumptions that organic farming can aid the soil in carbon sequestration (Müller-Lindenlauf 2009).

1.5.5 Carbon Sequestration

A review of studies comparing carbon sequestration in soil under organic and conventional management showed a higher soil carbon content with organic management, as compared to conventional management practices (Müller-Lindenlauf 2009). OA encourages agroforestry as well as the integration of landscape elements, leading to further carbon sequestration in plant biomass (e.g., IFOAM Norms 2002; East African Organic Standard 2007; Pacific Organic Standard 2008). Biomass burning, a major contributor to carbon dioxide emissions, is restricted in OA (Müller-Lindenlauf 2009). High carbon sequestration potential is also reported in grassland soils (Smith et al. 2007). Organic grassland farming could be a way to optimize carbon sequestration in grasslands (Liebig et al. 2005; Rice and Owensby 2001). Introducing grass and clover leys into the crop rotations as feedstuff for ruminants and diversifying the crop sequences, as well as reducing plow depth augment soil organic matter and contribute to carbon sequestration (Niggli 2009).

1.5.6 Reduction of Energy Use

A 12-year study of organic versus conventional management on energy use, energy output, and energy-use efficiency was carried out in Manitoba, Canada. Energy use was 50 % lower with organic than with conventional management (Hoeppner et al. 2006). Energy efficiency was highest in the organic and integrated rotations (Lynch and Truro 2009; Ziesemer 2007). Fossil energy inputs averaged 30 % lower for organic production systems than for conventional corn production (Lynch and Truro 2009). In a study between organic and conventional apple production in Washington State, it was found that the organic system allowed a 9 % reduction in energy inputs and was 7 % more efficient in energy use (Reganold et al. 2001).

According to Khanal (2009), OA is reported to be more efficient and effective in reducing GHG emissions mainly as a result of less chemical fertilizers and fossil fuel. OA is also reported to be resilient to climate change as it promotes the proper management of soil, water, and biodiversity. The level of atmospheric GHGs—mainly carbon dioxide (CO_2), nitrous oxide (N_2O), and methane (CH_4)—has been rapidly increasing since the onset of the industrial revolution. The increased level of GHGs has created a greenhouse effect which subsequently altered precipitation patterns and global temperatures around the world (Khanal 2009). Climate change is having great impacts on agriculture, forestry, water resources, biodiversity, desertification, human health, ecosystems' production and distribution of goods, and services worldwide (Fig. 1.1).

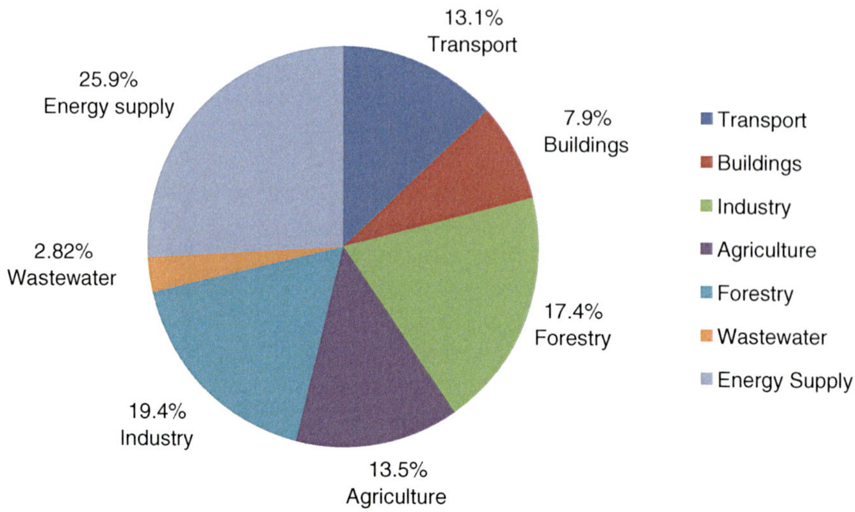

Fig. 1.1 Percentage of greenhouse gas emissions from different sectors (IPPC 2007) (*Credit* https://www.ipcc.ch/pdf/assessment-report/ar4/syr/ar4_syr.pdf)

1.5.7 Enhancing Biodiversity

OA has various ways of helping the environment, chiefly enhancing biodiversity (Hole et al. 2005; McNeely 2001). The cultivation of specific crops for the organic export market is not recommended because it makes farms defenseless against the unstable global market and it reduces farm biodiversity (Panneerselvam et al. 2012). Since organic weed, pest, and disease management depend on biodiversity (Niggli 2009), it is in the economic interest of organic farmers to promote diversity at all levels.

Biodiversity can be estimated at different levels of organization (e.g., genetic diversity within species, species diversity within taxa and trophic levels, functional diversity in communities) and at different spatial scales (Letourneau and Bothwell 2008). In a recent study carried out on crop fields and woody hedgerows of 16 conventional and 14 organic sites, clear differences were observed in species nutrient requirements and composition (Lynch and Truro 2009). Fields and woody hedgerows located in organic sites had more native and exotic plant species than those managed conventionally.

1.5.8 Soil Nutrient Management and Water Quality

Higher soil organic matter content increases the soil's water retention capacity and reduces the risk of soil erosion (Müller-Lindenlauf 2009). Avoidance of the use of chemical fertilizers lowers the leaching of nitrates and helps to improve the quality of water (Stolze et al. 2000). In a study carried out on a perennial orchard system in the USA, Kramer et al. (2006) discovered that after nine years, the organically managed soil exhibited not only greater soil organic matter and microbial activity, but also greater denitrification efficiency compared to conventionally managed or integrated orchard management systems (Lynch and Truro 2009). The emissions of N_2O were not significantly different among treatments, whereas emissions of N_2 were highest in the organic plots (Lynch and Truro 2009). There is abundant evidence from European, US, Australian, and African studies that organic farms and organic soil management enhance soil fertility (Niggli 2009).

1.5.9 Yield

A common question asked of the organic movement relates to its yields for example: Can OA feed the world? The public concern about what potential benefits OA might offer to small-scale or family farms often centers around the impact of adopting OA on yields (Goklany 2002). In a global view, the majority of scientific literature shows that organic yields are between 25 and 50 % lower than conventional yields, depending on whether the organic system has access to animal manure (Kirchmann et al. 2008; Wynen 1994; Stonehouse et al. 2001; Mendoza 2002). In less favorable crop-growing regions, organic yields tend to match conventional ones (Trewavas 2004). Yield also depends on many factors including the farmer's background, the farm's resourcefulness, and local and national support mechanisms (Kristiansen et al. 2006). Some of the main conclusions are that OA has consistently lower yields than conventional production and is thereby a less efficient method of land use; that environmental problems caused by processes such as nutrient leaching are not reduced by conversion to organic crop production; and that soil fertility status and microbial biodiversity are not improved by organic cropping (Bergström et al. 2008). Some other questions are related to sustainability.

1.6 Limitations of Organic Agriculture

The major challenge of organic farming is yield. Productivity in organic farming is limited by both nutrient shortages and high weed populations. Unlike conventional crop production, it can be cumbersome to increase yields through application of manures and the exclusive use of untreated minerals (Kirchmann et al. 2008). The main factors limiting organic yields are the availability of fewer nutrients, ineffective weed control measures, and limited possibilities to improve the nutrient status of infertile soils (Pattanapant 2009).

The progress in OA development has been slow, due to many barriers, which include the following: (a) lack of information and support from extension agencies;

(b) negative perception by some growers of OA; (c) improper management of weeds and pests; (d) decreases in yields; (e) lack of organic inputs; (f) insufficient labor supply; (g) insufficient research and development; (h) weak infrastructure; (i) complications in organic standards; (j) lack of awareness of existing standards and certification; (k) ineffective organic markets; (l) inadequate information on organic products on the part of the consumers; (m) pricing problems; and (n) the availability of crop cultivars bred and selected for organic production (Pattanapant 2009; Rigby and Cáceres 2001; Wynen 2003; Sittiwong and Varinrak 2004; Singpornpong 2005; Pornpratansombat 2006; Kerselaers et al. 2007; Wheeler 2008).

1.7 Benefits of Organic Agriculture

A growing demand for organically produced food in industrialized countries has the potential of providing access to premium prices and hence higher income (Willer et al. 2009). Providing healthy food for everyone is probably the most important survival issue for mankind in the future (Kirchmann et al. 2008). OA is promoted in a development context because of its possibility of improving livelihoods through increased incomes (UNCTAD 2006; Kilcher 2007; UNEP-UNCTAD 2008). Additionally, the adoption of OA in developing countries may provide economic, social, and cultural benefits (UNCTAD 2006). OA is generally considered to reduce external input costs due to the cessation of use of pesticides and mineral fertilizers and increase internal nutrient recycling using green manures, composts, and animal manures. However, given that labor use might increase following adoption, the reduction of external costs depends very much on labor substitutes for these inputs and whether farms supply their own labor. In developed countries, there is a consistent, growing market for organic products, driven by the rising consumer desire for healthy food and environmental protection (Willer et al. 2009).

1.8 Organic Agriculture Worldwide

OA is now practiced in about 160 countries (Willer and Kilcher 2011), and worldwide sales of organic products is about US $60 billion per annum (Biofach 2011). According to the (IFOAM), there has been an uphill progression in the development of organic agricultural land internationally (Fig. 1.2). The land devoted to OA worldwide has increased over the past decade from 15.8 million ha to 37.2 million ha, a compound rate of growth of 8.9 % per annum (Paull 2011a). Children consuming an organic food diet have reduced pesticide exposure and lowered the body burden of pesticides (Curl et al. 2003). Growth over the past decade is presented for 71 countries which taken together account for 35.3 million organic agricultural hectares, that is, 94.8 % of the total global OA area and 58.2 %

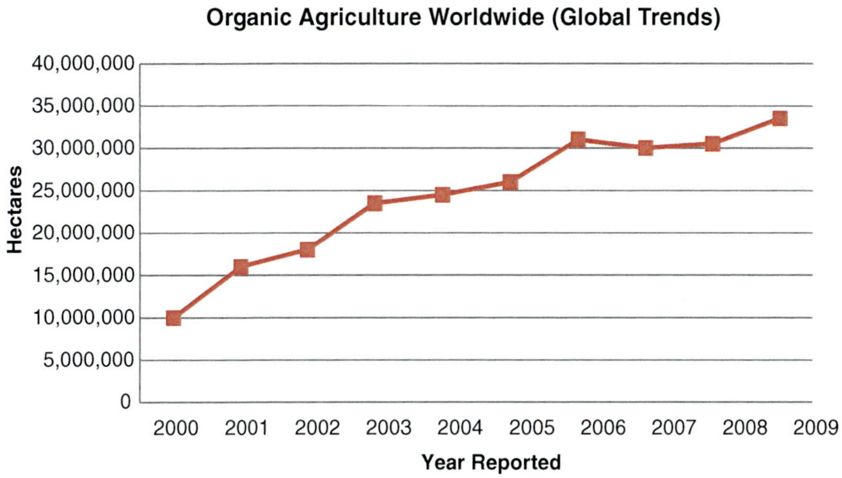

Fig. 1.2 *Source* The growth of organic agriculture worldwide (IFOAM 2000–2009)

of the total global agriculture area (Paull 2011a). This analysis reveals that underlying what appears at the global level is consistent increase in growth over the decade; growth is very uneven when disaggregated by country (Raynolds 2004; Willer et al. 2009).

1.8.1 Europe

At the onset of the introduction of organic products to the market, they were mainly sold by specialists (natural food shops). By the end of the 1990s, sales began to cross over to mainstream retailers (Pivato et al. 2008). Most European retail chains now sell organic products, compared with less than one-fifth in 1996, and supermarkets account for the sale of approximately one-half of all organic food and drink in Europe (Sahota 2004). OA has grown in the past two decades to be a significant sector within agricultural production, whereas in other countries it has remained at a relatively low level (Kirchmann et al. 2008). This rise of organic farming in Europe has been facilitated by considerable growth in the market for organic products within the EU (Haas et al. 2010).

The introduction of the Rural Environmental Protection Scheme (REPS) in Ireland to promote organic farming made it more economically attractive and caused a strong increase in the organic sector (Lapple 2010). About 80 % of the organic farms are dry stock farms, followed by horticultural enterprises accounting for about 10–15 %, with tillage, dairy, and poultry farms making up the remainder (DAFF 2002). Considering the fact that the Irish organic sector is less developed compared to other European countries, the sector is growing steadily, yet it remains small in comparison with other European countries (Lapple 2010). Ireland is currently self-sufficient in organic beef, and an export market, mainly to the UK, is emerging. A considerable part of organic produce is sold as conventional products, which reflects the lack of organic market development (Lapple 2010).

For about 20 years, certified organic farming was supported in Norway through a broad range of policies by means of legal, financial, and communicative conversion grants and support schemes (Koesling et al. 2012). However, in 1996, the food industry introduced organic premiums on several products, including milk and beef (Koesling et al. 2012). This resulted in a decline in growth of organic production in Norway, and most organic farmers started withdrawing from certification (Koesling et al. 2012) due to regulatory burdens and economic issues (e.g., Harris et al. 2008; Kaltoft and Risgaard 2006; Kirner et al. 2006; Sierra et al. 2008).

OA practiced in the Czech Republic is characterized by the prohibition of chemical fertilizers, chemical preparations, sprays, hormones and other artificial substances, and a positive relationship with animals, plants, and nature (Mala 2011). A significant deterrent of organic agricultural development in the Czech Republic is the high percentage of unfavorable localities for production. However, organic production is favored by subsidies and price allowances (Kroupová 2009).

1.8.2 Asia

OA has been identified by the Cambodian government as a priority sector (UNESCAP 2002), a means to achieve food security, diversify rural livelihoods, and gain access to value-added markets (Beban-France 2008). UNCTAD (2004) and the IMF (2004) have recently conducted studies into Cambodian agricultural markets and concluded that farmers lack the bargaining power necessary to achieve higher incomes due to a number of factors. These include limited long-term finance, uncertain property rights, limited access to markets, lack of information, lack of government support, low levels of trust, and lack of capacity to handle postharvest produce (Beban-France 2008).

Organic agriculture was first introduced in Turkey by European companies (Tate 1994), in the quest to supply Turkish organic products that cannot be grown in Europe (Demiryürek 2000). In tandem with the increase in demand from European countries, organic producers have increased since the mid-1980s (Rehber and Turhan 2002), developing rapidly between 1990 and 2006 (Demiryürek et al. 2008). Recent data showed that Turkey exports organic products to 37 countries (Güzel 2001; Demiryürek 2004; Kenanovglu and Karahan 2002; Babadovgan and Koç 2004; Olhan et al. 2005; Sayin et al. 2005; Aksoy and Engiz 2007).

OA started in Iranian universities through specific courses and lectures (Kledal et al. 2012). The University in Tehran established a master program in agroecology. In 2005, the Iranian Scientific Society of Agroecology (ISSA), an NGO, was established. Currently, its main focus is on research and education in the field of organic farming. In 2006, the Iranian Organic Association (IOA) was established to focus on marketing and trade. The ISSA and the IOA had active participation in the process of legislation of organic standards (Kledal et al. 2012).

India is globally recognized for having the largest number of organic farms (340,000) and increasing the export of organic products by 87 % over the past 3 years. Organic markets are expected to increase in India due to a strong demand for quality organic food by the rising income of the middle class (Panneerselvam et al. 2012). In India, the number of farmers converting to organic farming has increased in recent years despite the lack of government support in providing extension to the farmers (Panneerselvam et al. 2012).

1.8.3 Africa

OA has been adopted in few African countries (Kristiansen et al. 2006). Senegal and Burkina Faso have also established NGOs that set local certification standards to reduce external certification costs, as well as provide training in organic food processing, labeling, packaging, and storage. They have also established local and distant markets for selling organic produce (Anobah 2000). Many parts of Africa experience severe poverty and face difficult conditions for agricultural production (Kristiansen et al. 2006).

Much of the Sudanese agriculture is carried out under organic management by default, which simply means the farmers have no access to chemical fertilizers, pesticides, or other organically prohibited amendments for financial and other reasons (Alkhalifa et al. 2014). Most farming systems in Sudan depend holistically on natural methods of building soil fertility and fighting pests and diseases. However, most of these farms are not inspected or verified by any organic certification agency. The lack of a strategic plan to develop OA in Sudan has been the major problem which impedes the progress of organic farming there, along with the absence of an authorized body to register and certify the Sudanese organic products, albeit that most of the Sudanese products are produced traditionally and naturally using no or very low synthetic inputs (Alkhalifa et al. 2014).

Bakewell-Stone et al. (2008) reported that there are about 23 certified organic projects in Tanzania, including 16 export firms and seven projects for local markets. In addition to the prospects of improved incomes, organic producers in Tanzania are motivated by the high costs of hybrid seeds requiring chemical inputs, the ease of access to organic inputs, similarity of organic production systems to traditional practices, improved product taste and nutritional content, maintenance of soil moisture, heightened resistance to drought and diseases, improved handling qualities, and links between chemicals and health problems for people (Bakewell-Stone et al. 2008).

About three quarters of farmers in Nigeria practice OA by default because of the exorbitant prices of chemical fertilizers and other agrochemicals (Peace 2014). Unlike other African countries, Nigeria has yet to develop its potential in terms of organic farming, even though it is an agrarian country with a track record of being the world's leading producer of several crops at one time or another (Peace 2014). An organic farm in Nigeria currently sells organic lemon tea, turmeric, and other produce in the local market, a situation many regard as underutilization of the premium benefits of organic farming (Peace 2014).

1.8.4 South America

There has been a high level of adoption of OA in Central and South America in terms of certified land area and number of farms. Argentina has the second highest amount of land under organic production in the world, while Mexico has the greatest number of organic farms in Latin America (Kristiansen et al. 2006). However, socioeconomic constraints such as poverty and a land tenure system have greatly influenced the process of adoption and adaptation of OA in Central and South America (Parrott and Marsden 2002).

Vegetables, fruits, milk and milk products, honey and coffee are commonly sold in Mexico, Honduras, Nicaragua, Costa Rica, Peru, Bolivia, Brazil Uruguay, Chile and Argentina, and to a lesser extent in other countries (Garibay and Ugas 2009). In Costa Rica, more than half of the organic food sold is by supermarkets.

In Latin America, organic products are readily available in many supermarkets. Most Latin American countries feature specialized stores, or health food stores, which sell products from local organic farmers to an informed customer base. Many Latin American countries have been exporting their fruit to Europe and the USA (Garibay and Ugas 2009).

Willer and Yussefi (2007) reported that the amount of Mexican land set aside for organic crops has grown on average by 33 % yearly, employment in the sector by 23 %, and income generated by 26 %. Approximately 50 % of this production is accounted for by coffee, followed by herbs, vegetables, cacao, and other fruit crops (Nelson et al. 2009). About half of the organic producers are indigenous and 98 % are small scale, meaning they farm 30 ha or less (Nelson et al. 2009). The Mexican Network of Organic Markets has been a key player in the Mexican organic movement and has worked to develop local networks of organic production and consumption in Mexico (Nelson et al. 2009).

1.8.5 USA

In 2005, the USA recorded certified organic farmland in all 50 states. US producers dedicated over 1.6 million ha of farmland to organic production systems: 690,000 ha of cropland and 910,000 ha of rangeland and pasture. California remains the leading state in certified organic cropland, with over 89,000 ha, mostly for fruit and vegetable production (Gold 2007). The reason is that organic farming operations are well established in California and the USA and the data and materials on OA are obtainable. Organic growers in California and the USA often pinpoint the organic rules as their biggest challenge with increased documentation and tariffs for authorization and registration (Klonsky 2010). Thus, the persistent attempt of the NOP to define, amend, and report organic laws is key to the continuous enlargement of the organic industry (Klonsky 2010).

1.9 Organic Certification

The certification of farming practices as essential in OA provides a safe guarantee of organic principles and standards (Goh 2011). Non-certified OAcan be viewed from different perspectives. From the logic–poetic perspective, "non-certified OA" promises an alternative development path in rural areas of low-income countries (Halberg et al. 2006). And non-certified OA is less prone to the market pressures connected to growth and trade, which threaten to erode the standards and practices of certified OA. From a protest perspective, non-certified OA may play the role of opposition to both conventional and certified OA, whose growth in scale and markets may be anathema to some organic growers who also want it to be smaller in scale and catering to local markets (Alroe and Noe 2008). From the market perspective, non-certified OA is not even real and can hardly contribute any quota in the world market (Alroe and Noe 2008).

1.9.1 Process

The major steps involved in the certification process under the USDA NOP are highlighted below (Pamela 2012):

Step 1 Submission of an application to a certifier by a farmer: This application contains several documents such as organic system plan; map of the farm; field histories; operator agreement; and report of organic yields and sales.

Step 2 Reviewing the application by the certifier: The certifier will read through the application and assess whether the farm meets the regulations and specifications.

Step 3 The inspector visits the farm: Organic farms are usually inspected annually. However, an impromptu visit may be carried out, usually at the discretion of the certifier. At the end of inspection, the inspector reviews any areas of concern. A report is written by the certifier which is forwarded to the certification agency.

Step 4 Reviewing the inspection report: A decision is made after reviewing the report by the certifier whether it conforms to the standards and regulations.

Step 5 Issuance of the organic certificate.

It is worthy of note that organic farmers that gross less than $5000 from sales of organic products are exempt from the certification requirement (Klonsky 2010).

1.9.2 History

A sudden surge in the organic industry during the 1980s created a need for certified products. Different certifiers developed their own standards and certification processes (Pamela 2012). As a result, some certifiers did not accept the authenticity of organic certification by other certifiers. These discrepancies among certifier standards resulted in barriers to trade, which led many to believe that a single, consistent set of US standards for organic production, labeling, and marketing was needed. Congress passed the OFPA of 1990. This act resulted in the creation of the NOP, which is a regulatory program housed within the USDA Agricultural Marketing Service. (Pamela 2012).

1.9.3 Role of Organic Certification

When OA is regulated by the national government, firms can freely choose to apply organic practices in order to differentiate their product, but when they choose to produce organic food, they have to follow rules determined by the specific regulation. Some organic standard setters are beginning to refine their criteria so that organic products better match their ideals (Rosenthal 2011).

1.9.4 Organic Certification in Asia

Organic regulations are now in place in China, Japan, Korea, Israel, the Philippines, and Taiwan for domestic markets and international imports (Haas et al. 2010). The Indian regulation currently applies only to exports (Wai 2009). Thailand and Malaysia have published voluntary national organic standards, and they operate government certification programs as well (Haas et al. 2010). In April 2005, the China National Organic Product Standard (CNOPS) came into force (Haas et al. 2010). Out of a total of 157 CB listed for Asia in The Organic Standards (TOS) Certification Directory 2008 (Wai 2009), 140 are found in just five countries: Japan [60]; South Korea [32]; China [29]; India [13]; and Thailand [6], the countries in Asia with the largest market size and largest areas in organic production (Haas et al. 2010).

There are three major organic CB in Thailand, namely Northern Organic Standards Organization (NOSO), the Organic Agriculture Certification Thailand (ACT), and the Organic Crop Institute (Pattanapant and Shivakoti 2009). The first two are private organizations, while the latter is a government agency under the auspices of the Department of Agriculture. ACT and the Organic Crop Institute are nationally registered, and the organic products that are certified by these bodies are sold widely in domestic and foreign markets (Pattanapant and Shivakoti 2009).

Organic certification, which is yet to be practiced in Nepal, is too expensive for small farmers to pay for it. The consumers who understand this reality would be willing to pay more for certified and labeled organic vegetables, instead of wondering whether the food they consume is really organic. The organic farming and certification in National Agriculture Policy 2061 strengthens this sector (Bhatta 2009).

The Turkish Ministry of Agriculture has been supported by the Netherlands during the last few years in building up its organic certification (Van Leeuwen et al. 2008). Many people have their own vegetable gardens and do not realize that they are producing organic vegetables. The supermarkets in the big cities have started to sell organic products (Van Leeuwen et al. 2008). These organic products are more expensive than conventional products. The organic market in Turkey is very small, but has a good potential to grow according to Turkish companies.

CB in Thailand fall into 3 major categories: Thai government bodies, Thai private entities, and foreign entities. About half of the organic farmlands were certified by foreign companies in 2004. The Department of Agriculture offers a free certification service, but there is currently only one Thai-owned private certification body. Organic accreditation has been offered by the National Bureau of Food and Commodity Standards since 2004 Ellis et al. 2006. Thailand's organic sector is small but has also grown very rapidly over the past decades in line with global trends Ellis et al. 2006. In August 2004, ACFS received the first application for organic accreditation from Organic ACT. The Organic ACT, which was founded in 1995, is an independent private certification body. Its members are producer organizations, consumer groups, NGOs, environmentalists, and academics Ellis et al. 2006. ACT was the pioneer and is still the only organic certification body offering internationally recognized organic certification services that belong to Thailand.

The Japanese Ministry of Agriculture, Forestry and Fisheries (MAFF) established the Japanese Agricultural Standard (JAS) System in 1950. This organization governs all the agricultural and forestry products, with the exception of liquors, drugs/quasi-drugs, and cosmetics (Haas et al. 2010). The JAS standards for organic plants and organically processed foods are based on the FAO guidelines for the production, processing, labeling, and marketing of organically produced foods.

1.9.5 Europe

Within Europe, Austria is an organic pioneer and hosts one of the most developed consumer markets worldwide of organic produce (Haas et al. 2010). The per capita consumption of 56 euro (US$ 74) is one of the highest worldwide as is the market share of organic food, which is about 6 % of total food sales (Haumann 2007; Richter and Padel 2007). 87 % of Austrian consumers purchase organic food products at least occasionally (AMA Marketing 2007).

1.9.6 North America

In the city of Ontario, Canada, there are several CB that certify organic farms and food-processing enterprises. Inspection is usually carried out on farms in the year before certification, and operators must apply to their certification body at least 15 months preharvest for organic products. In the year in which certification is envisaged to be carried out, application should be made to the certification body in the early spring to allow ample time for the review of the applications and to allow the CB to arrange for inspections during the following planting season. Inspectors for organic certifications are independent from the CB, and most are members of the Independent Organic Inspectors Association (IOIA). Under the new regulations, all accredited CBs must strictly follow suit with the Canadian standard as the minimum requirement for all organic certification agencies in Canada.

The NOP is the federal regulatory body governing organic food in the USA. The regulations are governed by the USDA through the NOP under this act (Haas et al. 2010). It covers in detail all aspects of food production, processing, delivery, and retail sale. Producers with yearly sale not exceeding US $5000 are exempt and do not require certification. A USDA Organic seal identifies products with at least 95 % organic ingredients (Haas et al. 2010).

Haas et al. (2010) stated that there are about 56 US domestic certification agencies accredited by the USDA, among which are Organic Crop Improvement Association, California Certified Organic Farmers (CCOF), Quality Assurance International (QAI), and Indiana Certified Organic. About 41 accredited international agencies have also been recorded that offer organic certification services.

1.9.7 Challenges in Organic Certification in USA

California is ahead of all the states with the highest number of organic farms, land in organic production, and organic sales (Klonsky 2010). A survey carried out on a group of organic farmers indicated that 38 % felt that their most important challenges were regulatory problems. These included paperwork and record-keeping for certification, inspections, finding a certifier, and the cost of certification (Klonsky 2010). About one-fourth of organic farmers in California said that price issues (low premiums, lack of price information, or inconsistent prices) or market access (too much competition, not enough volume produced, or lack of buyers) were their greatest challenges.

1.9.8 South America

Certification organizations such as Organic Crop Improvement Association (OCIA) and Farm Verified Organic (FVO) from USA and Naturland, BCS Oeko-Garantie, and the Institute fur Marktoekologie (IMO) from Europe are very active in the region (Garibay and Ugas 2009).

Others are Ecocert, Control Union, and Ceres. Some national CB are very well developed, such as Argencert and Organización Internacional Agropecuaria (OIA, Argentina), Instituto Biodinamico (Brazil), Bolicert (Bolivia), and Biolatina (Peru and others). Other certification agencies include Ecológica from Costa Rica, BioNica from Nicaragua, Maya Cert from Guatemala, and CertiMex from Mexico. Uruguay has Urucert and Sociedad de Consumidores de Productos Biológicos (SCPB). Apart from the aforementioned, Argencert, Argentina, has more than 12 certification agencies, i.e., OIA, Mokichi Okada (MOA), Bio Letis, Food Safety, Ambiental, Fundación, and Agro Productores Organicos de Buenos Aires (APROBA) (Garibay and Ugas 2009).

1.10 Organic Certification Agencies

Organic food is produced under regulations and certification processes. There are different levels of regulations being implemented (Kahl et al. 2012). In the European Union, the process of organic production is regulated through EC Regulations 834/2007 and 889/2008. In addition, there are also regulations from local institutions (Bioland, Demeter, Naturland) and international institutions (e.g., Codex Alimentarius of FAO, World Health Organization, and the United Nations). Where a national regulation is not in existence or is not holistically implemented, organic certification may adhere to foreign rules (such as EU standards) or the standards set up by international organizations such as IFOAM. As a result of this situation, a certification body may be accredited by several agencies, thus being able to issue different types of organic certification. Despite the conventional purchase criteria (price, quality, and quantity), the organic food value chain pays specific attention to transport distances, reliability of certification, and the production itself (Haas et al. 2010).

1.11 Advantages of Organic Farming and Certification Process

If farmers can adhere to the production standards applied in OA, this development will pave the way for new marketing opportunities (Wollni et al. 2010). Given that organic production standards prohibit the use of synthetic fertilizers, it becomes imperative for farmers to follow alternative agricultural practices in order to maintain soil fertility, such as application of manure (Wollni et al. 2010). Conservation practices that restore soil functions and build soil organic matter also help to promote a gradual increase in organic production system productivity

(IFOAM 2006). Therefore, OA standards are likely to encourage the adoption of soil conservation practices. Bolwig et al. (2009) found in their study of organic smallholders in Uganda that certified organic farmers are more likely to apply various soil management techniques including mulching and manure applications (Wollni et al. 2010). In recent years, consumers in many regions have shown an increasing interest in the environmental and health-related aspects of food production, as shown in a substantial growth of markets for organic products (FAO 2000; Hobbs et al. 2001). Especially in major North American and European markets, annual growth rates of organic product segments have been as high as 20 % (Raynolds 2004). On the other hand, organic farmers may face specific challenges when implementing conservation practices (Wollni et al. 2010), especially during their initial establishment, as some conservation practices require more than one year. In most countries, there is a minimum of three years required using organic practices prior to organic certification (Wollni et al. 2010). Problems such as persistent weeding; require a substansial amount of labour input for manual weeding since farmers cannot apply synthetic weed killers are a particular challenge in organic farming (Wollni et al. 2010).

1.12 Stakeholders in Organic Agriculture—Knowledge Management

In organic farming, management knowledge is quite essential. Organic farming partially replaces external inputs by knowledge and information (Schmid et al. 2009). Attention is required to a broad range of factors such as consumer needs, producer's availability, potential processors, and retailers of organic food, but also knowledge of certifying bodies, governments, and other policy makers (Schmid et al. 2009). Local governments often support farmers' markets by providing the market infrastructure and advertising. It is important that long-term methods for the support of these efforts are developed (Panneerselvam et al. 2012). Therefore, grassroots organizations, governments, and research institutions need to cooperate in providing information, advice, training, and financial support during the conversion of small farms to organic. Also, marginal and small farmers should be encouraged to form organic farmers' associations, such as in the case area of Tamil Nadu, to facilitate the marketing of a variety of crops whether it is certified organic or not—at the local market. This could help to increase farm profit and reduces the susceptibility of market access for specific certified organic products. The type of external support offered to farmers either from contracting companies or from other stakeholders are major factors that determine the term of access to certified OA (Panneerselvam et al. 2012).

1.13 Conflicting Studies in Organic Agriculture

OA has often been advocated in the mass media as an alternative to supposedly unsafe and environmentally harmful conventional agriculture practices (Cahill et al. 2010). However, the available studies are quite conflicting to convince anyone who does not patronize organic, in that any differences are significant (Diver 2001). A report filed at the Institute of Food Technologists (IFT) says that the available scientific information is insufficient to prove that foodborne pathogens are killed during the process of pit composting and soil application (IFT 2002). The research to date supports the theory that the benefits of increasing consumption of fruits and vegetables far outweigh any negligible benefits that could be obtained through organic foods per se or any danger that may exist from residues of conventional agrochemicals (Ames and Gold 1997).

1.14 Conclusion

Organic farming and certification is receiving increasing attention in many parts of the world. There are many challenges that lie ahead in obtaining quality produce in compliance with the organic standards and regulations if the principles of health, environment, care, and economy would be sustained. Some of the constraints depend on the geographical region, while others affect organic farmers all around the world. Behind the billion dollars of markets and the millions of hectares in organic production today, there are growers and consumers who are deliberately opting for cleaner and safer goods that are produced with regard to the welfare of people and animals and with minimal impact on the environment.

In developing countries, the governing bodies play a major role in fostering the adoption of OA. With support from government and other stakeholders, researchers and extension workers are able to obtain funding and remain committed to OA research. An important factor that will enable OA to usefully contribute to food security is the attitude of decision-makers. OA must be discussed with an open mind, bearing the potential benefits and limitations. It is suggested that governing bodies should encourage objective debates on the potential benefits of OA and identifying the circumstances where it can be best applied. It is often believed that OA is easier to undertake under certain conditions, especially where the situation is good for agriculture in general, such as on fertile soils. Nevertheless, consideration should be given to the locations and circumstances that most suits the development of OA.

Acknowledgment Authors wish to thank Dr. Clinton Shock, Prof. George E Boyhan, Joshua Ibukun Raji, and Darleen Gunther for reviewing manuscript and valuable suggestions.

References

Aksoy U, Engiz M (2007) Country report on organic farming in Turkey. Unpublished report, May, General Directorate of Agricultural Production Development (TÜGEM), MARA, Ankara

Alkhalifa B, Azza E, Elhag A (2014) Organic farming in Khartoum. IFOAM Organic World Congress 2014 Istanbul, Turkey, 13–15 Oct 2014

Alroe HF, Noe E (2008) What makes organic agriculture move-protest, meaning or market? A polyocular approach to the dynamics and governance of organic agriculture. Int J Agric Resour Gov Ecol (IJARGE) 7(1/2):5–22

Altieri M (1987) Agroecology: the science of sustainable agriculture. Westview Press, Boulder

Altieri M, Nicholls CI (2004) Biodiversity and pest management in agroecosystems, 2nd edn. CRC Press, Boca Raton

AMA Marketing (2007) Roll AMA Motivanalyse-Bioprodukte. Available at: http://www.ama-marketing.at/home/groups/7/Konsumverhalten_Bio.pdf. Accessed July 2015

Ames BN, Gold LS (1997) Environmental pollution pesticides and the prevention of cancer: misconceptions. Fed Am Soc Exp Biol J 11(13):1041–1052

Anderson K, Nielsen CP (2000) GMOs food safety and environment: what role or trade policy and the WTO? Centre for International Economic Studies Policy Discussion Paper No 0034 Adelaide, SA. Available at: www.adelaide.edu.au/cies/0034.pdf. Accessed 12 Feb 2005

Anobah R (2000) Development of organic market in West Africa. In: Alföldi T, Lockeretz W (eds) IFOAM 2000—the World grows organic. Proceedings of the 13th international IFOAM scientific conference. Convention Centre Basel, Switzerland 28–31 August 2000. vdf Hochschulverlag an der ETH Zurich, Basel. p 433

Babadov gan G, Koç D (2004) Organic agriculture in Turkey 2004, Country Reports, Turkey, Organic-Europe, FiBL. http://www.organic-europe.net/country_reports/turkey/default.asp

Bakewell-Stone P, Lieblein G, Francis C (2008) Potentials for organic agriculture to sustain livelihoods in Tanzania. Int J Agric Sustain 6(1):22–36

Balfour E (1977) Towards a sustainable agriculture: the living soil. In: IFOAM conference Switzerland. http://www.soilandhealth.org/01aglibrary/010116balfourspeech.html. Accessed 24 July 2015

Beban-France A (2008) Organic agriculture: an empowering development strategy for small-scale farmers? A Cambodian case study. Institute of Development Studies, Master thesis, Massey University, Aotearoa New Zealand [Unpublished]

Benbrook C, Zhao X, Yanez J, Davies N (2008) State of science review: nutritional superiority of organic foods. The Organic Center, University of Arizona. http://www.organiccenter.org. Accessed 25 July 2015

Bergström L, Kirchmann H, Thorvaldsson G (2008) Widespread opinions about organic agriculture–are they supported by scientific evidence? In: Organic crop production–ambitions and limitations, Springer, Netherlands, pp 1–11

Bhatta GD, Werner D, Krishna Bahadur KC (2009) Potentials of organic agriculture in Nepal. J Agric Environ 10:1–14

Biofach (2011) Global organic sales approaching 60 Billion US Dollars. Biofach Newsletter, 239. 7 January

Blend JR, Van Ravenswaay EO (1999) Measuring consumer demand for eco labeled apples. Am J Agric Econ 81(5):1072–1077

Bolwig S, Gibbon P, Jones S (2009) The economics of smallholder organic contract farming in Tropical Africa. World Development (forthcoming). doi:10.1016/j.worlddev.2008.09.012

Burton S (2006) Contrasting organic and conventional food products: consumers' subjective perceptions and objective evaluations of nutrition and taste. J Int Soc Bus Disciplines Spring pp 37–47

Cahill S, Morley K, Powell DA (2010) Coverage of organic agriculture in North American newspapers: media: linking food safety, the environment, human health and organic agriculture. Br Food J 112(7):710–722

Chang H, Zepeda L (2005) Consumer perceptions and demand for organic food on Australia: focus group discussions. Renew Agric Food Syst 20(3):155–167

Conford P (2001) The origins of the organic movement. Floris Books, Glasgow

Curl CL, Fenske RA, Elgehun K (2003) Organophosphorus pesticide exposure of urban and suburban preschool children with organic and conventional diets. Environ Health Perspect 111 (3):377–382

Dainello, FJ (1998) Soil pH—how to make it work for you. Texas Gardener. Jan/Feb. pp 40–42

Defra (2004) Action plan to develop organic food and farming in England—two years on. Department of Environment Food and Rural Affairs, London. Department Food and Agriculture Organization of the United Nations, Rome, 2007

Demiryürek K (2000) The Analysis of information systems for organic and conventional hazelnut producers in three villages of the Black Sea Region, Turkey. PhD thesis, University of Reading, Reading

Demiryürek K (2004) Organic agriculture in the world and Turkey' (in Turkish). J Fac Agric Harran University 8(3–4):63–71

Demiryürek K, Stopes C, Güzel A (2008) Organic agriculture: the case of Turkey. Outlook Agric 37(4):261–267

Department of Agriculture, Fisheries and Food (DAFF) (2002) Census of Irish Organic Production 2002. Department of Agriculture and Food, Dublin, Ireland

Diver S (2001) Resource guide to organic and sustainable vegetable production. Appropriate Technology Transfer for Rural Areas (ATTRA) Fayetteville, AR. www.attra.ncat.org/horticultural.html. Accessed 3 May 2003

Durham CA, Andrade D (2005) Health vs. environmental motivation in organic preferences and purchases. American Agricultural Economics Association Annual Meeting Providence, Rhode Island July 24–27

East African Community (2007) East African organic products standard. Arusha, Tanzania

Ecobichon DJ (1996) Toxic effects of pesticides. In: Klaassen CD, Doull J (eds) Casarett and Doull's toxicology: the basic science of poisons, 5th edn. MacMillan, NewYork

Ellis W, Panyakul V, Vildozo D, Kasterine A (2006) Strengthening the export capacity of Thailand's organic agriculture. Geneva: International Trade Centre

FAO (Food and Agriculture Organization of the United Nations) (2000) Food safety and quality as affected by organic farming. FAO, Rome

FAO (Food and Agriculture Organization of the United Nations) (1999) Codex alimentarius commission guidelines for the production, processing, labelling and marketing of organically produced foods cac/gl 32–1999

Fearne A (2008) Organic fruit and vegetables—who buys what and why and do we have a clue? Kent Business School University of Kent, Canterbury

FSA (2009) Comparison of putative health effects of organically and conventionally produced foodstuffs: a systematic review. Food Standards Agency, London

Fuller R, Nortin L, Feber RE (2005) Benefits of organic farming to biodiversity vary among taxa. Biol Lett 1(4):431–434

Garibay SV, Ugas R (2009) Organic farming in Latin America and the Caribbean. The world of organic agriculture-statistics and emerging trends 2009. In: Willer H, Lukas K (eds) The world of organic agriculture—statistics and emerging trends 2010. IFOAM Bonn and FiBL, Frick, pp 176–185

Gliessman SR (2007) Agroecology: ecological processes in sustainable agriculture, 2nd edn. Lewis Publisher, Boca Raton

Goh KM (2011) Greater mitigation of climate change by organic than conventional agriculture: a review. Biol Agric Hortic 27(2):205–229

Goklany IM (2002) The ins and outs of organic farming. Science 298:1889

Gold MV (2007) Organic production/organic food: information access tools. Alternative farming systems information center. http://www.nal.usda.gov/afsic/pubs/ofp/ofp.shtml. Accessed 13 Aug 2015

Gold MV, Gates JP (2007) Tracing the evolution of organic/sustainable agriculture: a selected and annotated bibliography. http://www.nal.usda.gov/afsic/pubs/tracing/tracing.shtml. Accessed 20 Jan 2010

Gomiero T, Pimentel D, Paoletti M (2011) Is there a need for a more sustainable agriculture? Crit Rev Plant Sci 30:6–23

Gracia A, De Magistris T (2007) Organic food product purchase behavior: a pilot study for urban consumers in the South of Italy. Span J Agric Res 5(4):439–451

Green E (2004) Organic growing is best. Organic Consumers Association. http://www.organicconsumers.org/Organic/orgapples.cfm. Accessed 23 April 2015

Grunert KG, Jeppesen LF, Jespersen KR (2005) Market orientation of value chains. Eur J Mark 39:428–455

Güzel HT (2001) The production, export and development possibilities of ecological agricultural products in the world and Turkey (in Turkish), Publication No 2001–14. Istanbul Chamber of Commerce, Istanbul

Haas R, Canavari M, Slee B (2010) Looking east, looking west: organic and quality food marketing in Asia and Europe. Wageningen Academic Publishers, Wageningen, The Netherlands. ISBN 978-90-8686-095-1

Halberg N, Alrøe HF, Kristensen ES (2006) Synthesis: prospects for organic agriculture in a global context. In Halberg N, Knudsen KT, Alrøe HF. CABI Publishing

Harris F, Robinson GM, Griffiths I (2008) A study of the motivations and influences on farmers' decisions to leave the organic farming sector in the United Kingdom. In: Robinson GM (ed) Sustainable rural systems: sustainable agriculture and rural communities. Ashgate, Aldershot

Haumann B (2007) Organic farming in North America. In: Willer W, Yussefi-Menzler M, Sorensen N (eds) The world of organic agriculture. Statistics and emerging trends 2008. FIBL-IFOAM Report. IFOAM, Bonn; FiBL, Frick

Hobbs JE, Kerr WA, Phillips WB (2001) Identity preservation and international trade: signaling quality across national boundaries. Can J Agric Econ 49:567–579

Hoeppner JW, Entz MH, McConkey BG (2006) Energy use and efficiency in two Canadian organic and conventional crop production systems. Renew Agric Food Syst 21:60–67

Hole DG, Perkins AJ, Wilson JD (2005) Does organic farming benefit biodiversity? Biol Cons 122:113–130

Howard A (1943) An agricultural testament. Oxford University Press, London. http://journeytoforever.org/farmlibrary/howardAT/ATtoc.html

Hughner R, McDonagh P, Prothero A (2007a) Who are organic food consumers? A compilation and review of why people purchase organic food. J Consum Behav 6(2/3):94–110

Hughner RS, McDonagh P, Prothero A, Shultz CJ, Stanton J (2007b) Who are organic food consumers? A compilation and review of why people purchase organic food. J Consum Behav 6:1–17

IFOAM (2006) The four principles of organic farming. The International Federation of Organic Agriculture Movements Bonn, Germany. www.IFOAM.org. Accessed 6 May 2015

IFOAM (2011) Summary strategic plan. International Federation of Organic Agriculture Movements (IFOAM), Bonn

IFT (2002) IFT expert report on emerging microbiological food safety issues: implications for control in the 21st Century. Available at: www.ift.org/pdfs/microfs/webreport.pdf. Accessed 14 April 2003

IMF (2004) The fund's strategic direction—preliminary considerations. Report by The Secretary to the International Monetary Committee on the Fund's Strategic Direction, 30 September: 2

International Federation of Organic Agricultural Movements (IFOAM) (2002) Basic standards for organic production and processing approved by the IFOAM general assembly. Victoria, Canada

IPCC (Intergovernmental Panel on Climate Change) (2007) Synthesis report. In: Abdelkader A, Roxana B, Sandra D, et al (eds). An assessment report of the intergovernmental panel on climate change. https://www.ipcc.ch/pdf/assessment-report/ar4/syr/ar4_syr.pdfAccessed. 13Aug 2015

Jitsanguan T (2001) Development of sustainable agriculture. Paper presented at the policy meeting of Ministry of Agricultural and Cooperatives, Office of Agricultural Economics, Bangkok, 30 August
Jones GG (2012) Entrepreneurship in the natural food and beauty categories before 2000: global visions and local expressions. Harvard Business School Entrepreneurial Management Working Paper (13–024)
Kahl J, Baars T, Bügel S (2012) Organic food quality: a framework for concept, definition and evaluation from the European perspective. J Sci Food Agric 92(14):2760–2765
Kaltoft P, Risgaard ML (2006) Has organic farming modernized itself out of business? Reverting to conventional methods in Denmark. In: Holt G, Reed M (eds) Sociological perspectives of organic agriculture: from pioneer to policy. CAB International, Wallingford
Kenanovglu Z, Karahan O (2002) Policy implementations for organic agriculture in Turkey. Br Food J 104(3–5):300–318
Kerselaers E, De Cock L, Ludwig L, Huylenbroeck GV (2007) Modelling farm-level economic potential for conversion to organic farming. Agric Syst 94(3):671–682
Khanal RC (2009) Climate change and organic agriculture. J Agric Environ 10:116–127
Kilcher L (2007) How organic agriculture contributes to sustainable development. In: Willer H, Yussefi M (eds) The world of organic agriculture. Statistics and emerging trends 2007. FIBL-IFOAM Report 9th. ed IFOAM, FiBL, Bonn, Frick, pp 82–91
Kirchmann H, Thorvaldsson G, Bergström L (2008) Fundamentals of organic agriculture–past and present. In: Organic crop production–ambitions and limitations. Springer, Netherlands, pp 13–37
Kirner L, Vogel S, Schneeberger W (2006) Intended and actual behavior of organic farmers in Austria after a five-year commitment period. Renew Agric Food Syst 21:95–105
Kledal PR, Mahmoudi H, Damghani AMM (2012) Organic food and farming in Iran, The world of organic agriculture-statistics and emerging trends, pp 184–189
Klonsky K (2010) A look at California's organic agriculture production. Agric Resour Econ Update 14(2):8–11
Koch K (1998) Food safety battle: organic vs. biotech, vol 8, no 33. CQ Researcher, CQ Press, University of Maryland McKeldin Library, CollegePark, MD, pp 761–784
Koepf HH (1976) Biodynamic agriculture: an introduction. The Anthroposofic Press, Spring Valley, New York. Available at: http://continuousharvest.com/Documents/Bio-Dynamic%20Agriculture%20book.pdf
Koepf HH (2006) The biodynamic farm. Steiner Books, Dulles
Koesling M, Løes AK, Flaten O (2012) Farmers reasons for deregistering from organic farming. Organ Agric 2(2):103–116
Kramer SB, Reganold JP, Gloverf JD (2006) Reduced nitrate leaching and enhanced denitrifier activity and efficiency in organically fertilized soils. PNAS 103:4522–4527
Kristiansen P, Merfield C (2006) Overview of organic agriculture organic agriculture: a global perspective. Csiro Publishing, Collingwood, p 484
Kristiansen P, Taji A, Reganold J (2006) Organic agriculture. a global perspective. CSIRO Publishing, Collingwood
Kroupová Z (2009) Organic farming in less favoured areas. In Think Together [CDROM] Prague: Czech University of Life Sciences. ISBN 978-80-213-1906-6
Kuhar A, Juvancic L (2005) Modelling consumer preferences towards organic and integrated fruits and vegetables in Slovenia. In: 97th EAAE Seminar on 'The Economics and Policy of Diet and Health'. Reading, UK
Lairon D (2010) Nutritional quality and safety of organic food. A review. Agron Sustain Dev 30 (1):33–41
Lampkin N (2002) Organic farming. (Revised edition). Old Pond Publishing, Suffolk
Läpple D (2010) Adoption and abandonment of organic farming: an empirical investigation of the Irish drystock sector. J Agric Econ 61(3):697–714

Lea E, Worsley T (2005) Australians' organic food beliefs, demographics and values. In: McEachern MG, Willock, J Producers and consumers of organic meat: a focus on attitudes and motivations. British Food Journal 106(7):534–552

Letourneau DK, Bothwell SG (2008) Comparison of organic and conventional farms: challenging ecologists to make biodiversity functional. Front Ecol Environ 6(8):430–438

Liebig MA, Morgan JA, Reeder JD (2005) Greenhouse gas contributions and mitigation potential of agricultural practices in northwestern USA and western Canada, pp 25–52

Lockie S, Lyons K, Lawrence G (2004) Choosing organics: a path analysis of factors underlying the selection of organic food among Australian consumers. Appetite 43:135–146

Lockie S, Lyons K, Lawrence G, Mummery K (2002) Eating green: motivations behind organic food consumption in Australia. Sociologia Ruralis 42(1):23–40

Lotter DW (2003) Organic agriculture. J Sustain Agric 21(4):59–128

Loureiro ML, Mccluskey JJ, Mittelhammer RC (2001) Assessing consumer preferences for organic eco-labeled and regular apples. J Agric Resour Econ 26(2):404–416

Lynch D, Truro N (2009) Environmental impacts of organic agriculture: a Canadian perspective. Can J Plant Sci 87:1055–1066

Mader P, Fließbach A, Dubois D (2002) Soil fertility and biodiversity in organic farming. Science 296:1696–1697

Malá Z (2011) Efficiency analysis of Czech organic agriculture. E+M Ekonomika A Manage 1:14–28

McNeely JA, Scherr SJ (2001) Common ground, common future. How eco-agriculture can help feed the world and save wild biodiversity. IUCN and Future Harvest, May 2001

Meadows DH, Meadows DL, Randers J (1972) Limits to growth, a report of the club of Rome's project on predicament of mankind, Universe Books, New York, 205 p

Mendoza TC (2002) Comparative productivity, profitability and energy use in organic, LEISA and conventional rice production in the Philippines, vol 14, no 6. Livestock Research for Rural Development. http://www.cipav.org.co/lrrd/lrrd14/6/mend146.htm. Accessed 15 Mar 2015

Millock K, Wier M, Andersen LM (2004) Consumer's demand for organic foods-attitudes, value and purchasing behaviour. In: Selected paper for presentation at the XIII annual conference of European association of environmental and resource economics, Budapest, Hungary, 25–28 June

Muller A, Olesen JE, Davis J (2012) Reducing global warming and adapting to climate change: the potential of organic agriculture. Department of Economics School of Business, Economics and Law at University of Gothenburg Vasagatan 1, PO Box 640, SE 405 30 Göteborg, Sweden

Müller-Lindenlauf M (2009) Organic agriculture and carbon sequestration. Possibilities and constrains for the consideration of organic agriculture within carbon accounting systems. Natural Resources Management and Environment Department, Food and Agriculture Organization of the United Nations, Rome

Nelson E, Tovar LG, Rindermann RS (2009) Participatory organic certification in Mexico: an alternative approach to maintaining the integrity of the organic label. Agric Hum Values 27 (2):227–237

Niggli U (2009) Organic agriculture: a productive means of low-carbon and high biodiversity food production. Environment review, p 112

Norse D, Tschirley J (2003) Agriculture and the environment: changing pressures, solutions and trade-offs. In: Bruinsma J (ed) World agriculture: towards 2015/2030. An FAO perspective. Earthscan Publications and Food and Agriculture Organization of the United Nations, London, pp 331–356

Northbourne L (1940) Look to the land. Basis Books, London

Olhan E, Ataseven Y, Gün S (2005) Organic farming in Turkey. Pak J Biol Sci 8(3):505

Onyango B, Hallman W, Bellows A (2006) Purchasing organic food in U.S. food systems: a study of attitudes and practice. AAEA Annual Meeting, July 23–26

Pacific Organic Standard (2008) Secretariat of the Pacific Community Noumea, New Caledonia. http://www.ifoam.bio/sites/default/files/page/files/pacific_organic_standard.pdf. Accessed 13 August 2015

Paddock C (2007) Organic food is more nutritious say EU researchers. Medical News Today, 16 Apr 2009
Pamela C (2012) Guide for organic crop producers. National Center for Appropriate Technology (NCAT) Agriculture Specialist, 64pages. https://attra.ncat.org/attra-pub/summaries/summary.php?pub=67. ATTRA USDA. Accessed 16 April 2015
Panneerselvam P, Halberg N, Vaarst M et al (2012) Indian farmers' experience with and perceptions of organic farming. Renew Agric Food Systain 27(02):157–169
Parrott N, Marsden T (2002) The real green revolution. Greenpeace Environmental Trust, London, pp 11–16
Pattanapant A, Shivakoti GP (2009) Opportunities and constraints of organic agriculture in Chiang Mai province, Thailand. Asia-Pacific Develop J 16(1):115
Paull J (2008) The lost history of organic farming in Australia. J Organ Syst 3(2):2–17
Paull J (2010a) From France to the world: the international federation of organic agriculture movements (IFOAM). J Soc Res Policy 1(2):93–102
Paull J (2010b) The future of organic agriculture: otopia or oblivion? Innovative Sci Editions 1:11–14
Paull J (2011a) Attending the first organic agriculture course: Rudolf Steiner's agriculture course at Koberwitz, 1924. Eur J Soc Sci 21(1):64–70
Paull J (2011b) Biodynamic agriculture: the journey from Koberwitz to the world, 1924–1938. J Organ Syst 6(1):27–41
Paull J (2011c) The Betteshanger summer school: missing link between biodynamic agriculture and organic farming. J Organ Syst 6(2):13–26
Paull J (2011d) Organics Olympiad 2011: global indices of leadership in organic agriculture. J Soc Dev Sci 1(4):144–150
Peace (2014) USDA Foreign agricultural service. Global agricultural information. http://gain.fas.usda.gov/Recent%20GAIN%20Publications/Organic%20Agriculture%20in%20Nigeria_Lagos_Nigeria_6-5-2014.pdf. Accessed online 16 April 2015
Pearson D, Henryks J, Moffitt Z (2007) What do buyers really want when they purchase organic foods? Online J Organ Syst 2(1):1
Pearson D, Henryks J, Jones H (2011) Organic food: what we know (and do not know) about consumers. Renew Agric Food Syst 26(02):171–177
Pivato S, Misani N, Tencati A (2008) The impact of corporate social responsibility on consumer trust: the case of organic food. Bus Ethics: A Eur Rev 17(1):3–12
Pornpratansombat P (2006) Thai organic farming: explicit knowledge and empirical study. Kasetsart J Soc Sci 27(2):264–277
Raynolds LT (2004) The globalization of organic agro-food networks. World Dev 32(5):725–743
Reganold JP, Glover JD, Preston KA, Hinman HR (2001) Sustainability of three apple production systems. Nature 410:926–930
Rehber E, Turhan S (2002) Prospects and challenges for developing countries in trade and production of organic food and fibers: the case of Turkey'. Br Food J 104(3–5):371–390
Rice CW, Owensby CE (2001) Effects of fire and grazing on soil carbon in rangelands. In R. Follet et al (eds) The potential of U.S. grazing lands to sequester carbon and mitigate the greenhouse effect. Lewis Publishers, Boca Raton, pp 323–342
Richter T, Padel S (2007) The European market for organic food. In: Willer H, Yussefi-Menzler M, Sorensen N (eds), The world of organic agriculture—statistics and emerging trends 2007; IFOAM Bonn, Germany pp 143–154
Rigby D, Cáceres D (2001) Organic farming and the sustainability of agricultural systems. Agric Syst 68(1):21–40
Rodale JI (1942) Organic farming and gardening, vol 1, no 1. Rodale Press Emmaus, Pennsylvania, pp 1–16
Rodale JI (1945) Pay dirt: farming and gardening with composts. Devin-Adair Co., New York
Rosenthal E (2011) Organic agriculture may be outgrowing its ideals. New York Times, New York
Sahota A (2004) Changing retail landscape for organic products. Eur Reg Rev 44:50–53

Sayin C, Brumfield RG, Özkan B (2005) The organic farming movement in Turkey. Production and marketing report, HortTechnology 15(4):864
Schmid O, Padel S, Halberg N et al (2009) Strategic research agenda for organic food and farming TP Organics, Brussels, 116 p
Sharma AB (2006) Many India farmers returning to traditional ways of organic agriculture. Financial Express, New Delhi. www.organicconsumers.org/ofgu/traditional011306.cfm. Accessed 29 Sept 2006
Sharma AK (2001) A handbook of organic farming. Agrobios India, Jodhpur
Shepherd M, Bruce P, Bill C (2009) An assessment of the environmental impacts of organic farming. www.defra.gov.uk/farm/organic/policy/research/pdf/env-impacts2.pdf. 5 Apr 2009
Sierra L, Klonsky K, Strochlic R, Brodt S, Molinar R (2008) Factors associated with deregistration among organic farmers in California. California Institute for Rural Studies, Davis
Singpornpong S (2005) Farmers' attitudes towards organic rice production in the organic agriculture project in Sri Saket province. Department of Agricultural Extension Bangkok
Sittiwong N, Varinrak B (2004) 12 years of organic rice in the provinces of Chiang Rai and Pha Yao: from Tung Lao to the European Union and world markets. Bangkok, Top Organic Product and Supply Co. Ltd
Sloan AE (2002) The natural and organic foods marketplace. Food Technol 56(1):27–37
Smith P, Martino D, Cai Z, et al (2007) Agriculture. In: Metz B, Davidson OR, Bosch PR, Dave R, Meyer LA. (eds) Climate Change 2007: mitigation. Contribution of working group III to the fourth assessment report of the intergovernmental panel on climate change, Cambridge University Press, Cambridge, 497–540
Stolze M, Piorr A, Haring A et al (2000) The environmental impacts of organic farming in organic cropping systems for grain production and soil improvement. Agron J 99(5):1297–1305
Stonehouse DP, Clark EA, Ogini YA (2001) Organic and conventional dairy farm comparisons in Ontario Canada. Biol Agric Hortic 19(2):115–125
Sununtapongsak W (2006) Organic agriculture in Thailand. www.ldd.go.th/pldweb/tech/meet7/book1/f7.doc. Accessed on 29 Dec 2006
Tancho A (2006) Applied natural agriculture: ideas, principles, and techniques in Thailand. National Science and Technology Development Agency, Patum Thani
Tate WB (1994) The development of the organic industry and market: an international perspective. In: Lampkin NH, Padel S (eds) The economics of organic farming: an international perspective. CABI Publishing, Oxfordshire, pp 11–25
Texas A&M AGRI Life extension Texas A&M University (2015) Vegetable resources. Organic vegetable production guide. Managing pests. http://aggie-horticulture.tamu.edu/vegetable/guides/organic-vegetable-production-guide/managing-pests/. Accessed 16 Apr 2015
Thompson GD (1998) Consumer demand for organic foods: what we know and what we need to know. Am J Agric Econ 80(5):1113–1118
Thompson GD, Kidwell J (1998) Explaining the choice of organic produce: cosmetic defects, prices, and consumer preferences. Am J Agric Econ 80:277–287
Torjusen H, Lieblein G, Wandel M (2001) Food system orientation and quality perception among consumers and producers of organic food in Hedmark County, Norway. Food Qual Prefer 12:207–216
Trewavas A (2004) A critical assessment of organic farming-and-food assertions with particular respect to the UK and the potential environmental benefits of no-till agriculture. Crop Prot 23 (9):757–781
Tsakiridou E, Konstantinos M, Tzimitrakalogianni I (2006) The influence of consumer characteristics and attitudes on the demand for organic olive oil. J Int Food Agribusiness Market 18(3/4):23–31
UNCTAD (2004) Country Case Study on Environmental Requirements, Market access/entry and Export Competitiveness for Horticultural Produce from Cambodia. Paper presented at the subregional workshop on environmental regulation, Bangkok, 29 September–1 October 2004

UNCTAD (2006) Organic agriculture: a trade and sustainable development opportunity for developing countries, UNCTAD Trade and Environment Review 2006. UNCTAD/DITC/TED/2005/12. United Nations New York and Geneva, 161–223

UNEP-UNCTAD (2008) Organic agriculture and food security in Africa. United Nations Conference on Trade and Development. United Nations Environment Programme UNCTAD/DITC/TED/2007/15

UNESCAP (2002) Organic agriculture and rural poverty alleviation: potential and best practices in Asia. United Nations Economic and Social Commission for Asia and the Pacific, Bangkok

Ungoed-Thomas J (2007) Official: organic is really better. Times Online. http://www.timesonline.co.uk/tol/news/uk/health/article2753446.ece. Accessed 16 Apr 2015

University of Maryland, College Park (2009) Organic food and benefits. Department of English. http://www.english.umd.edu/interpolations/1793. Accessed 15 Mar 2015

Uphoff N (2002) The agricultural development challenges we face. In: Uphoff N (ed) Agroecological innovations: increasing food production with participatory development. Earthscan Publications Ltd, London, pp 3–20

Van Leeuwen MAE, Van Der Voort MPJ, Sukkel W et al (2008) Trade opportunities for organic fruit and vegetables. Applied Plant Research Arable farming and field production of vegetables PPO no. 32500743

Wai KO (2009) Organic Asia—from back to nature movement and fringe export to domestic market trend. Organ Stan 71:3–6

Wessells CR, Johnston RJ, Donath H (1999) Assessing consumer preferences for ecolabeled seafood: the influence of species, certifier, and household attributes. Am J Agric Econ 81 (5):1084–1089

Wheeler SA (2008) What influences agricultural professionals views towards organic agriculture? Ecol Econ 65(1):145–154

Willer H, Kilcher L (2011) The world of organic agriculture: statistics and emerging trends 2011: Bonn: International Federation of Organic Agriculture Movements (IFOAM). Switzerland Research Institute of Organic Agriculture (FiBL), Frick

Willer H, Rohwedder M, Wynen E (2009) Organic agriculture worldwide: current statistics. In: Willer H, Kilcher L (eds) The world of organic agriculture. Statistics and emerging trends 2009. FIBL-IFOAM Report. IFOAM, Bonn; FiBL, Frick; ITC, Geneva, pp 25–58

Willer H, Yussefi M (2007) The world of organic agriculture. Statistics and emerging trends. FIBL, Frankfurt

Winter CK, Davis SE (2006) Organic foods. J Food Sci 71:117–124

Wollni M, Lee DR, Thies JE (2010) Conservation agriculture, organic marketing, and collective action in the Honduran hillsides. Agric Econ 41(3–4):373–384

Wynen E (1994) Economics of organic farming in Australia. In: Lampkin NH, Padel S (eds) The economics of organic farming: an international perspective. CAB International, Wallingford, pp 185–199

Wynen E (2003) What are the key issues faced by organic producers? In: Organisation for economic cooperation and development (ed) Organic agriculture: sustainability, markets and policies. CABI Publishing, Wellingford

Zepeda L, Lin J (2007) Characteristics of organic food shoppers. J Agric Appl Econ 39(1):17–28

Ziesemer J (2007) Energy use in organic food systems. Natural Resources Management and Environment. http://www.fao.org/docs/eims/upload/233069/energy-use-oa.pdf. United Nations Food and Agriculture Organization. Accessed 13 Aug 2015

Zundel C, Kilcher L (2007) Organic agriculture and food availability. In: International conference on organic systems in an extreme climate year. American Journal of Alternative Agriculture, 18:146–154

Chapter 2
Organic Strawberry Production in Tennessee, USA, and Areas of Comparable Climate in China

Suping Zhou, Sarabjit Bhatti, Shu Wei and Fur-Chi Chen

Abstract Strawberry production and consumption is experiencing significant growth. Production in the USA has doubled since the 1990s. As the domestic demand increases, there is potential to expand production beyond California and Florida into other regions of the country. In the 1950s, strawberries accounted for 25.3 % of the total marketable value of all fruits and vegetables produced in Tennessee, before declining in the 1970s. Acreage in the last 2 decades has been on a slow upward trend. Several counties throughout Tennessee are well suited for the growing of high-quality and high-value strawberries. Efforts are being made to promote production and consumption of organic and naturally grown local strawberries in Tennessee. Strawberry production in parts of China that have a similar climate has also seen considerable growth. Hebei, Shandong, Anhui, and Liaoning are among the largest strawberry producing provinces in the country. In 2014, over 50,000 acres of strawberries were planted in Anhui Province, accounting for over 15 % of all strawberries grown in China. This chapter discusses the production systems, cultural practices including nutrient management, the selection of cover crops, integrated pest management, and weed and disease control in strawberry

S. Zhou (✉) · S. Bhatti · F.-C. Chen
Agricultural Biotechnology, Department of Agriculture and Environmental Sciences, College of Agriculture, Human and Natural Sciences, Tennessee State University, 3500 John a. Merritt Blvd, 37209 Nashville, TN, USA
e-mail: zsuping@tnstate.edu

S. Bhatti
e-mail: sbhatti@tnstate.edu

F.-C. Chen
e-mail: fchen1@tnstate.edu

S. Wei
State Key Laboratory of Tea Plant Biology and Utilization, College of Tea and Food Science and Technology, Anhui Agricultural University, 130 Changjiang Blvd West, Hefei 230036, Anhui, China

D. Nandwani (ed.), *Organic Farming for Sustainable Agriculture*,
Sustainable Development and Biodiversity 9,
DOI 10.1007/978-3-319-26803-3_2

production in Tennessee and China. Harvesting and postharvesting protocols, marketing channels, safety concerns, current constraints, and future potentials of organic strawberry production are also discussed.

Keywords Strawberry · Production · Market · Organic · USA–China

2.1 Introduction

Strawberry production and consumption is experiencing significant growth. According to the Food and Agriculture Organization (FAO) of the United Nations, world production of strawberries has exceeded 4 million tons since 2007. The USA is the world's largest producer of strawberries, producing over 36 billion pounds in 2012 and contributing approximately 29 % of the global supply (Naeve 2014). Production in the USA has doubled since the 1990s. Majority of this growth has occurred in California (89 %) and some in Florida (9 %), while there has been a decline in production for the other states with commercial strawberry industries (Rom et al. 2014). Year-round availability of fresh strawberries has significantly influenced their consumption by Americans. According to the USDA, Americans are eating more fresh strawberries grown in the USA, with annual per capita consumption doubling since 2002, to almost eight pounds in 2012. According to ERS's loss-adjusted food availability data (2012), strawberries rank the fifth most commonly consumed fresh fruit in the USA after bananas, apples, watermelon, and grapes. In terms of market value, they are the fourth most valuable fruit produced in the USA and considering only the fresh-market fruit, they are second only to apples in value. As the domestic demand increases, there is potential to expand production into other regions of the country.

Tennessee used to be among the largest strawberry producers in the USA. In the 1950s, strawberries planted on approximately 15,000 acres, accounted for 25.3 % of the total marketable value of all fruits and vegetables produced in Tennessee (Goble 1961). They were grown all over the state, particularly the Sale Creek/Soddy Daisy area near Chattanooga, Portland, and the surrounding area, and near Fruitland in Gibson County. Strawberries were celebrated as an important part of the local economy. The First Annual West Tennessee Strawberry Festival was held in Humboldt in 1934 and the Middle Tennessee Strawberry Festival in Portland goes back to the early 1940s when strawberries were the mainstay of this area. Portland, nicknamed the 'strawberry capital of Tennessee,' held that distinction until the 1960s when industrialization changed agriculture. By the mid-1970s, total acreage had dropped to just above 200 acres. Anthracnose infestation forced closure of several Arkansas and North Carolina nurseries which supplied plants (plugs) to the Tennessee growers. Acreage in the last 2 decades has been on a slow upward trend, though Tennessee does not rank in the top 10 states in strawberry production. The 2012 USDA agriculture census report lists the acreage harvested as 253 acres, up from 194 acres in 2007.

Strawberry production is scattered across the state of Tennessee. The top 5 counties in Tennessee in terms of strawberry acreage in 2002 were Unicoi, Washington, Rhea, Hawkins, and Bledsoe (Bost et al. 2003). Several other counties such as Wayne and Gibson in West Tennessee; Bedford, Coffee, Franklin, Montgomery, Moore, Robertson, Rutherford, Sumner, Warren, Williamson, and Wilson counties in Middle Tennessee; and Anderson, Bradley, Grainger, and Knox in East Tennessee; and many more throughout the state are well suited for the growing of high-quality and high-value strawberries. In East and Middle Tennessee, tobacco and row crop farmers have been looking for new ways to supplement farm income due to the uncertainty and changes occurring in tobacco production. These and other small farmers in Tennessee are open to new ideas that provide a better chance at making a living and keeping the farm in the family (Friedman 2007). Growing organic strawberries is one of their options.

US organic strawberry acreage is increasing in response to the market demands. For the past several years, strawberries have been on the 'dirty dozen' list released by the Environmental Working Group, a nonprofit organization which lists the top 12 types of conventionally grown fresh produce containing the highest amount of pesticide residues. Their recommendation is to look for organic strawberries to avoid exposure to a battery of toxins. Organic production has the potential to increase profits by providing consumers with locally grown, high-quality organic products in a rapidly growing market where the demand is greater than the supply.

Organic strawberries sold in grocery stores across Tennessee are mostly shipped from California, where organic strawberry sales have grown from $2 million in 1997 to over $63 million in 2011 (USDA/NASS). Organic certification is a vital part of ensuring that consumers are confident in the products they buy and trust that they meet USDA's organic requirements. Through the National Organic Program, USDA has helped organic farmers and businesses to achieve $35 billion annually in US retail sales. Efforts are being made to promote production and consumption of organic and naturally grown local strawberries in Tennessee. Although, the organic certification provides access to price premiums and specialty markets (Wszelaki et al. 2014), there are few growers that are certified organic. A higher percentage of farmers prefer to claim the 'naturally grown' term for their produce. They follow sustainable agricultural practices, minimizing or even avoiding the use of pesticides. As part of a project funded under the National Strawberry Sustainability Initiative (NSSI), 8 local strawberry producers in 5 Middle Tennessee counties were enlisted to participate in the initiative growing strawberries using organic practices. These farms agreed to follow sustainable strawberry production practices using preventive management to reduce problems with weeds, diseases, pests, and plant nutrition through integration of a variety of cultural, biological, and mechanical management practices. They provided 3 plots (12ft × 12ft) from which fresh strawberries were harvested for consumer preference evaluations and for studies dealing with enhancing food safety (Sect. 2.12).

In China, strawberries are cultivated on approximately 320,000 acres with a total output of over 260 million pounds in 2015. Hebei, Shandong, and Liaoning are the largest strawberry producing provinces in the country, each with nearly 25,000 acres of cultivation, followed by 20,000 acres in Jiangsu and 10,000 acres in

Anhui. Strawberry producing counties include Mancheng in Hebei, Dandong in Liaoning, Yantai in Shandong, and Changfeng in Anhui. At present, strawberries produced in China are mainly for domestic consumption and for frozen product export. Strawberry yield in China has great potential for improvement comparable to that in the USA, Japan, and Italy (Yan 2010). As of 2014, over 50,000 acres of strawberries are planted in Anhui Province, accounting for over 15 % of all strawberries grown in China (Liao et al. 2014). Changfeng with more than 30,000 acres (Yu et al. 2013) is the largest strawberry producer in Anhui.

Organic strawberry production in China began around 2000 in Liaoning Province. Although strawberries in Changfeng are largely certified as 'green food,' conventionally grown strawberries are still dominant in Anhui and other strawberry producing provinces. Until 2014, organic strawberries were planted on about 1300 acres, with total market value around $120 million. This high value per acre suggests great demands for organic strawberries in China. There were over 800 acres of organic strawberry production in Liaoning alone, which accounted for two-thirds of all organic strawberries in China (Liu 2014).

2.2 Cultural Management Systems

Since locally produced strawberries are an important cash crop in many Tennessee counties, it is imperative to maximize the yields and profits and be able to compete with the commercial growers. Selecting the best production system greatly affects the potential profit. There are basically two types of production systems utilized in Tennessee: plasticulture and the matted-row system. High Tunnel production has not gained much popularity among the small strawberry farmers, owing to the costs involved.

2.2.1 Plasticulture (Annual)

Decreased competition with weeds, less time to production, and an earlier harvest season are some of the reasons most producers would consider when using annual plasticulture. Organic as well as conventional farmers in California and Florida tend to be partial to this system (Guerena and Born 2007), even though it is quite intensive. Growers in Tennessee adopted the strawberry plasticulture production system in the late 1980s and it continues to be popular. Three-fourths of the producers participating in the project use this production system, which performs well in the milder areas of the southeast where temperatures rarely fall below 0 °F (Poling et al. 2005). It involves the planting of freshly dug bare-root plants or transplants started from runner tips (called plug plants) on raised beds covered with black plastic (Fig. 2.1a). Drip tape, buried at a depth of 2.5 in., is used for irrigation and supplemental fertilizing (fertigation).

Fig. 2.1 Strawberry production using plasticulture on a farm in middle Tennessee. **a** Raised bed plasticulture; **b** beds mulched and netted for winter. *Photo Credit* Tennessee State University

There are 2 types of raised beds used in plasticulture. In narrow beds, plants are spaced at about 12–14 in. in staggered double rows with one drip line running between the rows. The wide beds normally have two drip lines running between 4 rows of plants that are also spaced about a foot apart. Plants are set out at densities of 15,000–17,500 plants per acre (Poling et al. 2005). Strawberries grown using this system need to be intensely managed. It becomes extremely critical to control soilborne diseases and pests through soil treatments, crop rotation, green manure crops, and compost. The daily decision making regarding production and pest management strategies can have a major impact on yields and profit. There is also considerably more risk involved in terms of timing, frost and freeze strategies, and marketing using the plasticulture production system than the matted-row system.

Plug plants are planted through the plastic in late summer or early fall; dormant plants are planted in July. Lightweight plastic material with small holes is used for covering the beds for winter protection in the late fall (Fig. 2.1b). Growers in Tennessee prefer using the dual-purpose plastic mulch that allows penetration of soil warming radiation while eliminating light to prevent weed growth (Guerena and Born 2007). Raised beds provide good drainage; however, plants may be prone to freeze damage. Therefore, colder winter temperatures do create challenges for adopting this system. Strategies to best protect plants from winter injury may include using a thick row cover and applying the cover early. Rivard et al. (2014) reported that using a 1.2 oz/yd plastic as opposed to a 1 oz/yd cover resulted in a higher early yield. Both row cover thicknesses provided excellent protection to the crop even when temperatures fell below 10 °F; minimum temperatures under the covers were never less than 25 °F.

The raised beds allow for easier and faster picking and earlier harvesting (Poling et al. 2005). A north–south orientation of strawberry beds encourages a more uniform plant development and ripening on both sides of the double-row bed (Poling 1993). In most cases, berries can be harvested in 7–8 months after planting. Growers in Tennessee typically do not hold their plantings beyond the first year due to potential disease problems. If they choose to continue into a 2nd or 3rd year of production, runners must be removed each year. Higher yields and returns per acre are generally obtained using this production system. The majority of strawberries produced under annual plasticulture systems are June-bearing varieties, with 'Chandler' being the most commonly grown cultivar (Bost et al. 2003). In the northern part of TN, growers typically have a shorter harvest season (about 5 weeks) and lower potential yields per plant compared to the southern part (1 lb berries/plant vs. 1.5 lbs/plant). It is, however, still possible for them to make a profit with this yield which is equivalent to 15,000 pounds per acre (Safley et al. 2004).

2.2.2 Matted-Row System (Perennial)

A large percentage of strawberries grown in Tennessee are produced in matted-row production systems (Fig. 2.2). Franklin, Cheatham, Sumner, and Coffee counties in Middle Tennessee have historically been known for growing strawberries using this system. Though an old method of growing strawberries, it is still the predominant production method used by growers with great success. It is more commonly used in regions with a colder climate. The initial cost is low, since the matted-row system allows the plants to multiply into the rows to maximize production. In this production method, strawberry plants (crowns) are set out early to mid-spring on well-prepared soil at regularly spaced intervals (between 12 and 24 in. apart) within regularly spaced rows (36–52 in. apart). Planting in late spring causes fewer runners (stolons) to be produced, so the space between the plants must be reduced (Pritts 2002). From the initial planting, daughter plants from runners are established within the row and the intertwining network of runners creates a matted tangle of plants, giving the name 'matted row.' The daughter plants will produce their own strawberries for harvest the following spring. Weed management is the biggest challenge faced by farmers during the first year since there is much bare soil surface; frequent regular cultivation and hand weeding will greatly increase the life of the strawberry planting (Pritts 2002).

Fig. 2.2 Strawberry production using matted rows on a farm in Middle Tennessee. Matted rows with TSU experimental plots marked (*photo credit* Tennessee State University)

In the past, overhead irrigation was used more often, but currently matted-row growers are opting to use drip irrigation. During the spring, the overhead irrigation has a protective function to save the plants from frost damage. Drip irrigation users must use row covers for frost protection. The matted-row system takes a full year before a crop is harvested. Yields of anywhere from 12,000 to 25,000 lbs/acre are possible during the second year and onwards using this system. Each matted row is allowed to produce until the plants lose their productive ability (i.e., yields and berry size decline over time), which is usually between 3 and 5 years before rotation or replanting occurs. The most commonly grown cultivars in matted-row systems in Tennessee are 'Earliglow,' 'Allstar,' 'Delmarvel,' and 'Cardinal.'

One of the drawbacks of the matted-row system is that the strawberries require tending all year long. The strawberry renovation process must be done each year after harvest to prevent overcrowding caused by an excess number of runner plants rooting. This involves removing the foliage, narrowing the rows using a tiller, and then allowing runner plants to fill in the rows again.

An alternative to the matted row is a 'ribbon-row' high-density planting system, which has fewer weed problems and produces some fruit in the planting year and very high yields in the first fruiting year (up to 30,000 lbs/acre). This system is used if planting is done in late spring or early summer.

2.2.3 *High Tunnel Production*

A desire to extend the strawberry season has led to a great deal of interest in high tunnels. Also known as hoop houses, these greenhouse-like structures are covered with simple polyethylene, but they usually do not have exhaust fans or sources of heat. Temperatures and ventilation are controlled manually by opening and closing sides. Location and orientation of the high tunnel are very important for the success of this production system. The structure needs to be on a fairly level area protected from high winds, but on ground that is slightly higher than the surrounding area to help keep water away during heavy rainfall. High tunnel production is more expensive than open field production due to the costs involved in constructing and maintaining the structure. It still provides a reasonably low-cost investment considering that tunnels can extend the production season (both early and late in the season) and marketing window of strawberries in Tennessee. Late summer plantings in the tunnels allow the strawberries to produce a first crop in late fall and a second one in early spring (Wszelaki et al. 2013).

High tunnels provide crop protection from adverse weather conditions and better control of the growing environment. This significantly improves the survival rate of strawberry plants, as well as providing berries earlier (Wright 2012). Growers who are able to supply the customers with the earliest locally grown strawberries are often able to demand a premium price. As long as the plants are kept healthy and the environment in the high tunnels is appropriate, the strawberry harvest season will continue. High temperatures usually cause the harvest season to end (Wright 2012).

However, the tunnel environment does have limitations. The warm humid conditions are very conducive to pests such as white flies, aphids, and spider mites and *Botrytis*.

2.2.4 Cultural Management Systems in China

In China, national standards for organic strawberry production dictated by the Ministry of Agriculture (MOA) require no or minimum levels of chemicals present in the fruit (Liu 2014). There is tremendous variation in climate conditions in China. Therefore, strawberry cultivation systems differ significantly. Before the 1980s, strawberries were cultivated mainly in open fields. Since the 1990s, open field cultivation, small shed cultivation, normal greenhouse cultivation, and plastic greenhouse forced cultivation have all been in practice.

The climate of Anhui is fairly similar to that of Tennessee, with the average winter temperatures around 35–52 °F. There is a huge market for fresh strawberries in winter or early spring in China, when the popular Spring Festival is celebrated. This has led to the adoption of solar greenhouse production of strawberries all over the country, including Anhui during the winter months. The solar greenhouses are about 23 ft in width and 230–260 ft in length (Fig. 2.3b). The 20 in. wide beds

(a) (b) (c) (d)

Fig. 2.3 Production systems in China. **a** Glass greenhouse; **b** Plastic greenhouse; **c** Strawberries planted on bed in greenhouse; **d** Strawberrries planted in open field

allow for two rows of strawberries (Fig. 2.3a, c). Strawberry plantlets can be planted at densities of 72,000 plants per acre; after the first harvest the density can be reduced by 50 %. In this way, strawberry growers are able to significantly increase early yields (Lan et al. 2012).

In open field cultivation, strawberries are planted in the fall. Plug plants are grown on raised beds as shown in Fig. 2.3c, quite similar to plasticulture. Plants are set out at densities of 48,000–60,000 plants per acre (Zhang and Yang 2010), which is about 3 times that in the USA (Poling et al. 2005).

Regardless of the production system used, there are some key factors which cannot be ignored. The selection of a site for growing strawberries must consider size (to allow crop rotation), well-drained soil, adequate water supply for irrigation and also for frost protection, and good air movement (Carroll et al. 2014). Knowing soil conditions, such as pH, major and minor nutrients, infection by nematodes, and previous cropping history allows for better management decisions. For organic certification, the fields may not have been treated with prohibited products for 3 years prior to harvest. In the USA, a farm plan which describes production, handling and record-keeping is required by the USDA National Organic Program (NOP).

2.3 Cover Crops

Cover crops are essential to organic strawberry production to protect, maintain, and enrich the soil (Wszelaki et al. 2013). They have a beneficial effect on organic matter, encourage beneficial soil microbes, retain soil moisture, prevent erosion, and help control insects and diseases. Thus, the choice of cover crop (Table 2.1) depends on the specific goals of the production system. Growers in Tennessee usually plant these a year or two prior to establishing the strawberry crop so that they can succeed in suppressing weeds and in reducing nematode populations in the soil (Pritts 2002). There are two broad groups of cover crops commonly used in strawberry production in Tennessee: legumes and grasses. An emerging area of research has added a third group, the mustard family, which is being planted to suppress pests and also for its biofumigant activities. Cereal crops such as oats and rye have been used for their allelopathic properties and have the ability to suppress weeds and suppress pathogen and nematode pressure (Bhowmick et al. 2003). Oats are the most widely used cover crop in matted-row organic strawberry production in Tennessee.

Table 2.1 Common cover crops (lbs/acre) for strawberry production in Tennessee

Annual ryegrass	14–30	Temporary N tie-up when turned under, rapid growth, heavy N and moisture users, leaves behind heavy root system
Brassicas, e.g., mustards, rapeseed	5–10	Good cover and forage, cold hardy, mow or incorporate before seed formation, biofumigant properties
Buckwheat	35–134 60–70[a]	Use for site with low soil pH, matures in 70–90 days, low–moderate biomass, drought tolerant, will winter kill, incorporate shortly after bloom, scavenges nutrients, not a host for AMF
Cereal rye	60–200	Cold-tolerant cover crop, good catch crop, rapid germination and growth, incorporate before seed formation, temporary nitrogen (N) tie-up when turned under
Cowpea 'Iron clay'	100–130[a]	Moderate to good weed control, drought and heat tolerant legume (provides 100–150 lbs N/acre), good for beneficial insects
Hairy Vetch	30–40	Fast grower, mow or incorporate before seed formation
Marigold	5	Biofumigant properties, can be plowed under in 90 days, will winter kill
Pearl millet	30[a]	Excellent weed control, drought tolerant, does not get as tall as Sudan or Sorghum-Sudan, adapted to poor soils
Sorghum-Sudan	35–50 20–90[a]	Biofumigant properties, smothers weeds, tremendous biomass producer (4–6ft) in hot/dry weather; extra time to decompose thick stalks
Soybean 'Laredo'	110–130[a]	Moderate weed suppression, may need irrigation, more biomass than cowpea
Spring oats	60–100	Ideal quick cover crop, incorporate in early June if planted in spring, killed by successive frosts
Sudan	20–90	Excellent weed control, some drought tolerance, high biomass, adapted to poor soils
Velvet bean	80–100[a]	Very good weed control, good heat tolerance, very high biomass producer for legume, resistant to nematodes, seeds may be a bit more costly and less readily available than other legumes
Wheat	80–100	Mow or incorporate before seed formation

Adapted from Carroll et al. (2014) and McWhirt (2015)
[a]Rate used when planted with companion crops

No cover crops are used during strawberry cultivation in China. However, some cash crops (Fig. 2.4) are planted along with the berries to make better use of time or space (Table 2.2).

Fig. 2.4 Intercropping of strawberry and other crops in China. **a** Intercropping with vegetables; **b** intercropping with netted melon

Table 2.2 Common intercropping crops for strawberry production in China

Netted melon	7200 plants per acre	50 % net profit increase (Huo et al. 2008)
Rice		Soil improvement; decreased use of chemicals; 20 % increase in strawberry production
Cabbage		Planting May–Oct after strawberry is removed; early fruit; 20 % increase in strawberry production
Maize	1 row of corn per raised bed	Weed control; sun protection; pick corn and let the plant grow till Sep
Muskmelon	20–25 in. (row spacing)	Weed and insect control, soil improvement; exposed to solar for 40–50 days before strawberry plant
Tomato	Between 2 rows of strawberries, row spacing: 10 in.	Increased net profits (Zhang et al. 2011)
Luffa	row spacing: 5ft × 2ft	Planted March–July; increased net profits (Cheng 2007)
Watermelon	row spacing: 20 in.	15–20 % increase in net profit (Liu et al. 2008)

2.4 Variety Selection

Strawberry varieties are very sensitive to local conditions. A variety that performs well in one area may fail miserably in another area. Therefore, it is extremely critical to evaluate performance of the varieties under local conditions and select the right cultivars for optimum strawberry production. Selecting well-adapted varieties that are heat and humidity tolerant are very important in Tennessee. In organic production, the selected variety's relative resistance or susceptibility to diseases is extremely critical since there are a limited number of organic pesticides available for disease management (Carroll et al. 2014). Those varieties that exhibit disease

Table 2.3 Strawberry cultivars and their susceptibility to diseases in Tennessee

Variety	Leaf spot	Leaf scorch	Leaf blight	Anthracnose	Red stele	Verticillium wilt	Powdery mildew
Albion	M	–	–	M	R	R	M
Allstar	R	R	S	S	R	R	R
Annapolis	S	S	–	–	R	M	S
Cardinal	R	M	–	S	S	S	–
Chandler	S	M	S	S	S	–	R
Cavendish	R	R	–	–	R	M	S
Delmarvel	R	R	M	R	R	R	M
Earliglow	M	R	M	M	R	M-R	M
Honeoye	R	R	–	–	S	S	M
Idea	M	S	M	R	R	R	–
Latestar	M	R	–	–	R	R	–
Primetime	R	R	–	–	R	R	–
Redchief	S	R	S	S	R	M-R	R
Surecrop	M	M	–	M	R	R	R
Sweet Charlie	N	–	S	R	–	–	–

R Moderate to high resistance, *M* Moderate resistance to moderate susceptibility, *S* Moderate to high susceptibility, – Unknown
Modified from Bost et al. (2003)

and pest resistance (Table 2.3) for the area will have much better results. The market demand in terms of the qualities preferred also needs to be taken into consideration when selecting cultivars (Wszelaki et al. 2012a, b). Most Tennessee farmers grow strawberries for local consumption, so choosing varieties that have good shelf life to withstand shipping is not as critical.

Strawberry varieties are classified as either ‘June-bearing’ (short-day varieties) or ‘Ever-bearing’ (day-neutral varieties). ‘June-bearing’ start forming flower buds as the day-length gets shorter (in the fall) and temperatures get cooler. ‘Ever-bearing’ are insensitive to day-length and produce fruit throughout the season as long as nighttime temperatures drop below 60 °F (Strand 1993). June-bearers were developed after years of breeding for their attributes and are the type most often grown by commercial growers. They are preferred for their size and productivity and are recommended for early season production in a tunnel. June-bearing varieties are rated as early, midseason, or late according to when they bear fruit. Ever-bearing varieties are highly productive and favored for their flavorful berries. They bear fruit throughout the growing season, with three production peaks each year: June, mid-summer, and late summer to frost.

It is quite possible to grow 2–3 different cultivars using plasticulture in the southeast to extend the harvest season over a four-to-eight-week period based on the temperature during the season (Poling et al. 2005). ‘Sweet Charlie’ (early variety), ‘Chandler’ (early- midseason variety), and ‘Camarosa’ (mid-season variety) are all adapted for Tennessee. Four of the producers involved in this project grew 'Chandler' exclusively or as one of the varieties, along with 'Camarosa,' 'Ozark,' 'Haneoye,' and 'Sweet Charlie.' Others preferred 'Earliglow,' 'Allstar,' 'Santa Rosa,' 'Red Chief,' and 'Surecrop.' 'Chandler' is well liked by consumers for its attractive red color, size, and flavor (see Sect. 2.11). It is relatively cold hardy and does not generally require winter protection and also has high yields. However, row covers are recommended if temperatures dip below 10 °F for extended periods in the particular area, as this may cause significant flower damage and crown injury. 'Sweet Charlie' fills the early market niche. It can be ready anywhere from a week to two earlier than 'Chandler' and in some years, growers can have a second crop of 'Sweet Charlie' berries in the final week of the strawberry season (Poling et al. 2005). These are preferred for their high sugar to acid ratio. 'Camarosa,' bred at the University of California, is the most widely planted variety in the world. This variety has high yields and the fruit itself is large and very firm. This firmness makes it ideal for distant shipment or for long-term storage. For best flavor, 'Camarosa' is picked when the berries have taken on a darker color, just past the glossy bright red stage. If the color turns wine red, the berries are already too ripe. In Tennessee, 'Chandler' (Fig. 2.5a) and 'Camarosa' are typically grown in plasticulture production systems, while 'Earliglow' (early season), 'Allstar' (mid-/late season) (Fig. 2.5b), 'Delmarvel,' and 'Cardinal' are commonly grown in matted-row systems. 'Albion,' 'Tribute,' '300 Seascape,' and 'Festival' are also popular in Tennessee. 'Albion' has shown to be cold hardy in the Middle Tennessee area. A producer participating in the NSSI project harvested 'Albion' strawberries from his plasticulture production system in Middle Tennessee as late as the first week in November, before the first sub-freezing temperatures.

In China, varieties with large fruit were introduced in the early twentieth century. In the mid 1950s, many strawberry varieties were introduced from the former Soviet Union, Poland, and Yugoslavia. By the early 1980s, some excellent varieties had been screened and genetically improved and were being widely used for production. Among the cultivars grown, 'Cartl' and 'Elsanta' are usually used for

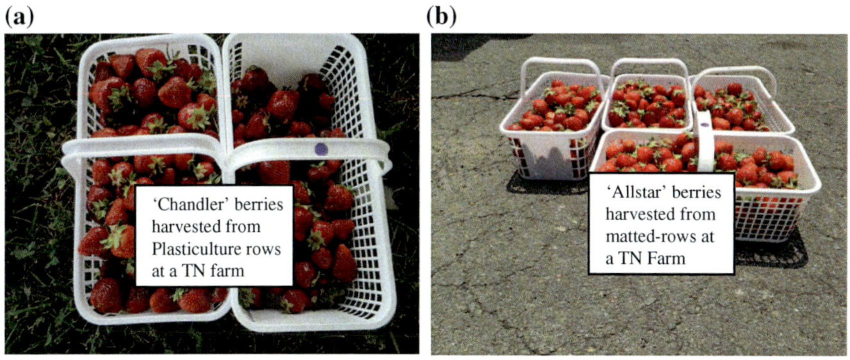

Fig. 2.5 Strawberries harvested from farms in Middle Tennessee in 2014

fresh consumption, while 'Honeye,' 'TuD La,' 'Darselect,' 'Senga Sengana,' and 'Darsanca' are used for processing (Liu 2014). The most widely planted cultivar in Changfeng is 'Confidante' (Lan et al. 2012).

More improved varieties are needed to increase production and capture a larger share of the local market. The University of Kentucky has been evaluating the performance of a new variety called 'Flavorfest.' This variety, which was developed at the USDA-ARS station in Beltsville, is showing great potential in terms of production, comparable to Chandler and is also exhibiting some disease and pest resistance (2015 news bulletin). Owing to similarity in growing conditions, 'Flavorfest' may show potential for growth in Tennessee in the coming years.

2.5 Soil Fertility/Nutrient Management

The need for balancing nutrition for strawberries can be greatly facilitated by improving soil organic matter content. Organic systems improve soil quality by increasing soil fertility, and promoting soil biological activity. Soil quality is further enhanced through management practices such as incorporating compost, animal manures, cover crops, and green manures (Wszelaki et al. 2012a, b). It is very critical to create a soil that is capable of holding nutrients and water since this is the basis of a good strawberry crop. The connection between soil, nutrients, pests, and weeds is so great that it can make a difference between the crop success and failure. Strawberry growers initiate soil preparation a year or two before establishing plants. Decomposing organic matter provides plant available nutrients in organically managed systems (Carroll et al. 2014).

Beneficial soil inoculants are added to the soil to introduce or re-establish 'good' soil microbes. These were initially put into practice to re-introduce 'good bacteria' back into the soil after fumigation. Now that chemical fumigation is being phased out, the soil inoculants continue to be popular. Two types that are commonly used are (1) vermicompost (compost made by earthworms), which has high microbial activity, and (2) arbuscular mycorrhizal fungi (AMF), which are fungi that form beneficial relationships with plant roots and help the plants find nutrients. Mycorrhizal cultures inoculated around strawberry plants (Fig. 2.6b) could improve phosphorus (P) availability in the soil. Compost applications encourage diverse soil microbe populations. Adding beneficial soil inoculants to the planting hole with the plug media have been shown to increase yields as compared to non-inoculated ones (McWhirt et al. 2015). Arancon et al. (2004) reported that vermicompost applications (2.02–4.05 ton/acre) increased strawberry growth and yields significantly. Welke (2004) reported a 20 % increase in yields with the foliar application of aerobically prepared compost tea, compared to the control.

Animal manure is the most traditional and widely recognized fertilizer used in organic systems. According to the US Department of Agriculture's (USDA) National Organic Program (NOP) standards, raw manure must be applied and incorporated into the soil at least 120 days prior to harvest of a crop that grows

Fig. 2.6 Weed Control and soil fertility practices. **a** Flame burning for weed control in a Robertson County farm; **b** inoculating biovam mycorrhizae around strawberry plant

close to the soil, such as strawberries. Due to food safety concerns, the use of raw manure is not recommended; the use of composted manure is preferred. However, manure-based compost can have high phosphorus or salt concentrations, leading to nutrient toxicities (McWhirt 2015).

Soil pH of 6.0–6.5, in the slightly acidic range, is recommended for strawberries to maintain nutrients at their optimum availability and help avoid micronutrient deficiencies (Carroll et al. 2014). A low pH or acid soil will have reduced availability of nitrogen (N), phosphorus (P), potassium (K), magnesium (Mg), and molybdenum (Mo), while a high pH or alkaline soil will have reduced availability of zinc (Z), boron (B), iron (Fe), manganese (Mn), and copper (Cu). Results of the soil analysis (Fig. 2.4c) determine whether pH needs to be adjusted by using either dolomitic lime (raise pH) or elemental sulfur (lower pH) during the soil preparation stage and will also determine the addition of nutrients. Generally, one-third of the N, half of the K, and all of the required P should be incorporated before planting (Poling 1993). Once the plantings are established, fertilizer applications are based on leaf analysis.

Many types of organic fertilizers (Tables 2.4 and 2.5) are available to provide supplemental nutrients during the growing season. However, the application timing is very critical. For instance, June-bearing strawberries (matted-row systems) must be fertilized in late summer, to give the organic fertilizer enough time to break down and release nutrients for the plants in the fall when the berries are setting buds for next year's crop (Guerena and Born 2007). It is important to balance the total plant nutrition taking into account competing characteristics of all fertilizer elements. Cover crops or compost alone might be inadequate to fulfill the late N demand of strawberry crop in organic production (Muramoto 2004). Due to the unpredictability of N mineralization from organic fertilizers and soil organic matter, providing the optimum amount of N to meet the needs of strawberries can be challenging in organic systems. Gaskell (2004) tested different sources of organic fertilizers such as guano, feather meal, liquid fish emulsion, fish meal, chicken manure, compost, and a green manure and found great variability in N availability.

Table 2.4 Available potassium in organic fertilizers

Sources	Available potassium
Sul-Po-Mag	22 % K_2O also contains 11 % Mg
Wood ash (dry, fine, gray)	5 % K_2O, also raises pH
Alfalfa meal (non-GMO)	2 % K_2O, also contains 2.5 % N and 2 % P
Greensand or granite dust	1 % K_2O
Potassium sulfate	50 % K_2O

Carroll et al. (2014)

Table 2.5 Available phosphorus in organic fertilizers

Sources	Available phosphorus
Bone meal	15 % P_2O_5
Rock phosphate	30 % total P_2O_5
Fish meal	6 % P_2O_5 (also 9 % N)

Carroll et al. (2014)

Though more N is utilized by strawberries in the mid to later growth stages than the early stage, it should be supplied over the entire production period to ensure optimum yields and high quality (May and Pritts 1990). Depending on the soil type, 150–220 lbs N/acre may be needed for new plantings and 75–120 lbs N/acre for established plantings. Excellent strawberry production requires about 1 lb N/acre per day during the harvest season.

Strawberries may require 100–150 lbs/acre of K, much of it being needed during flowering and fruit set, and continuing on into fruit maturation. The amount needed depends on the soil test results. The K:Mg ratio must be maintained at 4:1 to prevent magnesium deficiency, which is quite common in strawberries. Excess K affects Mg availability; recommendations call for providing 10–40 lbs/acre Mg in established plantings that are deficient in magnesium (Carroll et al. 2014). The most common source of Mg currently used to maintain soil Mg concentrations is dolomitic limestone, which contains a significant percentage of Mg, in addition to Ca. Epsom salt, also known as magnesium sulfate ($MgSO_4$), is another way to add Mg to the soil and may also be applied to plants as a foliar spray. Calcium (Ca) levels are usually adequate in the soil if the pH is in the appropriate range (6.0–6.2).

Tennessee soils contain sufficient amounts of phosphorus for good strawberry growth and as long as soil pH is within the 5.5–7.3 range, phosphorus is available for uptake. Supplemental amounts are generally worked into the soil prior to planting (25–30 lbs/acre).

The other important micronutrients for strawberries include boron (B) and zinc (Zn). Although B is often recommended as a nutrient supplement for strawberries, excessive levels can be toxic to the plants. Amounts applied to a field should not exceed one pound of actual boron per acre in any year. Zn is more available to plants under low soil pH; availability may be reduced with heavy applications of lime and/or phosphorus. It is typically applied as part of a fertilizer blend to the soil, or applied as a foliar spray. Elements such as iron, copper, and molybdenum, while still required for good growth, are needed in very small amounts. As soil pH has a strong influence on nutrient availability, it needs to be regularly monitored and adjusted as needed.

Fertilizers in China are mainly divided into 3 categories: (i) organic fertilizers which include animal manure, green manure, and straw and composted manure. (ii) microbial fertilizers which contain microbes that benefit plants, e.g., rhizobium and azotobacteria (Zhang and Yang 2010); and (iii) foliar nutrients such as amino acids, vitamins, sugar, and some trace elements (Liu 2014). Manure and the organic fertilizer may introduce insects, ovums, and weed seeds; therefore, manure must be kept at over 70 °C for over 5 days to kill them (Zhang 2012).

2.6 Weed Control

Weed management during the establishment of strawberry plants can have a major effect on optimal plant growth and yields. During the short-term production cycle of strawberries in Tennessee, it is extremely critical to control weeds within the rows, as weeds compete for water and nutrients and provide alternate hosts for pests. The use of cover crops helps to eliminate or suppress weeds (Sect. 2.3; Table 2.1). The preparation of the soil during the preplanting years also pays dividends by eliminating perennial weeds. Growers in Middle Tennessee occasionally use flame burning (Fig. 2.6a) to control weeds, though Wildung (2000) found that it is not as effective in new plantings as normal cultivation and hand weeding. Therefore, to prevent the weeds from going to seed, producers find manual cultivation to be the best approach. Weeder geese were once commonly used for weed control before the widespread popularity of herbicides. National Organic Program rules state that manure must be incorporated 120 days prior to strawberry harvest, so geese may only be employed to control weeds earlier in the production cycle.

Mulches provide weed control by smothering weed seedlings and blocking light from the soil surface, and preventing the germination of weed seeds. As an added benefit, they help to regulate soil temperature by shading soils in the warm summer months, insulating the soil during cool weather, retaining soil moisture by preventing water losses through evaporation, and protecting the soil from erosion from heavy rains. Mulches commonly used in organic production in Tennessee include straw, hay, sawdust, and wood chips. Some producers choose to use compost, grass clippings, plastic, biodegradable mulch, and landscape fabric as mulches.

Weed management in organic strawberry production can be carried out using commercially available lemongrass oil-based organic herbicides, containing the active ingredient citrus extract *d*-limonene. Rowley et al. (2011) evaluated weed control provided by mulches and organic herbicides alone or in combination and found that organically certified herbicides such as clove oil displayed 41–95 % weed control when applied without mulch. Some mulches and organic herbicide combinations provided weed suppression similar to conventional herbicide application.

For weed control in China, flame burning is done in late autumn prior to strawberry cultivation. After the soil is plowed, the beds are prepared and covered with black plastic to inhibit weed germination and growth; manual weeding must continue as needed (Liu and Zhu 2008). Moreover, as previously mentioned, corn and muskmelon intercropping can also help to control weeds (Table 2.2).

2.7 Insect Control

While appealing to humans, the strawberry fruit is also attractive to pests, rodents, and various fruit insects, and mites. In addition to reducing fruit quality and crop yield, these pests also are potential vectors for pathogens. Bost et al. (2003) listed the following pests commonly encountered in strawberry productions: strawberry crown borer (*Tyloderma fragariae*), strawberry leafroller (*Ancylis comptana fragariae*), strawberry rootworm (*Paria fragariae*), strawberry weevil (clipper) (*Anthonomus signatus*), lygus bugs (*Lygus lineolaris, Lygus Hesperus*), stink bugs (brown: *Euschistus servus*, green: *Acrosternum hilare*), whitefringed beetle (*Naupactus leucoloba*), spittlebugs (*Philaenus spumarius*, *P. leucophthalmus*), aphids (*Chaetosiphon Fragaefolii* and other genera), two-spotted spider mites (*Tetranychus urticae*), cyclamen mites (*Steneotarsonemus pallidus*), sap beetles (*Stelidota geminata*), flea beetles (pale-stripped *Systena balanda* and eggplant flea beetle *Epitrix fuscula*), root weevils (*Otiohynchus* spp.), and potato leafhopper (*Emposasca fabae*). Rhainds et al. (2002) measured the incidence of pests in strawberries under conventional and organic management systems through four fruiting seasons and found that the proportion of fruits damaged by plant bugs was higher in organic than in conventional plots, even higher than incidence of damage by gray mold or slugs.

Integrated pest management (IPM) involves the implementation of preventive practices before planting the crop (Rondon et al. 2003, revised 2009). The grower's first line of defense in a pest management program in organic strawberry production is to select varieties with pest and disease resistance (Wszelaki et al. 2012a, b). Plants must be inspected prior to transplant and only clean non-infested plants should be used. Other control measures include crop monitoring to detect pest presence and activity as well as identification of insects and diseases. The use of sticky traps is very beneficial for trapping of pests such as aphids. They can be used indoors such as in the high tunnels or in the fields hung from fence stakes.

Biological controls are most effective before the pests reach critical levels, so regular monitoring is very important (Guerena and Born 2007). Organic systems rely on populations of beneficial insects (Table 2.6) to maintain a natural balance between pest and predator species. It is possible to eliminate or reduce the use of pesticides on strawberries through early identification of arthropod problems and the use of living agents to suppress or destroy the undesirable pests (Rondon et al. 2003, 2009). Planting habitat for beneficial organisms may encourage populations of the 'good' insects (McWhirt et al. 2015).

Table 2.6 Some of the beneficial insects used in strawberry production

Pests	Biological agents	Common name	Category
Western flower thrips	*Neoseiulus cucumeris* *Orius insidiosus*	Cucumeris Minute pirate bug	Predator Predator
Cotton or melon aphid	*Hippodamia convergens* *Coleomegilla maculata* *Chrysoperla rufilabris* *Aphidius colemani*	Lady beetle Spotted lady beetle Chrysopa- green lacewing Aphidius wasps	Predator Predator Parasitoid Predator
Two-spotted spider mites	*Phytoseiulus persimilis* *Neoseiulus californicus*	Persimilis Californicus	Predator Predator
Caterpillars	*Bacillus thuringiensis (Bt)*	Dipel, MVP, MVP II and others	Pathogen derived
Strawberry leafroller (larvae)	*Macrocentrus ancylivrous* *Cremastes cookie*		Parasitoid Parasites
Lygus bugs	*Beauveria bassiana*	BotaniGard ES and Mycotrol O	Fungal pathogen
Cyclamen mites	*Amblyseius*		Mites
Whiteflies	*Beauveria bassiana*	BotaniGard ES, Mycotrol O	Fungal pathogen
White grubs (larvae of scarab beetles)	*Steinernema carpocapsae* *Heterorhabditus bacteriophora*		Nematodes Milky-spore bacteria
Strawberry root weevil	*Steinernema carpocapsae* *Heterorhabditus bacteriophora*		Nematodes Milky-spore bacteria

Modified from Rondon et al. (2003) and Bost et al. (2003)

Organic systems make great use of trap crops and companion planting. The principle of trap cropping is based on the pest's affinity for a particular plant which keeps it away from the main crop, thus eliminating the need to use insecticides (Wzselaki et al. 2012a, b). When designing a strawberry field to incorporate cover crops, it is helpful to know the target insect and its behavior.

There has been considerable interest in the use of certain crops as biological fumigants ahead of crop production to reduce the need for chemical fumigation. Advances in biopesticides with fumigant properties have been stimulated by the phase-out of methyl bromide, which was labeled for use in conventional strawberry production to target multiple pests. Plants in the mustard family, such as mustards, radishes, turnips, rapeseed, and sorghum species (sudangrass, sorghum-sudangrass hybrids), have shown the potential to serve as biological fumigants (see Sect. 2.3; Table 2.1). Plants from the mustard family produce chemicals called glucosinates, which are released from the roots or the foliage when the plant is cut. Glucosinates are further broken down into isothiocyanates that act like chemical fumigants. Sorghum also produces a syanogenic glucoside compound called dhurrin, which releases cyanide when the plant tissue is damaged.

Kirkegaard et al. (1999) identified compounds in Brassica roots and demonstrated their toxicity to fungal inoculum. Results from these agents have, however, been inconsistent in Delaware, often showing minimal benefits (Johnson 2009). Success with biofumigant crops depends on a number of factors. Firstly, using varieties bred for higher levels of active compounds will result in more effective chemical being released; secondly, plant material must be finely chopped and incorporated thoroughly for best results (Johnson 2009). However, even with these existing limitations, incorporating the biomass from the biofumigants adds rich organic matter to the soil. Sams and Kopsell (2011) found beneficial effects of different combinations of mustard meal and compost applications on vegetable production in terms of yield and quality, and protection from early blight. The application also increased the yield of strawberry plants and protected them against Anthracnose.

More recently, anaerobic soil disinfestation (ASD), also known as biological soil disinfestation, has been investigated as a potential fumigation alternative. This is a process whereby anaerobic soil conditions are created by incorporating organic soil amendments (substrate), covering with plastic mulch, and irrigating the soil to saturation for 2–6 weeks. The filling of the soil pores with water causes a reduction in soil oxygen, creating anaerobic conditions (Shrestha et al. 2014). Many soilborne pathogens and nematodes of concern showed susceptibility to ASD treatment. Muramoto et al. (2014) conducted anaerobic soil disinfestation using a range of carbon inputs and in field trials conducted at multiple sites in California and found that using rice bran as the carbon source (substrate) provided yields equivalent to those from preplant fumigation sites.

Insect control in China is primarily physical and biological. Physical control includes the removal of contaminated leaves by hand, trapping and killing using light, use of yellow boards and sex pheromones. Biological control is mainly protecting the predator (Liu and Zhu 2008).

Insect and mite populations vary from field to field and from year to year. When the pest population surpasses the established thresholds, chemical control measures may have to be used. It is still important to use pesticides that are compatible with the biological control agents. The reduction on chemical dependency is good for the environment, safe for people and in general good practice.

2.8 Disease Control

Diseases affecting the strawberry fruit cause a direct loss of the harvested product. A slight blemish by the pathogen can quickly engulf the entire fruit by the time it reaches the market. Anthracnose fruit rot (caused by the fungus *Colletotrichum*) affects not only the fruit, but also many other parts of the plant and may cause severe problems in perennial systems. The United States Department of Agriculture–Agricultural Research Service (USDA-ARS) initiated the development of anthracnose-resistant strawberry cultivars adapted to the southeastern USA in 1976 after an epidemic of anthracnose crown rot and this resulted in the release of 4 anthracnose-resistant breeding lines and one cultivar (Smith 2006).

Gray mold (caused by the fungus *Botrytis cinerea*) is the most commonly observed fruit contaminate in the market. Fuzzy brown to gray spores develop and cause fruit to rot. The gray mold fungus develops on dead plant material and is readily airborne. Local varieties such as 'Earliglow' and 'Delmarvel' are resistant to the gray mold. Leather rot (caused by *Phytophthora cactorum*) is not as devastating as the other two, but can be of concern in areas of poor drainage. Berries infested with leather rot appear normal, but have a sour odor and an unpleasant taste (Bost et al. 2003).

Good Agricultural Practices (GAP) demands that growers make smart choices for disease management. Planting disease-resistant varieties or less susceptible varieties will eliminate or greatly reduce the need for disease control. The first vulnerable area is the transplant nurseries, so disease-free healthy plants should be purchased from reputed nurseries that follow good propagation and cultural practices. Healthy soil with adequate organic matter will help maintain beneficial organisms that may suppress soilborne pathogens (Guerena and Born 2007). Other cultural controls include picking fruit frequently and removing infected fruits since pickers handling infected berries can spread the infection to healthy berries. If disease is spotted, sprinklers should be used only for frost protection, not for irrigation. Care must be taken to ensure that there is good air movement, proper drainage and no water retention between rows.

Field sanitation practices include removal, burning, or deep cultivation of crop residues to help prevent the spread of disease in a field. Diseased plant material can also be tilled into the soil to prevent the spread of spores in the wind and to hasten the breakdown of the disease pathogens by beneficial fungi, nematodes, and bacteria. It is good practice to remove winter-killed foliage before bloom, to eliminate a food base for fungi, and also to properly clean-up the matted rows after harvest. Additional sanitation practices include removing weedy habitat that may shelter pests and cleaning equipment to prevent the spread of disease or weed seed from field to field. Good sanitation practices can go a long way in preventing pest problems in organic strawberry production. However, practices such as deep plowing and burning may cause erosion, decrease soil organic matter, and reduce biodiversity. Therefore, these practices should be used with caution on a limited basis.

As discussed in Sect. 2.3, the use of biofumigant cover crops such as *brassica* can encourage beneficial soil microbes and help control insects and diseases. Pinkerton et al. (2002) demonstrated the potential of solarization in management of root diseases in strawberry production in hot and dry areas. Heat trapped under the clear plastic mulch, laid on moist soil, raises the soil temperature, and kills pests. Welke et al. (2004) reported a reduction in the incidences of *Botrytis* with the foliar application of aerobically prepared compost tea. Serenade, Mycostop, and Promot are some of the biorational products available commercially for *Botrytis* control (Guerena and Born 2007). If pesticide applications are needed prior to bloom, it is better to stick with biologicals (Table 2.7). Sprays should be used strategically and sparingly; 50 % of the applications are unnecessary (Schnabel and Peres 2015). Integrated pest management and weed control techniques can greatly reduce pesticide use in strawberry production.

Leaf diseases appearing on strawberries can cause significant damage to the plant causing it to be more susceptible to winter injury. Leaf blight, common leaf spot, and leaf scorch are caused by fungi and are best controlled by following the cultural practices described above. Angular leaf spot, caused by a bacterium, caused major problems in Tennessee strawberry production in the 1990s (Bost et al. 2003).

Table 2.7 Biological pesticides used for disease control in strawberries

Trade name	Active ingredient	Product rate	Type of control	Comments
Regalia Biofungicide	*Reynoutria sachalinensis*	1–3 qts/acre	Leaf blight, leaf spot, gray mold, anthracnose, red stele, black root rot, verticillium wilt	Start at first sign, then every 7–14 days
Actinovate-AG	*Streptomyces lydicus* WYEC-108	3–12 oz/acre	Powdery mildew, gray mold, anthracnose, leather rot, red stele, black root rot, verticillium wilt	Foliar application, apply before onset of disease, reapply at 7–14 day intervals
Double Nickel 55	*Bacillus amyloliquefaciens* str. D747	0.25–3 lb/acre	Powdery mildew, gray mold, anthracnose, leather rot, red stele, black root rot, angular leaf spot, verticillium wilt	Foliar application
Double Nickel LC	*Bacillus amyloliquefaciens* str. D747	0.5–6 qt/acre	Powdery mildew, gray mold, anthracnose, leather rot, red stele, black root rot, angular leaf spot, verticillium wilt	Foliar application

Carroll et al. (2014)

This disease is difficult to treat and prevention is still the best means of control. Red stele and Phytophthora crown rot are most often found in poorly drained areas of the field. Most current varieties are resistant to red stele, but none are known to be resistant to Phytophthora crown rot. Elemental copper and sulfur have been used by conventional and organic growers as pesticides for foliar bacterial diseases and powdery mildew, respectively (Guerena and Born 2007). A commercial formulation of *Bacillus pumilis* has been approved by OMRI (Organic Materials Review Institute) for the control of powdery mildew in strawberries. Table 2.8 lists some of

Table 2.8 Pesticides labeled for disease management in strawberries in Tennessee

Trade name	Active ingredient	Product rate	Type of control	Comments
Badge X2	Copper hydroxide, copper oxychloride	0.75–1.25 lb/acre	Leaf blight, leaf scorch, leaf spot, angular leaf spot	
Champ WG	Copper hydroxide	2–3 lb/acre	Leaf blight, leaf spot, angular leaf spot	May cause crop injury under certain conditions
CS 2005	Copper sulfate pentahydrate	19.2–25.6 oz/acre	Leaf blight, leaf Scorch, leaf spot	
Cueva Fungicide Concentrate	Copper octanoate	0.5–2.0 gal/ 100 gal	Leaf blight, leaf scorch, leaf spot, powdery mildew, gray mold, anthracnose, angular leaf spot	Applied as a diluted spray at 50–100 gal/A
Milstop	Potassium bicarbonate	2–5 lb/acre	Leaf blight, powdery mildew, gray mold, anthracnose	Not compatible with alkaline solutions
NuCop 50DF	Copper hydroxide	2–3 lb/acre	Leaf blight, leaf spot	Discontinue if phytotoxic
OxiDate 2.0	Hydrogen dioxide Peroxyacetic acid	32 floz-1 gal/100 gal water	Leaf blight, powdery mildew, gray mold, angular leaf spot	At planting and existing planting foliar application
PERpose Plus	Hydrogen peroxide/dioxide	1 fl oz/gal initial/curative 0.25–0.33 fl oz/gal weekly/ preventative	Leaf blight, leaf spot, powdery mildew, gray mold, anthracnose, leather rot, red stele, black root rot, angular leaf spot, verticillium wilt	Curative for 1–3 consecutive days Preventative every 5–7 days
Trilogy	Neem oil	0.5–1 % solution	Leaf blight, leaf spot, powdery mildew, gray mold, anthracnose, angular leaf spot	Apply in 25–100 gal water/A

Carroll et al. (2014)

the pesticides used in Tennessee to manage a broad spectrum of diseases in strawberries. The selection of disease-resistant varieties (Table 2.3) will minimize losses due to plant damage. There is effort placed in producing more disease-resistant varieties, but currently there are no major marketable varieties available with high levels of resistance to multiple pathogens (Mossler 2004).

The main strawberry diseases in China are white leaf spot, powdery mildew, and nematodes. In the early stage of white leaf spot of strawberry, peptaibol biological bactericide (200–300×) is used, but stopped 3 days before harvesting. For powdery mildew, the infected leaves and fruits are removed in the early morning. Eguenol (600×), a mixture of terramycin and streptomycin, is used to control the disease. Treating plants in hot water (35 °C and 45 °C each for 10 min) significantly reduced nematode population (Liu 2014).

2.9 Harvest and Postharvest Handling

In the State of Tennessee, harvest from plasticulture production systems typically begins during the third week of April to second week of June and matted row begins the second week of May and extends through mid-June. Harvest may be advanced by 1–2 weeks by the use of mulch covers and row covers. Strawberries are hand-picked in April–early June. Since the quality of strawberries does not improve after harvest, it is advisable to only pick fully colored strawberries at the peak of flavor; the fruit shoulders and tip should not be green or white (Wright 2012). The fruit must be firm and free from rot. When harvested at the right time and handled properly, strawberries remain in good condition for up to five days, in terms of appearance and taste. Strawberries are extremely perishable and have unusually demanding postharvest handling requirements. Proper handling will ensure the relative longevity of the fruit. Pelayo et al. (2003) found that the 3 strawberry cultivars that they investigated had a postharvest life (based on appearance) of 7–9 days when stored at 5 °C. However, the maximum period of storage during which the fruit maintained a flavor profile similar to freshly harvested fruit was shorter (5 vs. 7 days) for 'Diamante' and 'Aromas' and remained the same (9 days) for cultivar 'Selva.'

Strawberries remain alive and produce heat as a natural consequence of respiration even after they are harvested (Boyette et al. 1914). Fruits must be chilled rapidly by forced-air cooling (to 40 °F) within an hour of picking to remove the field heat and increase shelf life and must not be allowed to rewarm. Most strawberry producers pick the fruit early in the day, while temperatures are cool. This makes a significant difference in the shelf life of the berries, when combined with the postharvest cooling. This also keeps the fruit rots from developing or they develop more slowly. As the time difference between harvesting and cooling lengthens, more berries are lost to deterioration. Growers using the Pick-Your-Own (PYO) method of distribution do not need to be concerned about handling and storage since these are performed by the consumers. If they use other markets, such as grocery stores, restaurants and farmers' markets, they do need to maintain the quality of the product after harvesting with the right cooling, handling, and storage methods.

In China, strawberries are hand-picked when their surface is 70 % red. Picking is usually done at early morning and late afternoon to keep the fruit fresh, and picked fruits are cooled with wind (1 h at 1 °C after picking), then stored at low temperature till marketing.

2.10 Economics and Marketing

Strawberries, especially organic, are a high-value crop with special production requirements. They are highly perishable and time sensitive. The brief marketing season is very intense. So it makes sense to maximize the profit potentials. The initial investments in preparing the land, irrigation costs, and other equipment can range from about $2,000 per acre (for matted-row system) to $10,000 (for plasticulture systems). The yields obtained with plasticulture are almost double that of matted-row system and the harvest season is extended. The earlier harvest allows producers to receive higher 'beginning of season' prices. Since their production costs are higher, the berries must be sold at a premium in order to make a profit. In areas where the local market demand is fairly strong, prices tend to be higher (Guerena and Born 2007). Strawberry growers who sell direct to customers have great control over price, so usually they can set the price sufficiently high to make a profit. Consumers who are willing to pay higher prices for locally produced foods place importance on product quality, nutritional value, methods of raising a product and the effect of those methods on the environment, and support for local farmers (Martinez et al. 2010). Prices for fresh-market strawberries have been stable in recent years because of demand. With current yields and prices, strawberries continue to be a profitable crop to grow.

Organic strawberry growers support the idea of long-term land stewardship. However, they often choose production practices based on costs and yield potentials. Research has shown a direct link between incorporation of a practice and increases in strawberry yields, some resulting in immediate benefits, while others can result in yield benefits in the long run. The extension agents in the state provide support and resources for those who are currently farming and those who want to transition into organic production of this potentially lucrative specialty crop.

The organic strawberry market is seeing exponential growth. The number of organic strawberry growers continues to grow, with 160 registered with the California Organic Program as of 2004. Organic strawberries now rank sixth among all California organic fresh commodities. Even with growers in California, Florida, Oregon, and Washington producing 95 % of reported US output, there is immense opportunity for local growers to tap into the remaining 5 % of the market share. Growers outside the western USA will likely be called upon to offset shortages of strawberries caused due to the severe drought that has plagued California since 2013.

The surge in popularity of farmers' markets, and the CSAs (Community Supported Agriculture) is seeing an unprecedented growth. In 2009, USDA launched the 'Know Your Farmer, Know Your Food' initiative, an agency-wide effort to create new economic opportunities by better connecting consumers with local producers. The Tennessee Department of Agriculture started the 'Pick Tennessee Products' campaign in 1986 to encourage consumers to buy products grown or manufactured in the state. Their Web site maintains an active directory of farmers/producers who sell their products at farmers' markets, online, on the farm, or are strictly pick-your-own operations such as the strawberries growers. Recent studies show that more Tennessee farmers are selling directly to consumers. For smaller farms, direct marketing to consumers accounts for a higher percentage of their sales than for larger farms (Martinez et al. 2010).

Consumers want to know where their food is coming from. This has opened up the opportunity for the relatively small-scale producers who have had to contend with severe competition from the large companies. When it comes to perishable produce like strawberries, this gives the farmers a larger share of the local sales. In most locations, demand for locally produced strawberries exceeds the available supplies. Small-scale producers can thus receive higher returns from strawberries than from most other crops. Promoting local organic strawberry production means better access for shoppers to quality strawberries and better profitability for the farmers growing the crops. To avoid spoilage during shipping, most California and Florida grown strawberries are harvested before they are fully ripe. They cannot compete with the locally produced tasty and fresh berries which are picked at ripeness for the local market.

Assessment of the strawberry market potential in the area and the possible methods of marketing the fruit is a critical first step in the successful management of any strawberry farm (Himelrick et al. 2002). Most small farmers find great success from direct marketing through roadside farm stands and PYO operations. Bringing people to the farm for 'u-pick events' results in greater profits. In North Carolina, majority of the strawberry production is marketed toward u-pick customers (McWhirt et al. 2015). In recent years, more producers have been offering pre-picked strawberries for sale instead of PYO to maximize their profits. Other direct and niche marketing strategies can boost profits. The 'farm to table' concept, which is pushing more specialty restaurants to either partner with or buy locally grown foods directly from the producer, is proving to be another important channel for small strawberry growers. These types of restaurants typically open in places where consumers are highly supportive of the local foods movement (Martinez et al. 2010). Small, independent grocery retailers are better positioned to offer local food as they develop a strong relationship with local farmers and prefer products that have traveled a short distance.

2.11 Market Demand for Desirable Properties

Consumers, producers, and distributors all agree that freshness, good shelf life, and firmness are largely what determine the quality of strawberries. Consumers and retailers also attach importance to taste, while yields are important to the growers. Several studies, both national and on a smaller scale, have explored consumer preferences for locally produced food. In recent decades, public health promotion of healthier lifestyles has led to increased demand for fresh produce in many industrialized nations. The sustainability of the strawberry industry depends on its ability to satisfy the changing demands of its customers. So it is important for the producers to know their target audience, to explore consumers' preferences toward different sources of strawberries, specifically store-bought, industrially grown strawberries and locally grown, farm raised strawberries. In addition to raising awareness about the local production, it also helps in gathering public opinion. Surveys conducted on the campus of Tennessee State University and during the 2014 Middle Tennessee Strawberry Festival held in Portland revealed consumer preferences when it comes to choosing between organic vs. non-organic strawberries (Fig. 2.7). Five locally grown varieties were subjected to a blind taste test. Nine quality attributes were evaluated: color, freshness, size, appearance, smell, firmness, sweetness, juiciness, and overall quality. Locally grown and store-bought strawberries were judged to have similar appearance and firmness. Though the store-bought strawberries had a slight edge in terms of their size and color, the local varieties were picked for freshness, smell, sweetness, juiciness, and overall quality. Respondents rated the local strawberries as 'excellent' as compared to the store purchased ones which were rated 'very good.' People were willing to pay 33 % more for the local berries. 'Albion' was the most favored cultivar for color, size,

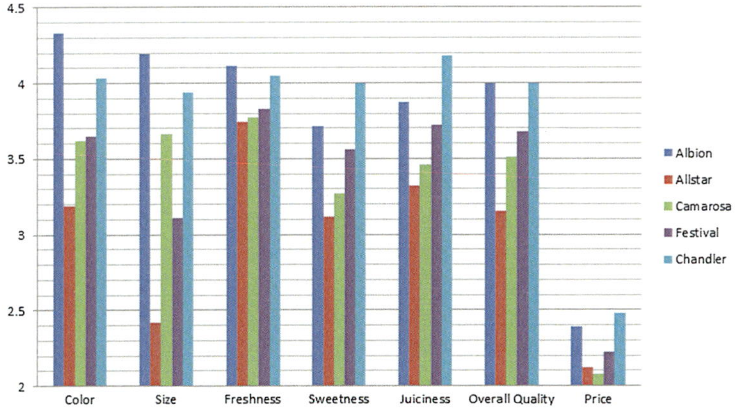

Fig. 2.7 Consumer preference of local varieties based on various attributes. Data from a Consumer Preference Survey conducted at Tennessee State University in Nashville (May 2014)

and freshness. 'Chandler' ranked number one in sweetness and juiciness. Both 'Albion' and 'Chandler' ranked highest for overall quality.

In China, size, color, and sweetness are preferred, as evidenced from the popularity of 'Confidante.'

2.12 Current Constraints and Future Potentials

While there continues to be a demand for locally produced organic berries, there are particular challenges related to growing strawberries in Tennessee. Many farms currently producing strawberries throughout the southeast have converted from intensive tobacco production to intensive strawberry production (McWhirt et al. 2015). This practice of using a single plot of land for annual production leads to a decline in nutrients and soil organic matter. Each year, plants are replanted in the same location, possibly due to land availability issues. Due to a lack of crop rotation, insect and pathogen populations increase, often resulting in an increased need for chemical inputs, whence destroying helpful biological life that contributes to soil health. The long-term viability of strawberry production needs to be maintained. Production systems must incorporate practices that not only improve the short-term productivity of the land but also contribute to long-term land stewardship (McWhirt et al. 2015).

The National Strawberry Sustainability Initiative administered by the University of Arkansas and funded by a grant from the Walmart Foundation has been supporting research at land-grant public universities across the country to study sustainable strawberry production that will benefit not only consumers, but also provide an economic boost for local farmers, from the production level through the supply chain to the market and finally to the consumer. Tennessee State University (Nashville) and the University of Tennessee (Knoxville) have strong agricultural research programs to solve problems facing farmers. The aim of research conducted is to expand the areas where crops such as strawberries can be grown, enabling shorter trips for the berries between farm and consumer. This helps provide the much needed boost for organic strawberry research to address critical issues facing the marketability of fresh strawberries. The county extension agents at these universities seek to move the science and technology developed in laboratories and experimental farms into the producers' fields.

In China, there is a need for virus-free strawberry nurseries to be established for providing growers with healthy plant varieties of excellent quality, high productivity, and strong stress resistance. Different cultivation procedures and protocols have to be established and standardized for major varieties, cultivation systems, and control of pests and diseases to further improve berry quality and yield.

2.13 Food Safety Concerns

Changing population structures and demographics along with a growing awareness of food safety issues have created a market for fresh strawberries that are free of microbial contamination, particularly those organisms that are known to cause diseases in humans. The increased consumption of fresh strawberries brings an higher risk of foodborne illnesses. Strawberries have been the culprit in numerous outbreaks of *Salmonella*, hepatitis A, norovirus, and *E. coli O157:H7* (Palumbo et al. 2013). An *E. coli* O157:H7 outbreak in 2011 occurring in Oregon resulted in 15 illnesses and two deaths (Palumbo et al. 2013). It was the first reported instance in which strawberries contaminated by deer feces were associated with an *E. coli* outbreak (Laidler et al. 2013). Ensuring the safety of fresh strawberry crop is of prime importance as every incidence of illness resulting from fresh fruits and vegetables contaminated with pathogenic microbes erodes consumer confidence and results in significant losses to the growers and retailers. The use of sanitizing solutions or vigorous washing is not possible for strawberries without causing mechanical damage to the delicate structure of the berries. Providing specific guidance and strategies to minimize potential contamination is important for all parties engaged in the production, marketing, and consumption aspects of the industry.

Consumers of fresh produce are always concerned about microbial populations whether organic or non-organic. A recent study on fresh produce-associated bacteria found that fruits and vegetables harbor diverse bacterial communities. Bacterial population from organically-grown produce is less complex than inorganic analogs. Thus, consumers are exposed to substantially different bacteria when eating conventionally and organically farmed varieties (Leff and Fierer 2013).

In March 2015, the North Carolina Strawberry Association along with North Carolina State University offered a workshop to provide farmers with strawberry-specific tools needed to identify potential food safety concerns. Strategies are needed to minimize potential contamination. Zhou and her group at Tennessee State University have developed an easy to use tool (dip-stick assay) to rapidly and reliably detect the presence of human pathogens on fresh strawberries, thus providing a guarantee on the safety of the produce. Gazula et al. (2014) studied the survival analysis and detection of pathogens (*Salmonella typhimurium, Listeria monocytogenes and Escherichia coli*) on mature strawberries from participating farms as part of a project conducted at Tennessee State University in Nashville and funded by the National Strawberry Sustainability Initiative (2013–2014). Results showed that the surface of strawberries did not support the growth of these pathogens (Unpublished data) suggesting that harvesting of strawberries following good agricultural practices reduces the risk of contamination. Having a science-based strategy and providing specific guidance on how to manage microbial contamination are very important for all parties engaged in the production, marketing, and consumption aspects of the industry. It eliminates the risk of foodborne diseases and the financial loss due to recall of contaminated foods.

2.14 Conclusion

Consumer preference for purchasing organic fresh strawberries is a growing trend in the fresh strawberry market. Therefore, it is highly desirable to develop an organic strawberry production system in Tennessee and neighboring states in the US, as well as in China as the living standard of the citizen continures to improve. The unique characteristics inherent to strawberry farming such as ability to harvest a high-yield crop on a small amount of land and heavy consumer demand support an environment for small farmers to operate successful businesses. Based on an assessment of available information, we have identified potential problems and possible solutions that will support the establishment of a sustainable strawberry industry, especially as an alternative crop for small farmers.

Acknowledgements This work was funded by a grant from the Walmart Foundation and administered by the University of Arkansas System Division of Agriculture Center for Agricultural and Rural Sustainability.

References

Arancon NQ, Edwards CA, Bierman P, Welch C, Metzger JD (2004) Influences of vermicomposts on field strawberries: 1 effects on growth and yields. Bioresour Technol 93(2):145–153

Bhowmik PC, Inderjit J (2003) Challenges and opportunities in implementing allelopathy for natural weed management. Crop Protect 22:661–671

Bost S, Buchanan J, Hale F, Hensley D, Straw A (2003) Crop profile for strawberries in Tennessee

Boyette MD, Wilson LG, Estes EA (1914) Postharvest cooling and handling of Strawberries. The Southern Region Small Fruit Consortium. Published by the North Carolina Agricultural Extension Service

Carroll J, Pritts M, Heidenreich C (2014) Production guide for organic strawberries. NYS-IPM Publication No. 226, July 2014. http://nysipm.cornell.edu/organic_guide/strawberry.pdf

Cheng YB, Jiang JH, Cheng YF (2007) Cultivation techniques of Strawberry interplanting gourd stereo. Mod Agric Sci 2 (in Chinese)

Friedman D (2007) Transitioning to organic production. Sustain Agric Res Edu Bull. National Institute of Food and Agriculture, U.S. Department of Agriculture. http://www.sare.org/Learning-Center/Bulletins/Transitioning-to-Organic-Production/Text-Version. Accessed 1 April 2015

Gaskell M (2004) Nitrogen availability, supply, and sources in organic row crops. In: Proceedings of California organic production and farming in the new millennium: A research symposium. Berkeley, CA, University of California Sustainable Agr. Res. and Educ. Program, University of California, Davis, pp 13–20, 15 July 2004

Gazula H, Yan Y, Bhatti S, Chen F, Zhou S (2014) Survival analysis and detection of human pathogens for organic strawberry production in Tennessee. In: Proceedings of the American society for horticultural science. Orlando FL, July 2014

Goble WE (1961) Tennessee's competitive position in producing and marketing strawberries. Univ Tenn Agric Exp Station Bull. http://trace.tennessee.edu/utk_agbulletin/274. Accessed 2 April 2015

Guerena, M, Born H (2007) Strawberries: organic production. A Publication of ATTRA- National Sustainable Agriculture Information Service. https://attra.ncat.org/attra-pub/summaries/summaryphp?pub=13. Accessed 23 March 2015

Himelrick, DG, Powell AA, Dozier Jr. WA (2002) Commercial production strawberry. Published by the Alabama Cooperative Extension System Reprinted, May 2002, ANR-0633

Huo HZ, Mei L, Li JF, Chen XP (2008) Cultivation techniques of relay intercropping Strawberry in greenhouse muskmelon. Jiangsu Agric Sci 6 (in Chinese)

Johnson G. (2009) Fumigation alternatives—Biofumigants. weekly crop update. University of Delaware Cooperative Extension. https://agdev.anr.udel.edu/weeklycropupdate/?p=837

Kirkegaard JA, Matthiessen JN, Wong PTW, Mead A, Sarwar M, Smith BJ (1999) Exploiting the biofumigation potential of Brassicas in farming systems. Tenth International Rapeseed Congress, Canberra Australia. http://www.regional.org.au/au/gcirc/2/125.htm. Accessed 1 April 2015

Laidler MR, Tourdjman M, Buser GL, Hostetler T, Repp KK, Leman R (2013) Escherichia coli O157:H7 infections associated with consumption of locally grown strawberries contaminated by deer. Clin Infect Dis 57(8):1129–1134. doi:10.1093/cid/cit468 Epub 2013 Jul 21

Lan W, Qian XQ, Run Z, Zhang DX, Ding WH, Xu SC (2012) Exploration on new culture model of “simplification,concentration and bees” for strawberry (*Fragaria gracilis Losinsk.*) in Fuyang area. J Fuyang Teach Coll (Nat Sci) 29(1) (in Chinese)

Leff JW, Fierer N (2013) Bacterial communities associated with the surfaces of fresh fruits and vegetables. PLoS ONE 8(3): e59310. doi:10.1371/journal.pone.0059310

Liao HJ, Jiang Q, Shen HY, Dong L, Ning ZY, Li WW, Xu XT (2014) Anhui seedling techniques of virus-free Strawberry elevated point. Fruit Trees 10 (in Chinese)

Liu YF (2014) Open field production mode for organic strawberry. Nongmin Zhifuzhiyou yuekan 18 (in Chinese)

Liu CG, Zhu YB (2008) Organic Strawberry production technology. China Veg 6:46–47 (in Chinese)

Liu YH, Lan Y, Wang Z, Zhang L, Qi CH (2008) Cultivation technique of Strawberry interplanting watermelon in solar greenhouse. Technol Innov 10 (in Chinese)

Martinez S, Hand MS, Pra MD, Pollack S, Ralston K, Smith T, Vogel S, Clark S, Lohr L, Low SA, Newman C (2010) Local food systems: concepts, impacts, and issues. Economic Research Report No. (ERR-97) 87 pp, May 2010

May G, Pritts M (1990) Strawberry nutrition. Adv Strawb Prod 9:10–24

McWhirt A, Fernandez G, Schroeder-Moreno M (2015) Sustainable practices for plasticulture strawberry production in the Southeast. Publication AG-796. North Carolina Cooperative Extension. http://content.ces.ncsu.edu/sustainable-practices-for-plasticulture-strawberry-production-in-the-southeast/. Accessed 23 March 2015

Mossler MA (2004) Florida crop/pest management profiles: strawberry. University of Florida Institute of Food and Agricultural Sciences, CIR 1239

Muramoto J, Gliessman SR, Schmida D, Shennan C, Swezey S (2004) Nitrogen dynamics in an organic strawberry production system. In: Proceedings of California organic production and farming in the New Millennium: a research symposium. The International House, University of California, Berkeley, pp 131–134

Muramoto J, Shennan C, Baird G, Zavatta M, Koike ST, Bolda MP, Daugovish O, Dara SK, Klonsky K, Mazzola M (2014) Optimizing anaerobic soil disinfestation for California strawberries. Acta Hortic 1044:215–220

Naeve L (2014) Commodity strawberry profile. Iowa State University Extension and Outreach. http://www.agmrc.org/commodities__products/fruits/strawberries/commodity-strawberry-profile/

Palumbo M, Harris LJ, Danyluk MD (2013) Outbreaks of foodborne illness associated with common berries, 1983 through May 2013 [Internet]. University of Florida, Florida Cooperative Extension Service. http://edis.ifas.ufl.edu/pdffiles/FS/FS23200.pdf. Accessed 1 April, 2015

Pelayo C, Ebeler SE, Kader AA (2003) Postharvest life and flavor quality of three strawberry cultivars kept at 5 °C in air or air +20 kPa CO_2. Postharvest Biol Technol 27:171–183

Pinkerton JN, Ivors KL, Reeser PW, Bristow PR, Windom GE (2002) The use of soil solarization for the management of soilborne plant pathogens n strawberry and red raspberry production. Plant Dis 86(6):645–651

Poling EB (1993) Strawberry plasticulture in North Carolina: II. Preplant, planting and postplant considerations for growing 'Chandler' strawberry on black plastic mulch. HortTechnology 3 (4):383–393
Poling EB, Krewer G, Smith JP (2005) Southeast regional strawberry plasticulture production guide. http://www.smallfruits.org/SmallFruitsRegGuide/Guides/2005culturalguidepart1bs1.pdf
Pritts M (2002) A future for the perennial matted row? Berry Basket 5(1):13
Rhainds M, Kovach J, English-Loeb G (2002) Impact of strawberry cultivar and incidence of pests on yield and profitability of strawberries under conventional and organic management systems. Biol Agric Hortic 19(4):333–353. doi:10.1080/01448765.2002.9754937 (Published online: 24 Apr 2012)
Rivard C, Gawron M, Oxley K, Smith J, Taylor M, Pryor D, Kennelly M (2014) Managing winter injury for annual strawberry production systems in the Great Plain. http://strawberry.uark.edu/Rivard_Poster_Managing_winter_injury_for_annual_strawberry_production_systems_in_the_Great_Plains.pdf
Rom C, Friedrich H, Freeman L, Malvar L, Lehovec B (2014) Moving the needle. Accomplishments of the national strawberry sustainability initiative 2013–2014
Rondon SI, Cantliffe DJ, Price JF (2003) revised in 2009. Biological control for insect management in strawberries. Document No. HS923. University of Florida, Institute of Food and Agricultural Sciences. http://edis.ifas.ufl.edu
Rowley MA, Ransom CV, Reeve JR, Black BL (2011) Mulch and organic herbicide combinations for in-row orchard weed suppression. Int J Fruit Sci 11:316–331
Safley CD, Poling EB, Wohlgenant MK, Sydorovych O, Williams RF (2004) Producing and marketing strawberries for direct market operations. HortTechnology 14(1) (Jan–March 2004)
Sams CE, Kopsell DA (2011) Brassica cover crops and seed meals as soil bio-fumigants in vegetable crop production. In: Proceedings for the vegetable, potato, flower, small fruits and general sessions. Mid-Atlantic Fruit and Vegetable Convention. Pennsylvania Vegetable Growers Association, Richfield, PA. pp 86–88. http://web.utk.edu/~tkarpine/EnhBiofum.html
Schnabel G, Peres NA (2015) Disease and resistance management in strawberry: top considerations for the coming Season. Small Fruit News 15(1):16–18
Shrestha U, Wszelaki AL, Butler DM (2014) Introduction to anaerobic soil disinfestation as a fumigant alternative. UT Extension SP 765-A
Smith BJ (2006) USDA-ARS strawberry Anthracnose resistance breeding program. Acta Hort (ISHS) 708:463–470
Strand LL (1993) Integrated pest management for strawberries. Pub 3351. University of California. p. 15
United States Department of Agriculture/National Agricultural Statistics Service (released 2012) 2011 California certified organic production survey. http://www.nass.usda.gov/Statistics_by_State/California/Publications/Other_Files/201210organics.pdf
USDA, Economic Research Service (2012) Loss-adjusted food availability data. http://www.ers.usda.gov/data-products/chart-gallery/detail.aspx?chartId=30486. Accessed 25 March 2015
Welke SE (2004) The effect of compost extract on yields of strawberries and the severity of *Botrytis cenerea*. Journal of Sustainable Agriculture 25(1):57–68
Wildung D (2000) Flame burning for weed control and renovation with strawberries. Greenbook 2000, Energy and Sustainable Agriculture Program. Minnesota Department of Agriculture
Wright S (2012) High tunnel strawberries. Cooperative Extension service, University of Kentucky College of Agriculture, Food and Environment. Updated April 2014
Wszelaki A, Lockwood D, Martin J (2013) High tunnel strawberry production in Tennessee. UT Extension, SP 754-A
Wszelaki A, Saywell D, Broughton S (2012a) Introduction to organic and sustainable agriculture practices: Defining the terms. UT Extension W235-A
Wszelaki A, Saywell D, Broughton S (2012b) Transitioning to organic farm systems. UT Extension W235-B
Wszelaki A, Saywell D, Smith B (2014) Organic certification in Tennessee. UT Extension W 215

Yan WZ (2010) Non-polluted cultivation technique of Strawberry. Shanghai Science and Technology Press, Shanghai (in Chinese)

Yu YX (2013) Changfeng brand Strawberry probe into the new mode of construction and development. Modern Commerce and Industry. 23 (in Chinese)

Zhang ZH (2012) Guide to safe production technique of Strawberry. China Agriculture Press (in Chinese)

Zhang W, Yang HQ (2010) Standardized production of strawberries full fine solutions. China Agriculture Press (in Chinese)

Zhang GL, Xie HZ and Ren MT (2011) Cultivation technique of Strawberry interplanting tomato in solar greenhouse. Chin Melons Veg 24 (in Chinese)

Chapter 3
Role of Plant Growth-Promoting Rhizobacteria (PGPR) as BioFertilizers in Stabilizing Agricultural Ecosystems

Peiman Zandi and Saikat Kumar Basu

Abstract Non-judicious and over applications of different toxic, synthetic chemical fertilizers lead to several environmental hazards, causing damages to human, animal, and ecosystem health and can even result in unfavorable economic turnaround. Residual chemical fertilizers in aquatic and/or rhizosphere zones could potentially disrupt the natural ecosystem balance severely hampering both agricultural productivity and initiate several critical health issues. To avoid such environmental, agricultural, and health crises, serious attention has now been shifted toward the production of environmentally friendly biofertilizers with higher economic returns and better financial gains in comparison with conventional synthetic chemical fertilizers. Under intensive agricultural practices, application of biofertilizers is of particular importance in increasing soil fertility and ensures right movement toward sustainable agriculture. To improve the agricultural productivity and yield stability, utilization of conducive terricolous microorganisms such as rhizobacteria, as biofertilizers, has been found to be of quite important under in case of modern agricultural management. The present review was aimed to elucidate firstly the main conceptions of rhizosphere and rhizobacteria, and secondly the direct/indirect functions of rhizobacteria-mediated plant growth promotion.

Keywords Plant growth-promoting rhizobacteria · Biofertilizer · Agriculture · Yield

Abbreviations

ABA Abscisic acid
ACC 1-aminocyclopropane-1-carboxylate

P. Zandi (✉)
Department of Agronomy, Takestan Branch, Islamic Azad University,
Takestan PO. Box 466 Iran
e-mail: z_rice_b@yahoo.com

S.K. Basu
Department of Biological Sciences, University of Lethbridge,
Lethbridge AB T1K 3M4, Canada

D. Nandwani (ed.), *Organic Farming for Sustainable Agriculture*,
Sustainable Development and Biodiversity 9,
DOI 10.1007/978-3-319-26803-3_3

AM	Arbuscular mycorrhiza
CK	Cytokinins
ET	Ethylene
GAs	Gibberellins
GB	Glycine betaine
HCN	Hydrogen cyanide
IAA	Indole-3-acetic acid
MHB	Mycorrhization helper bacteria
MOA	Mechanism of action
PC	*Pseudomonas chlororaphis*
PEG	Polyethylene glycol
PF	*Pseudomonas fluorescens*
PGPR	Plant growth-promoting rhizobacteria
ePGPR	Extracellular PGPR
iPGPR	Intracellular PGPR
PGRs	Plant growth regulators
PH	Plant hormones (phytohormones)
PSR	Plant systemic resistant

3.1 Introduction

The crop yield can be highly affected by symbiotic relations established between microorganisms in the rhizosphere (or rhizobacteria) and plant species. The term 'PGPR' refers to a group of bacteria that benefits plant growth and development (Antouni et al. 1998). The so-called various bacterial genera (or soil vital components) contribute to different biotic activities to bring the mobility (nutrient circulation) and stability (crop yield) to the soil ecosystem (Chandler et al. 2008). Until more recently, most of the studies were focused on Gram-negative genera (e.g., *Fluorescent pseudomonads* with the greatest strains) of PGPR (Antouni et al. 1998). However, there are many reports suggesting the close relationship of Gram-positive bacteria, such as *Bacillus*, to PGPR bacteria that can colonize roots. At first, the PGPR studies were concentrated on root crops such as sugar beet (*Beta vulgaris* L.), tomato (*Solanum tuberosum* L.), and radish (*Raphanus sativus* L.); however, in later stages, a broad array of other hosts such as legumes, non-legumes, and trees were also covered (Antouni et al. 1998). PGPR can promote plant growth directly through biosynthesis of numerous phytohormones (or plant growth regulators), increase in uptake capability of mobilized nutrients (Antouni et al. 1998; Tank and Saraf 2010), inciting the plant systemic resistant (PSR) against microbial pathogens (Antouni et al. 1998; Russo et al. 2008; Braud et al. 2009), modifying soil structure (Ahemad and Kibret 2014), and bioremediation of contaminated soil through degrading environmentally damaging xenobiotic substances, e.g., pesticides (Ahemad and Khan 2012a) and isolating toxic heavy metals (Hayat et al. 2010) through their detoxifying potentials (Wani and Khan 2010). Being versatile in transformation, mobilization, solubilization, and recycling of soil nutrients, the rhizobacteria inhabiting in rhizoplane are determinant in fortifying soil fertility (Glick 2012; Ahemad and Kibret 2014).

The antagonistic impacts of PGPRs over root pathogens have been approved by earlier works (Braud et al. 2009; Hayat et al. 2010; Ahemad 2012; Ahemad and Kibret 2014). The indirect function of PGPR often occurs when microbial balance in rhizosphere, by which, is modified. In other words, PGPR by preventing the activity of other deleterious microorganisms surrounding the roots stimulates the plant growth and development (Ahemad and Kibret 2014). Root pathogens can be influenced by a series of mechanisms usually implemented by rhizobacteria such as producing antibiotics, hydrogen cyanide (HCN), or siderophores (Antouni et al. 1998; Russo et al. 2008; Jahanian et al. 2012). *Rhizobium* bacterium or its partner *Bradyrhizobium* both are able to form N_2-fixing nodules on legume roots. Not only legumes such as beans or fenugreek (*Trigonella foenum-graecum* L.) are under the exposure of being nodulated by their relevant symbionts (Zandi et al. 2014) but even the roots of non-legume crops have the potential or enough talent to be affected by other nodule producing/inducing bacteria (Ahemad 2012). The colonization process is regarded as the first step of interaction between beneficial bacteria and plants (Antoun and Prévost 2005). The formation of hypertrophies or root hair curling (nodule-like structure) induced by PGPR bacteria has also been observed in case of non-host legumes (lupine—*Lupinus albus* L.; pea—*Pisum sativum* L.) (Trinick and Hadobas 1995) and non-legumes (corn—*Zea mays* L.; oilseed rape—*Brassica napus* L.; asparagus—*Asparagus officinalis* L.; wheat—*Triticum aestivum* L.; oat—*Avena sativa* L.; rice—*Oryza sativa* L.; *Arabidopsis thaliana* (L) Heynh.) (Antouni et al. 1998).

Recently, there has been an increasing demand for the over application of symbiotic [*Rhizobium* (e.g., *R. oryzae*), *Bradyrhizobium* (e.g., *B. japonicum*), *Mesorhizobium* (e.g., *M. loti*)] and non-symbiotic [*Azomonas* (e.g., *A. agilis*), *Pseudomonas* (e.g., *P. fluorescens*), *Klebsiella* (e.g., *K. pneumonia*), *Bacillus* (e.g., *B. thuringiensis, B. subtilis*), *Azotobacter* (e.g., *A. chroococcum*), *Azospirillum* (e.g., *A. brasilense, A. amazonense*)] rhizobacteria as bioinoculants, under diverse biotic/abiotic stress conditions caused by herbicides, insecticides, fungicides, heavy metals, salinity, and draught stress (Wani and Khan 2010; Ahemad and Khan 2011a, b, 2012b). Although rhizobacteria have been well known to enhance plant growth, the actual process of microbial interaction mediating plant growth is still under active investigation (Zaidi et al. 2009).

3.2 Rhizosphere: A Confident Habitat for Rhizobacteria

The term rhizosphere is referred to a narrow zone in top soil encompassing the root system, while a group of heterogeneous bacteria in the rhizosphere with root-colonizing capability are known as rhizobacteria (Ahemad and Kibret 2014).

In other words, the part of the soil influenced by plant roots, their exudates, and root hairs is defined as rhizosphere (Dessaux et al. 2009). The rhizosphere (soil), rhizoplane (root surface with soil-adhering particles), and the root itself are known as three principal components of the soil (Barea et al. 2005). Irrespective of their mechanical backing and water/nutrient uptaking facilities, the roots individually are able to synthesize, gather, and exude a diverse array of chemical compounds that are mainly attractant to a large number of soil microbial communities (Walker et al. 2003). The root secretions modify physicochemical properties in the soil zone and, hence, help to modulate the structural basis of soil microbial community in the rhizosphere (Kang et al. 2010).

The exudates are also responsible for promoting the plant–bacterial beneficial symbiotic associations and inhibiting the growth trend in competing plant species (Nardi et al. 2000; Dakora and Phillips 2002). In return, the soil microbial activity impacts the nutrient supply to the roots and root template and, thus, modifies the root secretion's quality and quantity. Secretions are plant-derived small organic molecules in a way that their composition is deeply dependent on plant species, plant physiological status, and type of microbes (Kang et al. 2010). Among which, there are some with attractant properties to ensnare the microbes, and others are microbial repellants (Nardi et al. 2000). After releasing the root exudates, a fraction of them [i.e., nitrogen and carbon ($\sim$5–21 %) sources] are further metabolized by root neighboring microbes, and then, some of the molecules are reabsorbed by roots for growth and development (Marschner 1995). Endophytes and other microbes have the ability to colonize the root system within the rhizosphere layer (Barea et al. 2005). The root colonization is identified as colonization of root tissue or rhizoplane, while rhizosphere colonization goes back to the soil volume colonized under the influence of root secretions (Ahemad and Kibret 2014).

3.3 Plant Growth-Promoting Rhizobacteria (PGRP)

The plant growth-promoting rhizobacteria (PGPRs) must possess three main features: (1) the proficiency to colonize the root surface, (2) their survival, multiplication, and competition with residual microbiota when necessary, and (3) their promoting and protective roles (Ahemad and Kibret 2014). Among the total rhizobacteria reintroduced into the soil by the action of bioinoculation (comprising of a vast array of competitive microflora), only 2–5 % of them are reported to demonstrate their beneficial effects over plant growth and are known as PGPRs (Kloepper and Schroth 1978). The beneficial effect of soil rhizobacteria greatly depends on the soil and plant conditions, type of bacterial strains, and the environment in which they promote plant growth (Şahin et al. 2004). The PGPRs burgeoning in rhizosphere niche, where the plant tissues are located, stimulate the plant growth by direct/indirect multiple mechanisms (Vessey 2003). According to Antoun and Prévost (2005), they can be classified into four main groups based on their functional activities which include (a) an increased availability of nutrients to plants or biofertilizers, (b) enhanced plant growth through phyto-based hormones or phytostimulators, (c) an abated rate in organic pollutants or rhizoremediators, and (d) the controlled capacity of disease through production of antifungal and/or antibacterial metabolites or pesticides. Generally, the PGPR associations, which are mainly related to the degree of bacterial vicinity to the roots and the intimacy of an established association, are divided into extracellular (ePGPR) and intracellular (iPGPR) types.

The 'ePGPR' associations can be found in rhizosphere and rhizoplane or around the cortex cells of roots, while the 'iPGPR' associations are created inside the root cells in nodular structures (Figueiredo et al. 2011). Some of the promoting substances secreted by PGPRs are as follows: IAA (indole-3-acetic acid/indole acetic acid), siderophores, HCN (hydrogen cyanide), ammonia, exo-polysaccharides, phosphate solubilization, heavy metal mobilization, nitrogenase activity, ACC (1-aminocyclopropane-1-carboxylate) deaminase, biocontrol potentials, antifungal activity, N_2 fixation, induced systemic resistance, Zn solubilization, Zn resistance, Pb and Cd resistance, antibiotic resistance, gibberellin, kinetin, metal resistance, and cytokinin (Dakora and Phillips 2002; Ahemad and Khan 2012c; Bhattacharyya and Jha 2012). Furthermore, *Chromobacterium, Agrobacterium, Bacillus, Arthrobacter, Azospirillum, Caulobacter, Erwinia, Micrococcous, Azotobacter, Burkholderia, Pseudomonas*, and *Flavobacterium* belong to the 'ePGPR' groups, while *Bradyrhizobium, Rhizobium, Azorhizobium, Allorhizobium*, and *Mesorhizobium* have been reported to be represented by the 'iPGPR' groups (Bhattacharyya and Jha 2012; Ahemad and Kibret 2014). In connection with microbial communities existing in the rhizosphere niche, we should note that there are numerous soil-inhabiting actinomycetes with beneficial features affecting plant growth (Merzaeva and Shirokikh 2006; Ahemad and Kibret 2014). Among these, *Thermobifida* sp., *Micromonospora* sp., and *Streptomyces* spp. have the potential to act as biocontrol factors against the fungal pathogens in the root environment (Russo et al. 2008; Franco-Correa et al. 2010; Salcedo et al. 2014).

3.3.1 Nitrogen Biofixation

Nitrogen is considered as a key nutrient in producing agricultural crops and one of the main constituents of protein synthesis, nucleic acids, and the other essential cellular components (Franche et al. 2009; Martínez-Viveros et al. 2010). Although nitrogen constitutes 78 % of the air volume, it is still of the most limiting factors for plant growth in which to address its deficiency, nitrogen fertilizers are applied, but this issue, by itself, also causes the cost of production to be increased (Franche et al. 2009). Under these conditions, the utilization of fixed atmospheric nitrogen attained by symbiotic relationships in leguminous cover crops (affected by *Rhizobium* species) and/or non-symbiotic mutualism in non-leguminous crops (affected by free-living molecular N_2-fixing microorganisms) mainly through nitrogen biofixation process supplies the required nitrogen input for arable soils and helps with the replacement of depleted soil nitrogen reserves as an appropriate option (Martínez-Viveros et al. 2010). Different estimations have defined the contribution rate process of nitrogen biofixation in supplying soil nitrogen as 44–200 kg/ha annually and in average 140 kg/ha of pure nitrogen per annum (Franche et al. 2009). Nitrogen biofixation is considered as the first proposed mechanism to enhance plant growth by the plant growth-promoting rhizobacteria. Although native populations of these bacteria are found in the soil, they likely may not be capable of fulfilling the assumed performance in terms of nitrogen fixation. So the efficient and effective strains of such bacteria are applied in the form of biofertilizers.

Most bacterial species, such as *Azotobacter*, *Azospirillum*, *Bacillus*, *Beijerinckia*, *Clostridium*, *Enterobacter*, *Pseudomonas*, and *Spirillum lipoferum*, are capable of fixing nitrogen. Enhancement in the yield of different cereals was achieved by inoculation via N_2-fixing bacteria in several studies (Ozturk et al. 2003; Şahin et al. 2004). The N_2 fixation can help restore nitrogen balance in the plant and stimulate nitrogenase activity in the inoculated roots. Recent researches have shown that the N_2-fixing process via microorganisms is an energy-requiring reaction which is provided by accessible organic carbon (it is used to break down the linkage among nitrogen atoms) (Martínez-Viveros et al. 2010). For this reason, application of various green, organic, and some other chemical fertilizers affect the efficacy and activity of *Azotobacter* (Wani et al. 1988). Free-living and symbiotic N_2-fixing bacteria are regarded as the most important non-leguminous inoculants of plants, especially for the cereals. The so-called free-living bacteria can fix to about 15 kg/ha pure nitrogen annually and the root mutualistic (or symbiotic) bacteria to 30 kg/ha pure nitrogen per year. The amount of nitrogen fixed through the action of such bacteria can be effective in a long-term preservation of soil fertility (Şahin et al. 2004).

3.3.2 Increased Absorption and Availability or Nutrient-Solubilizing Ability in the Area Around the Roots

Different reports suggest that PGPR bacteria increase plant growth through facilitation in uptaking of nutrients and minerals [$K(K^-)$, $P(H_2O4^-)$, $N(NO_3^-)$], as well as micronutrients (Barber 1985). There are several discussions about the PGPR mechanisms in the absorption of mineral elements. Many studies have shown that rhizosphere organisms increase the capability of mineral absorption by plant roots, but there is lack of proper interpretation for accepting this. On the other hand, it is reported that increase in mineral absorption by plants is mainly because of increase in the extension of root system, increase in root numbers, thickness (fresh and dry weight), and root length and is not relevant to any specific mechanism for increasing ion absorptions (Barber 1985; Biswas et al. 2000). For example, more absorption of 'Fe' and 'K' with thicker roots and higher absorption of 'P' are in association with increase in root hair growth (Barber 1985).

Other studies show that growth-promoting bacteria are capable of changing morphology and successfully increasing the root absorption area. These changes due to inoculation with bacterium increase the mineral element absorption by the plants. In one study, it has been shown that inoculation with *Azospirillum brasilense* influences membrane activity following H^+ extrusion from the roots, likely caused by releasing a signal (Bashan 1991). Emission (passing out) of protons (H^+) from root cells that acidifies rhizosphere zone has been proposed as the main mechanism involved in the increased mobility of minerals (Marschner et al. 1986). Production of organic acids by both the plants and bacteria in the rhizosphere decrease the soil pH and eventually increases the access to mineral elements such as P, Ca, Fe, and Mn (Ahemad and Kibret 2014).

3.4 The Biosynthesis of Plant Growth-Regulating Substances (PGRS)

Fairly recently, several evidences have demonstrated that PGPRs besides their ability of fixing atmospheric nitrogen can be effective in the biosynthesis of different phytohormones (PH). PHs are known as plant growth regulators (PGRs) that have shown to have important role in plant growth and development. PGRs are organic matters that influence plant physiological processes at very low concentrations (Arshad and Frankenberger 1998). For the reason that concentration of hormonal secretions is a diagnostic indicator for regulating the physiological processes within the plant, any positional change in the level of plant hormones may lead to some variations in the characteristics of growth and development of plants (Ahemad and Kibret 2014). Among important plant hormones, auxins, gibberellins (GAs), cytokinins (CK), abscisic acid (ABA), and ethylene (ET) can be mentioned.

Many studies done in recent years suggested that inoculation of cereal grasses with *Azospirillum brasilense* improves the plant growth and productivity in many cases. After many investigations, it was specified that such bacteria by producing auxin around both sides (under and/or upper) of the root stimulate plant growth (Quispel 1991). Also, auxin biosynthesis has been already detected by many *Azotobacter* strains. This hormone especially involves in several developmental processes such as root elongation (lengthwise), increase in root volume, influencing the number of root hairs, and permeability of root cells and finally root exudates (Ahemad and Kibret 2014).

3.4.1 The Importance and Regulation of Ethylene Level in Plants

Another mechanism by which the plant growth improvement has been authenticated is mainly associated with ethylene (ET). Ethylene is a potent plant growth regulator that is effective in many aspects of plant growth development and senescence. ET promotes lateral root growth, root hair structure, and germination. It also releases the primary and secondary seed dormancy (KeÇpczyński and KeÇpczyńska 1997). Depending on the concentrations, the type of physiological processes, and the specific plant growth stage, ET can stimulate or inhibit plant growth (Martínez-Viveros et al. 2010). The synthesis of ET at lower levels increases primary growth and root extension but at higher levels leads to inhibition of root elongation (Mattoo and Suttle 1991). Growth-promoting bacteria can increase plant growth through mechanism of lowering ET levels in plant. This process is mainly in association with the enzymatic activity of 1-aminocyclopropane-1-carboxylate (ACC) deaminase (Ahemad and Kibret 2014). For instance, *Pseudomonas putida* CR12-2 bacterium having the enzyme ACC deaminase hydrolyzes ACC substrate. ACC is a precursor and primary substrate responsible for ET biosynthesis in plants (Ahemad and Kibret 2014).

This model describes that, at first, this bacterium attaches to the plant seed cover and surrounds the seed. Next, by hydrolyzing ACC as the only source of nitrogen for its growth converts it to ammonium cation and α-ketobutyric acid, thereby decreasing the ET level (Glick et al. 1998; Martínez-Viveros et al. 2010). The amount of ET around the seedling roots does not go higher than a specific level. On the other hand, it has been demonstrated that most of the *Pseudomonas* strains make auxin, and production of auxin in these bacteria causes the activation of ACC deaminase synthesis pathways (Martínez-Viveros et al. 2010). Therefore, it can efficiently decompose the ET precursor. Some bacteria such as *Pseudomonas putida* containing the ACC-decomposing enzymes enhance the plant growth through hydrolyzing the ACC contents available in the seed sprouts. Shaharoona et al. (2006) demonstrated that inoculation with specific *Pseudomonas* strains in corn (*Zea mays* L.) leads to significant increase in plant height, root weight, and total biomass compared to the control. It seems that by decreasing the inhibitory functions of ET in the roots, these strains support crop yield through improved root growth. Recently, it has been found that seed yield and root growth in plants inoculated with growth-promoting rhizobacteria are improved as a result of elevated activity of the ACC deaminase (Belimov et al. 2002).

3.4.2 Siderophore Biosynthesis

Microbial siderophores are recognized as relatively medium-sized organic molecules with molecular mass ~1000–1500 Da (daltons) that have high affinity in binding or chelating with soluble ferric ions (Fe^{+3}) (Milagres et al. 1999; Ahemad and Kibret 2014). The transmission of these iron-chelating compounds (or Fe^{+3}-binding agents) into plant cells is made feasible by specific receptors through active absorption. It has been reported that *Pseudomonas fluorescens* under conditions in which the rhizosphere lacks adequate Fe^{+3} (or encounters Fe^{+3} deficiency) expose the pathogens to severe Fe deficiency and make them passive by producing a large number of specific siderophores (*Pseudobactin*, *Pyoverdine* (Pvd), *pyochelin* (Pch), etc.) that cannot be utilized by the pathogens (Rajkumar et al. 2010; Ahemad and Kibret 2014). Thus, siderophores serve as a solubilizing factor (they reduce the insoluble Fe^{+3} to Fe^{+2} on bacterial membrane and release them into the root cells) forming stable complexes of Fe^{+3}-siderophore under Fe starvation conditions (Rajkumar et al. 2010; Ahemad and Kibret 2014). It is essential to note that the solubility coefficient of Fe compounds is very low so that with every unit increase in the soil pH, its rate decreases 1000 times. Hence, in soil conditions with pH scales greater than 4, these compounds are often observed in their non-absorbent forms (Rajkumar et al. 2010).

3.4.3 The Biosynthesis of Vitamins

Plants under perfect growth conditions usually synthesize adequate amount of vitamins. But due to abiotic stresses (such as drought, temperature fluctuations, and mineral deficiency), often the stress-induced plants with vitamin deficiency suffer from poor yield performance. Vitamins are able to neutralize the negative consequences of lacking adequate minerals, and under natural conditions, plant growth and yield after applying vitamins are increased (Martínez-Toledo et al. 1996). Sometimes, the vitamins are added to the list of compounds produced by PGPR bacteria that have recently been found to play an important role in promoting plant growth.

The researchers reported that *Azospirillum*, *Azotobacter*, and *Rhizobium* strains are capable of synthesizing some or all of the water-soluble B vitamins, including niacin (vitamin B_3), pantothenic acid (vitamin B_5), thiamine (vitamin B_1), riboflavin (vitamin B_2), cyanocobalamin (vitamin B_{12}), pyridoxine (vitamin B_6), and biotin (vitamin H) in defined medium (Martínez-Toledo et al. 1996; Revillas et al. 2000; Ahemad and Kibret 2014). Evidences show that roots are able to absorb vitamin B from exogenous sources that create positive impacts on root extension, stem length, and dry matter production and in the absorption of nutrients (Martínez-Toledo et al. 1996).

3.4.4 The Production of Antibiotics

The antibiotics play important roles in preventing numerous plant diseases. An important rhizobacteria capable of antibiotic production is *Pseudomonas fluorescens* (PF). Phenazine derivatives have been isolated from this bacterium as the first identified antibiotics in biocontrol (Martínez-Viveros et al. 2010). The bacterium also produces pyoluteorin (an antifungal antibiotic) which can cause the inhibition of oomycetes such as the soil-born plant pathogen, *Pythium ultimum* Trow. Moreover, the effect of phenazine produced by *Pseudomonas chlororaphis* (PC) to cope with all types of fungi has been approved by some researchers (Chin-A-Woeng et al. 2000).

3.4.5 Cyanide Biosynthesis

The indirect method of function in PGPR bacteria is the biological control of phytopathogenic factors. Various accessions of biocontrolling agents may exhibit different responses to specific media for diffusing distinctive secondary metabolites (Martínez-Viveros et al. 2010). As mentioned earlier, metabolites produced by biocontrol agents are often harmful to target organism and in most cases do not cause toxicity to the host plant. However, such a case typically traces back to the content and kind of metabolites produced, sensitivity of target organism against metabolite, and tolerance level in host plant (Martínez-Viveros et al. 2010). For example, the hydrogen cyanide (HCN) excreted by bacterium PF eradicates tobacco black root rot that is caused by the fungus *Thielaviopsis basicola* (Berk. & Br.) (Martínez-Viveros et al. 2010). Nevertheless, overproduction of HCN (having antifungal features) may be lethal to host plant. For producing HCN enough Fe is required. It has been hypothesized that some siderophore-producing plants increase yield by reducing the synthesis of cyanide by decreasing available Fe (a prerequisite factor in the biosynthesis of cyanide).

3.5 Improvement of Plant Resistance Against Abiotic Stresses

In 1980s, some experiments were conducted in order to examine the impact of *Azospirillum* bacteria on the stress-imposed plants. Sarige et al. (1988) reported that sorghum (*Sorghum bicolor* L Moenech) plants inoculated with *Azospirillum brasilense* were less affected by drought stress. Their research results suggested that the inoculated crops had higher water content in the vegetation canopy in comparison with non-inoculated crops. It was also reported that the application of inocula leads to increase in the water potential of the leaves and decrease in the canopy temperature as compared to non-inoculated ones. The inoculated plants with *A. brasilense* absorbed more moisture from the soil. In this experiment, employment of inoculant caused the water to be drawn out from deeper layers of the soil profile. Therefore, increase in sorghum yield in inoculated plants was dependent on the improved and efficient utilization of soil moisture. In another study, the effect of *A. brasilense* inoculation on the water relations of two wheat varieties increased wheat seedling growth in saline and dark conditions (Creus et al. 1998). Indeed, inoculation with *Azospirillum* bacteria significantly increased the water relation parameters in intercellular (apoplastic) parts of plant tissues. In this study that was conducted under hydroponic conditions, the plants did not expose to (or receive) any nutrient, and hence, their improved water condition and increase in their growth trend were not correlated with increased absorption of nutrients (Creus et al. 1998). In their more recent studies, Creus et al. (2004) found that despite regular osmotic adjustment occurring during drought stress in both non-inoculated and *Azospirillum*-inoculated wheat cultivars, the inoculated cultivars had more improved water relations, lower grain yield loss, and higher grain reserves of Ca, Mg, and K.

Also, in a hydroponic system with no application of nutrient solution, *A. lipoferum* mitigated the negative effects of drought stress on the wheat seedlings (Bacilio et al. 2004). To maintain the balance of environmental osmotic stress and cellular protection, *Azospirillum* species can accumulate organic compounds (e.g., proline, reduced sugars, and ions) which are called organic osmolytes (osmotic regulators). All osmolytes accumulated in several strains of *A. brasilense* have been identified by imposing an osmotic stress of NaCl. In *Azospirillum* species, trehalose, glycine betaine (GB), glutamate, and proline have been characterized as dominant and compatible osmolytes (small solutes) that seem to have an important role in adaptation to saline fluctuations (osmoadoptation) (Bacilio et al. 2004).

It has been proved that in the genus *Azospirillum* resistance to higher doses of sodium chloride (salt), sucrose or polyethylene glycol (PEG) is increased in *Azospirillum amazonense*, *A. lipoferum*, *A. brasiliense*, and *A. halopraeferens*, respectively, which implies the existence of an osmoregulatory mechanism as a result of increase in osmolyte biosynthesis and/or its absorption from soil environment (Hartman 1989). *A. halopraeferens* as an osmotolerant species can absorb choline and convert it into the osmoregulator/osmoprotectant 'GB.' In some of the salt-tolerant plants, it has been recognized that choline absorption exists in root exudates. In *A. halopraeferens* and *A. brasiliense*, 'GB' promotes plant growth and N_2 fixation under salt stress conditions (Hartman 1989). These bacteria can improve wheat resistance to saline conditions through nitrogen fixation and producing hormones such as auxin. For these reasons, utilization of their strains as inoculant is recommended for improving plant yield, especially grasses in arid regions (Saatovich 2006).

3.6 PGPRs Function Through a Series of Mechanisms

Studies indicate that PGPRs operate through multiple mechanisms and/or mutual interactions. In this regard, the *Azospirillum* bacterium can be referred as having the concurrent ability to process atmospheric nitrogen (bio fixation), to solubilize P, and to produce plant growth-regulating substances (Banerjee et al. 2006; Zaidi et al. 2009). Also, it can be inferred that even those group of bacteria with only one type of function or mechanism of action (MOA) can increase growth in the host plant by integrating their separate effects (synergistic effects).

3.6.1 Association with Its Surrounding Environment: Interaction with Other Soil Microorganisms

Irrespective of the individual effects of PGPRs, the promoting process can be modified by means of a double inoculation with other microorganisms through synergistic or additive effects (Ahemad et al. 2009). The combined/simultaneous/dual inoculation of *Rhizobium* alongside *Azotobacter* or *Azospirillum* leads to an increase in dry matter production, seed yield, and higher amount of nitrogen in different leguminous crops as compared to single inoculation (Rodelas et al. 1999). Similar results were also reported by Tilak et al. (1982) on sorghum and maize and by Rai and Gaur (1988) on wheat. Positive results in conjunction with dual inoculation of legumes have been attributed to faster nodulation, increase in the number of nodes, more nitrogen fixation, and increase in root development. Although the concomitant application of *Azospirillum* and *Rhizobium* does not always lead to an increase in nodulation, even in some cases nodulating ability of *Rhizobium* is restricted in host plant (Ahemad et al. 2009). Moreover, increase or inhibition in nodulation depends upon the concentration of bacterium and time of inoculation (Ahemad et al. 2009). Experiments conducted in hydroponics system showed that inoculation with *Azospirillum brasilense* increased the excretion of flavonoids from the roots of bean seedlings (Burdman et al. 1996). Flavonoids with three important functions in higher plants (i.e., N_2 fixation through symbiosis, UV filtration, and pigmentation for flower coloration) lead to the expression of genes involved in root nodule formation by rhizobia in the roots (Burdman et al. 1996). Therefore, increase in flavonoid production by root can be regarded as another contributing factor in stimulating nodule formation by growth-promoting bacteria. Although most of the studies had *Azotobacter* as the bacterium with the potential to establish synergistic associations with various soil microbes, however, some reports imply that *Azotobacter* has an antagonistic relationship too. For example, growth control in some fungal species by *Azotobacter* can be pointed. Some strains of *Azotobacter* by synthesizing antibiotic substances inhibit fungal growth (e.g., *Alternaria, Helminthosporium* and *Fusarium*) on synthetic media (Subba Rao 1988).

3.6.2 The Role of PGPRs in the Efficiency of Arbuscular Mycorrhizal Fungi

Nowadays, it has been elucidated that a group of rhizobacteria or [mycorrhization helper bacteria (MHB)] help to increase colonization of mycorrhizal fungus in the root zone (Wu et al. 2005; Rigamonte et al. 2010). The synergistic effect of arbuscular mycorrhiza (AM), *Azotobacter*, and/or *Azospirillum* on infected/inoculated roots of maize, barley (*Hordeum vulgare* L.), and rye (*Secale cereale* L.) has been observed (Subba Rao et al. 1985). The 'AM' fungi can increase plant growth, especially in less fertile soils, by boosting uptake of phosphates and other necessary nutrients. Indeed, the N and P content in inoculated plants with *Azospirillum brasiliense* and AM with no fertilizer application was equal to those of non-inoculated plants that had received N and P fertilizers (Gadkar et al. 2001). The growth-promoting bacteria (e.g., diazotrophs) supplying vitamins to the rhizosphere help to increase the mycorrhizal effects due to the dependence of these fungi on certain vitamins (Ahmad et al. 2008). So inoculation with mycorrhizal fungi along with vitamin-producing PGPRs can efficiently increase and improve plant growth.

3.7 Conclusion

Soil is a living system, highly sensitive and vulnerable, and biofertilizers are considered as an important natural and environment-friendly approach for keeping soil essential systems alive and active. Certainly application of biofertilizers is also accompanied with challenges such as difficulty in application, failure to respond quickly and economically, susceptibility of living organisms to their environment, complex relationship among different microbial communities, and antagonistic reaction. But at the same time, the exudates excreted by these inoculating bacteria (PGPRs) located on the root surface are found to be directly/indirectly involved in promoting plant growth and development. The PGPRs assist in plant growth via three approaches: regulating hormonal level, nutrient resource acquisition, and biocontrol of different pathogens. Their application to different plant species under stressed and non-stressed conditions has shown to be a panacea in terms of plants' health and development and in bettering yield. It can be concluded that in future the application of PGPRs as biofertilizers is expected to increase across the globe and could play an important role in stabilizing agricultural ecosystems by reducing the indiscriminate use of synthetic agrochemicals that have long-term negative impact on the local ecosystems and the environment.

References

Ahemad M (2012) Implications of bacterial resistance against heavy metals in bioremediation: a review. IIOAB J 3(3):39–46

Ahemad M, Khan MS (2011a) Response of Greengram [*Vigna radiata* (L.) Wilczek] grown in herbicide-amended soil to inoculation with *Bradyrhizobium* sp. (*vigna*) MRM6. J Agric Sci Technol 13(6):1209–1222

Ahemad M, Khan MS (2011b) Insecticide-tolerant and plant growth promoting *Bradyrhizobium* sp. (vigna) improves the growth and yield of greengram [*Vigna radiata* (L.) Wilczek] in insecticide-stressed soils. Symbiosis 54:17–27. doi:10.1007/s13199-011-0122-6

Ahemad M, Khan MS (2012a) Ecological assessment of biotoxicity of pesticides towards plant growth promoting activities of pea (*Pisum sativum*)-specific Rhizobium sp. strain MRP1. Emirates J Food Agric 24(4):334–343

Ahemad M, Khan MS (2012b) Productivity of greengram in tebuconazole-stressed soil, by using a tolerant and plant growth-promoting *Bradyrhizobium* sp. MRM6 strain. Acta Physiol Plant 34:245–254. doi:10.1007/s11738-011-0823-8

Ahemad M, Khan MS (2012c) Evaluation of plant growth promoting activities of rhizobacterium *Pseudomonas putida* under herbicide-stress. Ann Microbiol 62(4):1531–1540. doi:10.1007/s13213-011-0407-2

Ahemad M, Kibret M (2014) Mechanisms and applications of plant growth promoting rhizobacteria: current perspective. J King Saud Univ Sci 26(1):1–20. doi:10.1016/j.jksus.2013.05.001

Ahemad M, Khan MS, Zaidi A, Wani PA (2009) Remediation of herbicides contaminated soil using microbes. In: Khan MS, Zaidi A, Musarrat J (eds) Microbes in sustainable agriculture. Nova Science Publishers, New York

Ahmad F, Ahmad I, Aqil F et al (2008) Diversity and potential on nonsymbiotic diazotrophic bacteria in promoting plant growth. In: Ahmad I, Pichtel J, Hayat S (eds) Plant-bacteria interactions: strategies and techniques to promote plant growth. Wiley-VCH Verlag GmbH & Co. KGaA, Weinheim, Germany. doi: 10.1002/9783527621989.ch5

Antoun H, Prévost D (2005) Ecology of plant growth promoting rhizobacteria. In: Siddiqui ZA (ed) PGPR: biocontrol and biofertilization. Springer, Dordrecht, pp 1–38

Antouni H, Beauchamp CJ, Goussard N et al (1998) Potential of *Rhizobium* and *Bradyrhizobium* species as plant growth promoting rhizobacteria on non-legumes: effect on radishes (*Raphanus sativus* L.). Plant Soil 204:57–67. doi:10.1023/A:1004326910584

Arshad M, Frankenberger WTJ (1998) Plant growth-regulating substances in the rhizosphere: microbial production and functions. Adv Agron 62:145–151. doi:10.1016/S0065-2113(08)60567-2

Bacilio M, Rodriguez H, Moreno M et al (2004) Mitigation of salt stress in wheat seedlings by a gfp-tagged *Azospirillum lipoferum*. Biol Fertil Soils 40(3):188–193. doi:10.1007/s00374-004-0757-z

Banerjee MR, Yesmin L, Vessey JK (2006) Plant-growth promoting rhizobacteria as biofertilizers and biopesticides. In: Rai MK (ed) Handbook of microbial biofertilizers. Food products. Academic Press, New York, pp 137–181

Barea JM, Pozo MJ, Azcón R, Azcón-Aquilar C (2005) Microbial co-operation in the rhizosphere. J Exp Bot 56(417):1761–1778

Barber SA (1985) Potassium availability at the soil-root interface and factors influencing potassium uptake. In: Munson RD (ed) potassium in agriculture American Society of Agronomy-CSSA-SSSA, Madison, USA, pp 309 324. doi:10.2134/1985.potassium.c11

Bashan Y (1991) Changes in membrane potential of intact soybean root elongation zone cells induced by Azospirillum brasilense. Can J Microbiol 37(12):958–963. doi:10.1139/m91-165

Belimov AA, Safronova VI, Mimura T (2002) Response of spring rape (*Brassica napus* var. oleifera L.) to inoculation with plant growth promoting rhizobacteria containing

1-aminocyclopropane-1-carboxylate deaminase depends on nutrient status of the plant. Can J Microbiol 48(3):189–199

Bhattacharyya PN, Jha DK (2012) Plant growth-promoting rhizobacteria (PGPR): emergence in agriculture. World J Microb Biotechnol 28(4):1327–1350. doi:10.1007/s11274-011-0979-9

Biswas JC, Ladha JK, Dazzo FB (2000) Rhizobia inoculation improves nutrient uptake and growth of lowland rice. Soil Sci Soc Am J 64(5):1644–1650. doi:10.2136/sssaj2000.6451644x

Braud A, Jézéquel K, Bazot S, Lebeau T (2009) Enhanced phytoextraction of an agricultural Cr-, Hg- and Pb-contaminated soil by bio augmentation with siderophore producing bacteria. Chemosphere 74:280–286. doi:10.1016/j.chemosphere.2008.09.013

Burdman S, Volpin H, Kigel J et al (1996) Promotion of nod gen inducers and nodulation in common bean (*Phaseolus vulgaris*) roots inoculated with *Azospirillum brasilense* Cd. Appl Environ Microbiol 62(8):3030–3033

Chandler D, Davidson G, Grant WP et al (2008) Microbial biopesticides for integrated crop management: an assessment of environmental and regulatory sustainability. Trends Food Sci Technol 19(5):275–283. doi:10.1016/j.tifs.2007.12.009

Chin-A-Woeng TF, Bleomberg GV, Mulders IH et al (2000) Root colonization by phenazin-1-carboxamide-producing bacterium *Pseudomonas chlororaphis* PCL1391 is essential for biocontrol of tomato foot and root rot. Mol Plant Microbe Interact 13(12):1340–1345

Creus CM, Sueldo RJ, Barassi CA (1998) Water relations in *Azospirillum*-inoculated wheat seedling under osmotic stress. Can J Bot 76(2):238–244

Creus CM, Sueldo RJ, Barassi CA (2004) Water relations and yield in *Azospirillum* inoculated wheat exposed to drought in the field. Can J Bot 82:273–281. doi:10.1139/B03-119

Dakora FD, Phillips DA (2002) Root exudates as mediators of mineral acquisition in low-nutrient environments. Plant Soil 245:35–47. doi:10.1023/A:1020809400075

Dessaux Y, Hinsinger P, Lemanceau P (2009) Rhizosphere: so many achievements and even more challenges. Plant Soil 321(1–2):1–3. doi:10.1007/s11104-009-0063-5

Figueiredo MVB, Seldin L, Araujo FF, Mariano RLR (2011) Plant growth promoting rhizobacteria: fundamentals and applications. In: Maheshwari DK (ed) Plant growth and health promoting bacteria. Springer-Verlag, Berlin, Heidelberg, pp 21–42

Franche C, Lindström K, Elmerich C (2009) Nitrogen-fixing bacteria associated with leguminous and non-leguminous plants. Plant Soil 321:35–59. doi:10.1007/s11104-008-9833-8

Franco-Correa M, Quintana A, Duque C et al (2010) Evaluation of actinomycete strains for key traits related with plant growth promotion and mycorrhiza helping activities. Appl Soil Ecol 45 (3):209–217. doi:10.1016/j.apsoil.2010.04.007

Gadkar V, David-Schwartz R, Kunik T, Kapulnik Y (2001) Arbuscular mycorrhizal fungal colonization. Factors involved in host recognition. Plant Physiol 127(4):1493–1499. doi:10.1104/pp.010783

Glick BR (2012) Plant growth-promoting bacteria: mechanisms and applications. Scientifica vol 2012 (Article ID 963401):1 15. doi:10.6064/2012/963401

Glick BR, Penrose DM, Li J (1998) A model for the lowering of plant ethylene concentrations by plant growth-promoting bacteria. J Theor Biol 190(1):63–68. doi:10.1006/jtbi.1997.0532

Hartman A (1989) Ecophysiological aspects of growth and nitrogen fixation in *Azospirillum* Spp. Plan Soil 110(2):225–238. doi:10.1007/BF02226803

Hayat R, Ali S, Amara U, Khalid R, Ahmed I (2010) Soil beneficial bacteria and their role in plant growth promotion: a review. Ann Microbiol 60(4):579–598. doi:10.1007/s13213-010-0117-1

Jahanian A, Chaichi MR, Rezaei K et al (2012) The effect of plant growth promoting rhizobacteria (pgpr) on germination and primary growth of artichoke (*Cynara scolymus*). Int J Agric Crop Sci 4:923–929

Kang BG, Kim WT, Yun HS, Chang SC (2010) Use of plant growth-promoting rhizobacteria to control stress responses of plant roots. Plant Biotechnol Rep 4(3):179–183. doi:10.1007/s11816-010-0136-1

KeÇpczyński J, KeÇpczyńska E (1997) Ethylene in seed dormancy and germination. Physiol Plant 101:720–726. doi:10.1111/j.1399-3054.1997.tb01056.x

Kloepper JW, Schroth MN (1978) Plant growth-promoting rhizobacteria on radishes. In: Proceedings of the 4th international conference on plant pathogenic bacteria, vol 2, Station de Pathologie Végétale et de Phytobactériologie, INRA, Angers, France, pp 879–882

Marschner H (1995) Mineral nutrition of higher plants. Academic Press, London

Marschner H, Römheld V, Horst WJ, Martin P (1986) Root-induced changes in the rhizosphere: importance for the mineral nutrition of plants. J Plant Nutr Soil Sci 149(4):441–456. doi:10.1002/jpln.19861490408

Martínez-Toledo MV, Rodelas B, Salmeron V et al (1996) Production of pantothenic acid and thiamine by Azotobacter vinelandi in a chemically defined medium and a dialysed soil medium. Biol Fertil Soils 22(2):131–135. doi:10.1007/BF00384444

Martínez-Viveros O, Jorquera MA, Crowley DE et al (2010) Mechanisms and practical considerations involved in plant growth promotion by rhizobacteria. J Soil Sci Plant Nutr 10 (3):293–319

Mattoo AK, Suttle CS (1991) The plant hormone ethylene. CRC Press, Boca Raton, p 337

Merzaeva OV, Shirokikh IG (2006) Colonization of plant rhizosphere by actinomycetes of different genera. Microbiology 75(2):226–230. doi:10.1134/S0026261706020184

Milagres AM, Machuca A, Napoleão D (1999) Detection of siderophor production from several fungi and bacteria by a modification of chrome azurol S (CAS). agar plate assay. J Microbial Meth 37(1):1–6. doi:10.1016/S0167-7012(99)00028-7

Nardi S, Concheri G, Pizzeghello D et al (2000) Soil organic matter mobilization by root exudates. Chemosphere 41(5):653–658

Ozturk A, Caglar O, Sahin F (2003) Yield response of wheat and barley to inoculation of plant growth promoting rhizobacteria at various levels of nitrogen fertilization. J Plant Nutr Soil Sci 166(2):262–266. doi:10.1002/jpln.200390038

Quispel A (1991) A critical evaluation of the prospects for nitrogen fixation with non-legumes. Plan Soil 137(1):1–11. doi:10.1007/BF02187425

Rai SN, Gaur AC (1988) Characterization of *Azotobacter* SPP. and effect of *Azotobacter* and *Azospirillum* as inoculant on the yield and N-Uptake of wheat crop. Plant Soil 109(1):131–134. doi:10.1007/BF02197592

Rajkumar M, Ae N, Prasad MN, Freitas H (2010) Potential of siderophore-producing bacteria for improving heavy metal phytoextraction. Trends Biotechnol 28(3):142–149. doi:10.1016/j.tibtech.2009.12.002

Revillas JJ, Rodelas B, Pozo C et al (2000) Production of B-group vitamins by two Azotobacter strains with phenolic compounds as sole carbon source under diazotrophic and adiazotrophic conditions. J Appl Microbiol 89(3):486–493. doi:10.1007/s00726-004-0153-x

Rigamonte TA, Pylro VS, Duarte GF (2010) The role of mycorrhization helper bacteria in the establishment and action of ectomycorrhizae associations. Braz J Microbiol 41(4):832–840

Rodelas B, González-López J, Martínez-Toledo MV et al (1999) Influence of *Rhizobium/Azotobacter* and *Rhizobium/Azospirillum* combined inoculation on mineral composition of faba bean (*Vicia faba* L.). Biol Fertil Soils 29(2):165–169. doi:10.1007/s003740050540

Russo A, Vettori L, Felici C et al (2008) Enhanced micropropagation response and biocontrol effect of *Azospirillum brasilense* Sp245 on *Prunus cerasifera* L. clone Mr. S 2/5 plants. J Biotechnol 134(3–4):312–319. doi:10.1016/j.jbiotec.2008.01.020

Saatovich SZ (2006) *Azospirilli* of Uzbekistan soil and their influence on growth and development of wheat plants. Plant Soil 283(1–2):137–145. doi:10.1007/s11104-005-5690-x

Şahin F, Çakmakçi R, Kantar F (2004) Sugar beet and barley yields in relation to inoculation with N_2-fixing and phosphate solubilizing bacteria. Plant Soil 265(1–2):123–129. doi:10.1007/s11104-005-0334-8

Salcedo LDP, Prieto C, Correa MF (2014) Screening phosphate solubilizing actinobacteria isolated from the rhizosphere of wild plants from the Eastern Cordillera of the Colombian Andes. Afr J Microbiol Res 8(8):734–742. doi:10.5897/AJMR2013.5940

Sarige S, Blum A, Okon Y (1988) Improvement of the water status and yield of field grown grain sorghum by inoculation with *Azospirillum brasilense*. J Agric Sci 110:271–277. doi:10.1017/S0021859600081296

Shaharoona B, Arshad M, Zahir ZA et al (2006) Performance of *Pseudomonas spp.* containing ACC-deaminase for improving growth and yield of maize (*Zea mays* L.) in the presence of nitrogenous fertilizer. Soil Biol Biochem 38(9):2971–2975. doi:10.1016/j.soilbio.2006.03.024

Subba Rao NS (1988) Biofertilizers in agriculture. Oxford and IBH publishers, New Delhi

Subba Rao NS, Tilak KVBR, Singh CS (1985) Synergistic effect of vesicular-*arbuscular mycorrhiza* and *Azospirillum brasilense* on the growth of barley in pots. Soil Biol Biochem 17 (1):119–121. doi:10.1016/0038-0717(85)90101-4

Tank N, Saraf M (2010) Salinity-resistant plant growth promoting rhizobacteria ameliorates sodium chloride stress on tomato plants. J Plant Interact 5(1):51–58. doi:10.1080/17429140903125848

Tilak KVBR, Singh CS, Roy VK, Rao NSS (1982) *Azospirillum brasilense* and *Azotobacter chroococcum* inoculum: effect on yield of maize (*Zea mays* L.) and sorghum (*Sorghum bicolor*). Soil Biol Biochem 14(4):417–418. doi:10.1016/0038-0717(82)90016-5

Trinick MJ, Hadobas PA (1995) Formation of nodular structures on the non-legumes Brassica napus, *B. campestris*, *B. juncea* and *Arabidopsis thaliana* with *Bradyrhizobium* and *Rhizobium* isolated from *Parasponia* spp. or legumes grown in tropical soils. Plant Soil 172(2):207–219. doi:10.1007/BF00011323

Vessey JK (2003) Plant growth promoting rhizobacteria as biofertilizers. Plant Soil 255(2):571–586. doi:10.1023/A:1026037216893

Walker TS, Bais HP, Grotewold E, Vivanco JM (2003) Root exudation and rhizosphere biology. Plant Physiol 132:44–51. doi:10.1104/pp.102.019661

Wani PA, Khan MS (2010) Bacillus species enhance growth parameters of chickpea (*Cicer arietinum* L.) in chromium stressed soils. Food Chem Toxicol 48(11):3262–3267. doi:10.1016/j.fct.2010.08.035

Wani SP, Chandrapalaih S, Zambre MA, Lee KK (1988) Association between N_2-fixing bacteria and pearl millet plants: responses, mechanisms and persistence. Plant Soil 110(2):289–302. doi:10.1007/BF02226810

Wu SC, Cao ZH, Li ZG et al (2005) Effects of biofertilizer containing N-fixer, P and K solubilizers and AM fungi on maize growth: a greenhouse trial. Geoderma 125(1–2):155–166. doi:10.1016/j.geoderma.2004.07.003

Zaidi A, Khan MS, Ahemad M, Oves M (2009) Plant growth promotion by phosphate solubilizing bacteria. Acta Microbiol Immunol Hung 56(3):263–284. doi:10.1556/AMicr.56.2009.3.6

Zandi P, Basu SK, Bazrkar Khatibani L et al (2014) Fenugreek (*Trigonella foenum-graecum* L.) seed: a review of physiological and biochemical properties and their genetic improvement. Acta Physiol Plant 37:1714. doi:10.1007/s11738-014-1714-6

Chapter 4
Safety, Nutrition, and Health Aspects of Organic Food

Ying Wu, Li Wang and Ankit Patras

Abstract With the growing popularity of organic food, more and more attention is being paid to the safety, nutrition, and health aspects related to organic food. Consumers are having higher expectations for organic foods for their benefits to the environment, animal welfare, worker safety, and the safety, nutrition, and health benefits of the products. This chapter discusses the differences between organic and conventional food regarding their safety, nutrition, and health perspectives. This chapter covers products from both plant and animal origins such as fruits, vegetables, seafood products, meat, and dairy products. Microbiological and chemical hazards are addressed as safety indicators for the comparison of organic and conventional products. The levels of nutritional components and their related health benefits are also compared between organic and conventional products. Due to the limited database, it is premature to conclude that either system is superior to the other. At present, no information is available yet to ascertain whether the differences in the levels of certain chemicals between organic and conventional foods are of biological significance. More data is needed to advance the knowledge on the safety, nutritional quality, and health benefits of organic foods versus conventional foods.

Keywords Organic food · Safety · Health benefit · Nutritional quality

---

Y. Wu (✉) · L. Wang · A. Patras
Department of Agricultural and Environmental Sciences, College of Agriculture, Human and Natural Sciences, 3500 John A. Merritt Blvd., Nashville, TN, USA
e-mail: ywu@tnstate.edu

L. Wang
e-mail: lwang2@my.tnstate.edu

A. Patras
e-mail: apatras@tnstate.edu

D. Nandwani (ed.), *Organic Farming for Sustainable Agriculture*,
Sustainable Development and Biodiversity 9,
DOI 10.1007/978-3-319-26803-3_4

4.1 Introduction

Consumption of organic food is becoming increasingly popular. Compared to conventional food products, organic food is facing various challenges especially on chemical and microbial safety issues. Some investigators try to compare safety, nutrition, and health benefits of organic food with conventional grown products. However, due to the scarce amount of data on organic products, investigators are not able to provide any conclusions yet.

According to a 2014 Organic Industry Survey conducted by the Organic Trade Association (OTA), sales of organic products in the USA rose up to $35.1 billion in 2013, of which nearly 92 % came from organic food sales. The increasing rate of organic food sales has kept around 10 % since 2010, which is much higher than the average annual growth of 3 % in total food sales during that same period. In 2012, organic food sale first broke through $30 billion and now it consists of more than 4 % of the annual food sales in the USA. The fruit and vegetable category stands on the top of the sector with $11.6 billion in sales, up 15 %. More than 10 % of the fruits and vegetables sold in the USA are organic. New sales from organic fruits and vegetables contributed 46 % of the new organic sales. The sales for organic bread, organic meat, poultry and fish, and organic packaged and prepared food sectors are at 15, 12, 11, and 10 %, respectively (Electronic code of federal regulations; PRNewswire-USNewswire 2014). This rapid growth may be traced to the increased consumer confidence in organic food as well as to the concern about possible health risks and environmental impacts of conventional food production methods.

Recent food crises, for example, mad cow disease and foot-and-mouth disease, have lessened consumer confidence in food in general and especially in conventionally produced food that may use pesticides, antibiotics, and other chemicals in food production (Dreezens et al. 2005). Surveys have indicated that many consumers buy organic food because of the perceived health and nutrition benefits of organic products.

This chapter will discuss critical food safety issues, nutritional values, and health benefits of organic food products based on the information obtained from the currently available sources.

4.1.1 What Is "Organic Food"?

The term "organic" is not defined by law or regulations FDA enforces (FDA). "Organic" covers the food itself including how it was produced, and how the source animals and plants were raised. Food labeled organic must subject to the regulations of the National Organic Program. They must be raised, grown, and processed using organic farming methods that recycle resources and promote biodiversity. Crops must be grown without using any synthetic pesticides, bioengineered genes, petroleum-based fertilizer, or sewage sludge-based fertilizers. Regulations on "organic food" by the US Department of Agriculture (USDA) are summarized as such that crops or animals must be grown without using synthetic pesticides, bioengineered genes, petroleum-based fertilizer, nonagricultural substances, ionizing radiation, or sewage sludge-based fertilizers. Some non-synthetic substances such as lead salts and strychnine are also prohibited. Organic livestock must be provided with some kind of welfare, such as cage-free. In the absence of illness, other than vaccinations, animal drugs are not allowed. Any hormones for growth promotion, and most synthetic substances are not allowed (Electronic code of federal regulations).

4.1.2 Organic Production and Handling Requirements

Organic products have strict production and labeling requirements. Products on the shelf labeled as "100 % organic," "organic," or "made with organic (specified ingredients or food group(s)" must be certified by USDA with corresponding requirements (Electronic code of federal regulations). Otherwise, organic claims should not be made on the principal display panel nor use the USDA organic seal anywhere on the package (Electronic code of federal regulations). However, under certain circumstances, some operations can be exempted. For example, distributors and traders who only handle products in closed containers can get exemption; also, a handling operation only identifies organic ingredients on the information panel, and it can also get exempted (Electronic code of federal regulations).

For plants, the land to produce the plants has to meet certain requirements such as having distinct, defined boundaries and buffer zones such as runoff diversions to prevent the unintended applications of a prohibited substance coming out of organic management (Electronic code of federal regulations). The producer is responsible to select and implement tillage and cultivation practices to maintain or improve the soil condition and minimize soil erosion. The producer must use organically grown seeds, annual seedlings, and planting stock. The producer must also plan the strategies for crop rotation and pest control, as well as weeds and diseases control (Electronic code of federal regulations).

Organic livestock products must be grown under continuous organic management. The poultry for poultry products must have been under continuous organic management beginning no later than the second day of life. Milk or milk products must be from animals that have been under continuous organic management beginning no later than 1 year prior to the production of the milk or milk products. However, livestock used as breeder stock may be brought from a nonorganic operation onto an organic operation at any time. During the raising procedure, livestock must be fed with organically produced agricultural products. Preventive practices for livestock health care must be established and maintained. The producers also need to provide living conditions to accommodate the health and natural behavior of animals (Electronic code of federal regulations).

4.1.3 Labeling

The National Organic Program strictly regulates the labeling of organic food. Basically, there are four types of labeling: (1) "100 % organic": The raw or processed agricultural products must contain 100 % organically produced ingredients (by weight or volume). (2) "Organic": must contain no less than 95 % organically produced raw or processed agricultural products. The USDA organic seal can only be shown on the products under these two circumstances. (3) "Made with organic": specified ingredients or food group(s), which indicates that it must contain at least 70 % organically produced ingredients. (4) "Less than 70 % organic ingredients: It may only be claimed in the ingredient statement displaying the product's percentage of organic contents on the information panel (Electronic code of federal regulations). Figure 4.1 (USDA) elaborates the differences among the 4 types of organic labeling.

4.1.4 Organic Food Related Issues

USDA has established Organic Standards for farmers and processers to cover the topics from farm to table, including soil and water quality, pest control, livestock practices, and rules for food additives. USDA also oversees organic farmers and businesses to make sure that organic food is produced with organic methods. Each year, organic farmers update a farm plan and complete an inspection to confirm that their practices match their records. The farmer must correct any issues to continue certification. Organic food processors meet similar requirements (USDA).

The number of organic farmers has increased steadily. Similarly, the growth of the US organic industry has increased by approximately 20 % per year for more

Fig. 4.1 Understanding the organic label

than 10 years. Consumer demands for organic food also continue to grow. As a consequence, organic products are available in ~20,000 natural food stores and about 3 out of 4 conventional grocery stores. The US sales of organic products were estimated $28.4 billion in 2012 (over 4 % of total food sales) and will reach $35 billion in 2014 (USDA-Economic Research Service).

Organic farming aims to produce high-quality food along with benefits such as sustainable agriculture and ecosystems, balanced animal life, and food crops. Organic food from plant origins should avoid the use of artificial fertilizers, synthetic chemicals, and genetically modified organisms. Organic foods of animal origins are prohibited from using antibiotics or growth hormones (Fabiansson 2014).

These regulations may cause some challenges when producing these organic products. As a consequence, quality, safety, nutrition, and health benefits of organic food have been drawing increasing attentions in recent years. However, relevant systematic scientific evidence is very limited. Thus, the conclusion of safety, nutrition, and health benefits remains tentative in the lack of sufficient evidence. In the following sections, safety, nutrition, and health benefits of various organic products will be discussed.

4.2 Safety Aspect of Organic Foods

Many people choose organic food to avoid risks associated with the pesticides, herbicides, and other chemicals used in conventional farming. Meanwhile, high expectations of product quality and safety are reported among certain consumer groups (Piqué et al. 2013). Nevertheless, the safety of organic food is still unclear and needs to be thoroughly evaluated. Magkos et al. (2006) indicated that food safety of organic products is facing challenges by both chemical and biological hazards.

4.2.1 Chemical Safety

Organic farming is restricted from using pesticides, herbicides, fertilizers, fungicides, synthetic veterinary drugs (antibiotics, and growth hormones), auxiliaries, and synthetic preservatives. However, some compounds such as pesticides, nitrates or nitrites, and naturally occurring toxins are still detected in some organic products. This poses questions regarding the chemical safety of the organic products. In this section, the chemical safety of organic food will be discussed by comparing them with conventionally grown products.

Generally speaking, organic products present some clear advantages over the conventional products on the well-known toxicants (pesticides and nitrates), and natural toxins (Pussemier et al. 2006; Hoefkens et al. 2010).

Vegetables: Araújo et al. (2014) compared the physical–chemical composition and pesticide residue content of lettuce, peppers, and tomatoes grown in organic and conventional systems. They found that all three of the organic vegetables contained a higher total dietary fiber. The composition of the minerals and heavy metals varied between all three vegetables. Contamination by pesticide residues was found in both conventional peppers and organic tomatoes.

Fruit: Patulin is a mycotoxin mainly present in rotten apples and apple-based products. Piqué et al. (2013) analyzed the content of patulin in apple juices and purees derived from organic and conventional production systems on 93 apple-based products with 49 from conventional and 44 from organic farming. Their results showed a significantly higher concentration of patulin in the organic apple purees and juices compared to in conventional ones.

Maize: Maize is traditionally used for bakery. One of the risks of cereal consumption is mycotoxin contamination. Mycotoxins are dangerous for health and might be present in any grain depending on genotypes and environments. de Galarreta et al. (2015) assessed the natural levels of mycotoxins, fumonisin, and deoxynivalenol (DON), in nine maize varieties grown in four different locations, under organic or conventional conditions during two years. They have found that locations and varieties are the two major reasons that contribute to the fumonisin contamination but not for DON content. Therefore, they warned producers of the danger of natural contamination with mycotoxins for some varieties in specific environments. They found no differences between organic and conventional environments.

Wheat and corn flour: Aflatoxins (AFs) are mycotoxins produced by certain species of *Aspergillus*. Aflatoxin B1 (AFB1) is the most toxic, consistently carcinogenic and genotoxic. Armorini et al. (2015) studied 90 different samples of organic and conventional flours (20 conventional wheat flour, 20 organic wheat flour, 42 conventional corn flour, and 8 organic corn flour). AFB1 was found in 13 samples of corn flour, specifically 4 organic and 9 conventional. These results confirm a higher incidence of contamination in corn compared to wheat, as reported in literature. No significant differences were observed comparing conventional corn flour samples and organic corn flour samples (Armorini et al. 2015).

Oats: Kuzdraliński et al. (2013) studied the mycotoxin levels in oats from both organic and conventional farming systems for 3 years. Only one significant difference occurred between organic and conventional farming systems—the concentration of diacetoscirpenol (DAS) was higher in samples from conventional farms. Among the mycotoxin-positive samples, the concentrations of deoxynivalenol (DON), nivalenol (NIV), and aflatoxins were slightly higher in samples from conventional farming but not statistically significant.

Ewe's and goat's milk: Not many studies on chemical safety of organic animal products were found in the literature. Malissiova et al. (2013) conducted the study of Aflatoxin M1 (AFM1) in milk and dairy (in ewes and goats raw milk in Greece) and identified possible risk factors comparing organic and conventional milk. In their study, 39 organic and 39 conventional farms participated in this study and 243 samples were collected, during a lactation period. There were no conventional samples found over the maximum limit for AFM1 (0/117), while 4/117 (3.4 %) organic samples exceeded 50 ng/kg. They found that organic milk samples were contaminated with AFM1 and with higher contamination in comparison with conventional milk. The authors believed this contamination were associated with season, feed storage practices, and feeding pea.

4.2.2 *Microbiological Safety*

The use of animal manure as fertilizer presents potential microbiological risks if the manures have not been properly composted: They can contaminate foodstuffs. While both conventional and organic agriculture frequently use animal manure for fertilization, the use of manure is more widespread among organic production since organic producers cannot use synthetic fertilizers. Interestingly, organic standards require that animal manures be composted according to specific procedures or applied more than 90 d before harvest; conventional food production does not have such requirements (Winter and Davis 2006).

Due to the method of cultivation and processing, organic products may present increased risks to public health than conventional production. However, very few scientific evidence is available to support this assumption (McMahon and Wilson 2001). Organic produce is more exposed to microbiological contamination than conventional produce, since organic fertilizers often consist of manure, which may harbor pathogenic microorganisms such as *Salmonella spp.*, *Listeria monocytogenes* and *Escherichia coli* O157:H7 (McMahon and Wilson 2001). The restricted use of chemicals or medicines in organic farming can eliminate the residue and antibiotics in products and in the environment; however, the risk of microbiological contamination increased in foodstuff. Therefore, it is a long-time debating issue on how safe organic food can be. Compared to conventional food production which has a proven system and an abundant amount of data to investigate, the data for organic food is scarce.

Antibiotic resistance: Antibiotic resistance is posing a serious safety issue in the world. Fernández-Fuentes et al. (2012) studied biocide and antibiotic sensitivity in a collection of 378 isolates derived from 36 organic foods. Most isolates were sensitive to low concentrations of biocides. The study indicated that organic food may act as reservoirs for antibiotic-resistant bacteria and suggest that high levels of biocide tolerance could facilitate the prevalence of antibiotic-resistant strains. The prohibition of antibiotic use in organic animal production also appears to be responsible for the lower incidence of antimicrobial resistance in bacterial isolates from organically raised food animals compared with conventionally raised food animals. This has been demonstrated in several studies and is concisely summarized in an IFT expert report (Winter and Davis 2006).

Vegetables: A number of reports try to demonstrate that organic produce poses a greater risk of transmitting foodborne diseases than does conventional produce (Avery 2002). Oliveira et al. (2010) investigated 72 lettuce samples from 18 farms. Their results showed that the consumption of organically produced lettuce does not represent an increasing risk of a foodborne disease. Mukherjee et al. (2004)

compared microbiological safety of organic and conventional produce on 476 organic samples and 129 conventional samples in Minnesota and analyzed for *Escherichia coli*, *Salmonella*, and *E. coli* 0157:H7. Their results showed that no samples contained the pathogen *E. coli* 0157:H7, and only 2 samples (1 from organic lettuces and 1 from organic green peppers) contained *Salmonella*. Their results clearly indicated that certified organic produce was not at a higher microbiological risk than conventional produce.

McMahon and Wilson (2001) examined a range of commercially available organic vegetables for the presence of *Salmonella*, *Campylobacter*, *Escherichia coli*, *E. coli* O 157, *Listeria,* and *Aeromonas* spp., to provide information on the occurrence of such organisms in organic vegetables in Northern Ireland. No *Salmonella*, *Campylobacter*, *E. coli*, *E. coli* O 157, or *Listeria* was found in any of the samples examined. Maffei et al. (2013) studied 130 samples of different organic and conventional vegetable varieties sold in Brazil for mesophilic aerobic bacteria, yeasts and molds, total coliforms, Escherichia coli, and Salmonella spp. Salmonella spp. was not found in any sample. Some organic varieties have greater bacterial counts. No significant difference was observed between organic and conventional products.

Animal products: Organic animal producers are prohibited from using antibiotics, which may result in increased pathogen levels and elevated microbiological safety risks. However, research findings are inconsistent. For example, *Campylobacter* sp. isolated from bovine feces was 26.7 % in organic farms and 29.1 % in conventional farms (Sato et al. 2004), while in another study, 100 % of 22 organic broiler-flock samples were positive for *Campylobacter* spp. compared with 36.7 % of 79 conventional broiler-flock samples (Heuer et al. 2001). Rosenquist et al. (2013) studied Danish organic broiler meat carcasses at the end of processing after chilling. They reported that the yearly mean prevalence was 54.2 % for organic and 19.7 % for conventional carcasses. *Campylobacter jejuni* was the most frequently isolated species. The result showed a higher risk of illness from organic broiler carcasses compared with conventional broiler carcasses. Miranda et al. (2008) carried out a study on 30 samples of organic chicken meat, conventional chicken meat, and conventional turkey meat to assess differences in contamination. The study examined the total bacterial count for *Enterobacteriaceae*, and the antibiotic resistance including the resistance to ampicillin, chloramphenicol, cephalothin, doxycycline, ciprofloxacin, gentamicin, nitrofurantoin, and sulfisoxazole. Their results showed that the bacterial count in organic samples was significantly higher than those from the conventional meat samples. However, the antibiotic resistance from organic chicken meat was less than those from conventional meat samples. This result showed that although organic chicken meat contains more bacterial count, organic farming contributes to the elimination of antibiotic resistance.

4.3 Nutritional Quality of Organic Food

Huber et al. (2011) reviewed the nutritional differences between organic and conventional food as summarized in the following paragraphs. According to their review, it is generally believed that organic products had a higher dry matter, lower nitrate content, higher vitamin C content, and less pesticide residues (Woese et al. 1997; Worthington 2001). For example, Worthington (2001) did a meta-analysis on the level of vitamin C and the results showed significantly higher vitamin C in organic plant food than in conventional ones. A higher carotenoid content was also found in organic sweet peppers, yellow plums, tomatoes, and carrots (Chassy et al. 2006). However, some conflicting results showed lower or similar contents of carotenoids in organic blanched carrots and tomatoes (Rossi et al. 2008; Stracke et al. 2008). Barrett et al. (2007) explained that the inconsistency may be due to the soil type, genotype, fertilizers, and pesticides used. Some studies have shown that the content of phenolic compounds is higher in organic products (e.g., Chassy et al. 2006), whereas other studies (Chassy et al. 2006; Lombardi-Boccia et al. 2007) have found similar or lower contents of phenolic compounds in organic products. In most studies comparing conventionally with organically grown cereals, higher levels of proteins and amino acids were found in the conventionally produced grain (reviews by Heaton 2001; Worthington 2001; Benbrook et al. 2008). The higher N-fertilization rate in conventional production systems is very likely to explain this difference. Some studies also observed that the quality of the amino acids was higher in the organic products than in the conventional products, meaning that more essential amino acids were available in the organic grains (Maeder et al. 2007).

Wheat flour: Some researchers evaluated the nutritional content of wheat flours from organic and conventional production systems. Their study revealed that organic agriculture has the potential to yield products with high-quality proteins and higher microelements contents (Vrček et al. 2014).

Soybean product: Tofu was developed using organic soybean. Li et al. (2015) prepared organic tofu using organic compatible coagulants of magnesium chloride and three polysaccharides including carrageenan, guar gum, and gum Arabic. These organic compatible coagulants did not affect most of the protein structure. The overall-acceptability of organic tofu prepared with $MgCl_2$ and guar gum or gypsum was almost the same as conventional tofu except having a beany-flavor. Among these organic coagulants, tofu made from 0.6 g guar gum and $MgCl_2$ mixture was the most similar to conventional tofu.

Potato: Carillo et al. (2012) compared primary metabolites in potato (*Solanum tuberosum* L., cultivar Agria) grown under organic and traditional farming systems and evaluated the influence of heat processing (for producing potato powder) on nutritionally important compounds such as essential amino acids, proteins, and carbohydrates. Their study found that the potato powders both from conventional and organic farming were very similar in nutritional value, color, and consistency when rehydrated.

Oil: Samman et al. (2008) compared the fatty acid composition of edible oils from certified organic and conventional agricultural methods in Sydney. No consistent difference in the fatty acid was found in their study. Therefore, their study does not support that organic food have a higher nutritional quality than conventional food.

Milk: Bergamo et al. (2003) studied milk and dairy products produced by organic and conventional systems by comparing their fatty acid and fat-soluble vitamin concentrations. They found significantly higher *cis*-9 *trans*-11 $C_{18:2}$ (CLA), linolenic acid (LNA), *trans*-11 $C_{18:1}$ (TVA), and α-tocopherol (TH) in organic buffalo milk and mozzarella cheese. Similarly, all organic samples contained significantly higher CLA, TVA, LNA, TH, and β-carotene concentrations than that in conventional dairy foods.

Meat: Kamihiro et al. (2015) compared meat and fat quality of sirloin steaks from organic and conventional farms. They found little difference in meat quality (pH, shear force, and color), but the fat profiles varied considerably between the production systems and the season. Meat fat from organic and summer finished cattle contained more conjugated linoleic acid, vaccenic acid, omega-3 fatty acids, and had a lower ratio of omega-6 to omega-3 fatty acids compared with conventional and winter finished cattle, respectively.

Sea bass: Little data is available for organic seafood. One study on sea bass was carried out by Trocino et al. (2012) with 80 specimens. In their study, they compared the biometric and nutritional traits of European sea bass from organic or semi-intensive conventional production systems at two commercial sizes (small and medium). Their result indicated that the biometric traits and the texture were not affected by the production system but by the fish size.

Rabbit: Pla (2008) compared the meat of conventional and organic rabbits with 50 rabbits slaughtered at 63 or 90 days, respectively. Organic rabbits had a higher carcass length to circumference ratio. Also, organic carcasses were leaner and had a lower meat-to-bone ratio than in conventional rabbits. Organic carcass had less saturated fatty acids, less monounsaturated fatty acids, but more polyunsaturated fatty acids. The ratio of polyunsaturated/saturated fatty acid was higher compared to its conventional counterparts. The proteins in the organic meat were richer in methionine and cystine.

4.4 Health Benefit of Organic Food

Health benefit is a motivating factor for consumers. Huber et al. (2011) summarized several in vitro studies comparing health-related properties of organic versus conventional foods in the following paragraphs.

It is believed that conventionally produced food are more likely associated with higher contents of pesticide residues which are known to exert genotoxic, carcinogenic neuro-destructive, endocrine, and allergenic effects. Scientific evidence has shown that dietary exposure of children to organophosphorus pesticides, measured as the level of pesticide metabolites in urine, is much lower on an organic than on a conventional diet (Curl et al. 2003). Therefore, Huber et al. (2011) concluded that consumption of organic food provides protection against exposure to organophosphorus pesticides commonly used in agricultural practices (Lu et al. 2006). As summarized by Huber et al. (2011), a few observational studies compared the health effects of organic and conventional food on humans. According to one of these studies on 14,000 children in 5 European countries, children consuming biodynamic and organic food were found to have less allergies and a lower body weight compared with a group consuming conventional food (Alfven et al. 2006). At the same time, a KOALA Birth Cohort Study in the Netherlands (about 2700 newborns) indicated that children at the age of 2 years with the consumption of organic dairy products are associated with a lower eczema risk (Kummeling et al. 2008). It is also found that organic dairy consumption resulted in higher CLA levels in breast milk of their mothers (Rist et al. 2007). Rembiałkowska et al. (2008) assessed the health status and found that consumers eating organic food are significantly healthier than consumers eating non-organic food although this may attribute to other factors, i.e., nutritional pattern, living conditions physical activity, and ways to manage stress.

Fruits: Some studies indicated that there is no significant difference between organic and conventional fruits on health-promoting compounds. For example, Cardoso et al. (2011) compared the concentration of ascorbic acid (AA), dehydroascorbic acid (DHA), and carotenoids (lycopene and β-carotene) between three fruits, acerola, strawberry, and persimmon, produced by organic and conventional farming. They found that the AA content was significantly higher in organic acerola compared to its conventional production. Conversely, the AA content was significantly higher in conventional strawberries. The DHA and β-carotene contents were significantly higher in the conventional fruits. Lycopene was only detected in persimmons and no significant difference was found.

Many studies have shown that organic products provide more health-promoting compounds. Vrček et al. (2011) evaluated the antioxidant capacity, the polyphenol, and the metal content in conventional and organic wines using similar winemaking processes. The values of antioxidant activity, as evaluated by two free radical methods, ABTS and DPPH, were found systematically higher in organic wines compared to conventional ones. The phenol concentrations (chlorogenic acid, ferulic acid, catechin, *trans*-resveratrol, hydroxybenzoic acids, and flavonols) were higher in the organically produced wines. No apparent trend was found in the metal contents of the wines. Uckoo et al. (2015) studied the levels of phytochemicals in organically and conventionally cultivated Meyer lemons (Citrus meyeri Tan.) grown in south Texas under similar climatic conditions, using organic and conventional cultivation practices. Mature fruits were harvested in two seasons, stored at market-simulated post-harvest conditions for four weeks, and periodically evaluated for levels of phytochemicals, including flavonoids, amines, organic acids, and minerals. They found that organically grown lemons contain significantly higher levels of hesperidin, didymin, and ascorbic acid than those cultivated in conventional systems. Phenolic content was higher in organic lemons, whereas levels of

Table 4.1 Summary of recent studies comparing organic and conventional foods with respect to nutrient values

Food Matrix	Phyto-nutrients	Findings	References
Apples	Phenolics	Phenolics are higher in organic apple pulps than in conventional ones, no difference between organic and conventional apples with respect to phenolics in apple peels	Veberic et al. (2005)
Grapes	Diphenolase enzymes	Diphenolics activity is 2 times higher in organic grapes than in conventional ones.	Nunez-Delicado et al. (2005)
Marionberries, corns, strawberries	Phenolics and ascorbic acid	Phenolics and ascorbic acid are higher in organic than in conventional; highest levels of phenolics and ascorbic acid in crops grown "sustainably"	Asami et al. (2003)
Black currants	Flavonols	No consistent differences are noted between flavonol levels in organic and conventional	Mikkonen et al. (2001)
Qing-gen-cai, Chinese cabbage	Flavonoids	Organic food generally contain higher levels of flavonoids	Ren et al. (2001)
Spinach, welsh onion, green pepper	Flavonoids	Organic food exhibit higher levels of flavonoids	Ren et al. (2001)
Tomatoes	Vitamin C	Organic tomatoes have higher levels of vitamin C	Caris-Veyrat et al. (2004)

citric acid and amines were higher in conventionally cultivated lemons. These results suggest that organically grown Meyer lemons are a good source of enhanced levels of flavonoids and ascorbic acid. A summary of the findings from the various research studies on phytonutrient of organic and conventionally grown fruits and vegetables is presented in Table 4.1.

Green vegetables: One of the studies analyzed the antioxidant an anti-mutagenic activity of organic and conventional green vegetables (qing-gen-cai, Chinese cabbage, spinach, Wels onion, and green pepper) (Ren et al. 2001). The authors found much higher antioxidant activity in the organic vegetables than that in the conventional ones. Moreover, organic vegetable juices exhibited significantly stronger suppressive effects against mutagens. Another study compared the effects of extracts from organic and conventional strawberries on the proliferation of colon and breast-cancer cells (Olsson et al. 2006). The extracts from organic strawberries showed higher anti-proliferative activity on both types of cancer cells. Therefore, these results suggest a possible health benefit organic foods by reducing human cancer risks.

Tomato juice: (Vallverdú-Queralt et al. 2012) compared the phenolic and hydrophilic antioxidant profiles of organically and conventionally produced tomato juices. The results demonstrated statistically higher levels of phenolic compounds in organic tomato juices. This increase corresponds not only with the increased amount of soil organic matter in organic plots but also with reduced manure application rates once soils in the organic systems had reached equilibrium levels of organic matter. The authors indicated that the phenolic compounds and the hydrophilic antioxidant capacity were responsible for the differentiation between organic and conventional tomato juices. Thus, these appear to be genuine differences in the bioactive components of organic and conventional tomato juices not previously reported.

Winter Wheat: Mazzoncini et al. (2015) studied the effects of organic versus conventional cropping systems on yield and the phenolic composition of winter wheat cv. "Bologna." The results showed that the organic wheat yielded less than that of the conventional wheat mainly due to the nitrogen shortage. The cultivation system did not affect the total amount of phenolic compounds. Phenolic composition and quantity were significantly affected by the milling fraction (bran or white flour): Phenolic compounds were more concentrated in the bran. Under the conditions in their study, the organic cropping system can maintain or increase the health properties of the milled wheat products with an acceptable reduction in grain yield.

Buckwheat: Some studies indicated that the growing conditions can significantly affect the flavonoid content in conventional and organic food products. Kalinova and Vrchotova (2011) compared the level of flavonoids in conventionally and organically grown common buckwheat (*Fagopyrum esculentum* Moench) under the same environmental conditions. The level of rutin, epicatechin, catechin, and epicatechin gallate in buckwheat groats (hulled achenes) were quantified using high-pressure liquid chromatography (HPLC). Rutin and epicatechin gallate were significantly higher in organic groats. However, this difference may attribute to the environmental conditions in the given year and variety.

Spices: Lv et al. (2012) investigated conventional and organic cinnamon and peppermint for their phenolic profile, anti-proliferative, anti-inflammatory, and antioxidant properties. They extracted the samples using 75 % acetone. They found no significant difference between conventional and organic spices in phenolic composition. All conventional and organic peppermint and cinnamon extracts exhibited strong anti-proliferative and anti-inflammatory properties.

4.5 Summary

This chapter illustrates that tradeoffs exist between organic and conventional food production. Organic fruits and vegetables rely upon far fewer pesticides than conventional fruits and vegetables do, which results in fewer pesticide residues, but may also stimulate the production of naturally occurring toxins. The popularity of organic food continues to grow dramatically (Organic Trade Assn. 2015). Consumers purchase organic food with the expectations for perceived benefits to the environment, animal welfare, worker safety, and the safety, nutrition, and health benefits. This chapter discusses the differences between organic and conventional food with respect to food safety, nutrition composition, and potential health benefits in order to make clear of several qualitative differences. Organic fruits and vegetables possess fewer pesticide residues and lower nitrate levels than conventional fruits and vegetables do. In some cases, organic food may have higher levels of health-promoting compounds such as antioxidants, which may be beneficial, but may also be of potential health concern when considering naturally occurring toxins. Some studies have suggested potential increased microbiological hazards from organic produce or animal products due to the restricted antimicrobial use, while many other studies have not reached the same conclusion. Bacterial isolates from food animals raised organically appear to show less resistance to antimicrobial agents than those from food animals raised conventionally (Winter and Davis 2006). Although many studies demonstrated these qualitative differences between organic and conventional food, it is premature to conclude that either food system is superior to the other. Pesticide residues, naturally occurring toxins, nitrates, and phenolic compounds exert their health risks or benefits on a dose-related basis. At present, no information is available yet to ascertain whether the differences in the levels of certain chemicals between organic and conventional food are of biological significance. More data is needed to advance the knowledge on the safety, nutritional quality, and health benefits of organic food versus conventional food.

References

Alfven T, Braun-Fahrlander C, Brunekreef B, von Mutius E, Riedler J, Scheynius A, van Hage M, Wickman M, Benz MR, Budde J, Michels KB, Schram D, Ublagger E, Waser M, Pershagen G (2006) Allergic diseases and atopic sensitization in children related to farming and anthroposophic lifestyle—the PARSIFAL study. Allergy 61:414–421

Araújo DFDS, Silva AMRBDS, Lima LLDA, Vasconcelos MADS, Andrade SAC, Sarubbo LA (2014) The concentration of minerals and physicochemical contaminants in conventional and organic vegetables. Food Control 44:242–248

Armorini S, Altafini A, Zaghini A, Roncada P (2015) Occurrence of aflatoxin B1 in conventional and organic flour in Italy and the role of sampling. Food Control 50:858–863

Asami DK, Hong YJ, Barrett DM, Mitchell AE (2003) Comparison of the total phenolic and ascorbic acid content of freeze-dried and air-dried marionberry, strawberry, and corn grown using conventional, organic, and sustainable agricultural practices. J Agric Food Chem 51:1237–1241

Barrett DM, Weakley C, Diaz JV, Watnik M (2007) Qualitative and nutritional differences in processing tomatoes grown under commercially organic and conventional production systems. J Food Sci 72:441–450

Benbrook C, Zhao X, Yanez J, Davies N, Andrews P (2008) New evidence confirms the nutritional superiority of plant-based organic foods. State Sci Rev www.organic-center.org

Bergamo P, Fedele E, Iannibelli L, Marzillo G (2003) Fat-soluble vitamin contents and fatty acid composition in organic and conventional Italian dairy products. Food Chem 82:625–631

Cardoso PC, Tomazini APB, Stringheta PC, Ribeiro SMR, Pinheiro-Sant'Ana HM (2011) Vitamin C and carotenoids in organic and conventional fruits grown in Brazil. Food Chem 126:411–416

Carillo P, Cacace D, Pascale SD, Rapacciuolo M, Fuggi A (2012) Organic vs. traditional potato powder. Food Chem 133:1264–1273

Caris-Veyrat C, Amiot MJ, Tyssandier V, Grasselly D, Buret M, Mikolajczak M, Guilland JC, Bouteloup-Demange C, Borel P (2004) Influence of organic versus conventional agricultural practice on the antioxidant microconstituent content of tomatoes and derived purees; consequences on antioxidant plasma status in humans. J Agric Food Chem 52:6503–6509

Chassy AW, Bui L, Renaud ENC, Van Horn M, Mitchell AE (2006) Three-year comparison of the content of antioxidant microconstituents and several quality characteristics in organic and conventionally managed tomatoes and bell peppers. J Agric Food Chem 54:8244–8252

Curl CL, Fenske RA, Elgethun K (2003) Organophosphorus pesticide exposure of urban and suburban preschool children with organic and conventional diets. Environ Health Perspect 111:377–382

de Galarreta JIR, Butrón A, Ortiz-Barredo A, Malvar RA, Ordásb A, Landac A, Revillab P (2015) Mycotoxins in maize grains grown in organic and conventional agriculture. Food Control 52:98–102

Dreezens E, Martijn C, Tenbult P, Kok G, Vries N (2005) Food and values: an examination of values underlying attitudes toward genetically modified- and organically grown food products. Appetite 44:115–122

Electronic code of federal regulations, Title 7 (Agriculture), subtitle B (Regulations of the Department of Agriculture (Continued), Chapter I (Agricultural Marketing Service (Standards, Inspections, Marketing Practices), Department of Agriculture (Continued), Subchapter M (Organic Foods Production Act Provisions), part205 (National Organic Program). Subpart 100–101, 105, 202–206, 236–239, 300–305, 602–604

Fabiansson SU (2014) Safety of food and beverages: safety of organic foods. Encycl Food Saf 3:417–421

FDA, http://www.fda.gov/AboutFDA/Transparency/Basics/ucm214869.htm

Fernández-Fuentes MA, Morente EO, Abriouel H, Pulido RP, Gálvez A (2012) Isolation and identification of bacteria from organic foods: Sensitivity to biocides and antibiotics. Food Control 26:73–78

Heaton S (2001) Organic farming, food quality and human health. Soil Association Report, Bristol, p 87

Heuer OE, Pedersen K, Andersen JS, Madsen M (2001) Prevalence and antimicrobial susceptibility of thermophilic Campylobacter in organic and conventional broiler flocks. Lett Appl Microbiol 33:269–274

Hoefkens C, Sioen I, Baert K, Meulenaer BD, Henauw SD, Vandekinderen I, Devlieghere F, Opsomer A, Verbeke W, Camp JV (2010) Consuming organic versus conventional vegetables: the effect on nutrient and contaminant intakes. Food Chem Toxicol 48(11):3058–3066

Huber M, Rembiałkowska E, Średnicka D, Bügel S, van de Vijver LPL (2011) Organic food and impact on human health: assessing the *status quo* and prospects of research. NJAS—Wageningen J Life Sci 58(3–4):103–109

Kalinova J, Vrchotova J (2011) The influence of organic and conventional crop management, variety and year on the yield and flavonoid level in common buckwheat groats. Food Chem 127(2):602–608

Kamihiro S, Stergiadis S, Leifert C, Eyre MD, Butler G (2015) Meat quality and health implications of organic and conventional beef production. Meat Sci 100:306–318

Kummeling I, Thijs C, Huber M, van de Vijver LP, Snijders BE, Penders J, Stelma F, van Ree R, van den Brandt PA, Dagnelie PC (2008) Consumption of organic foods and risk of atopic disease during the first 2 years of life in the Netherlands. Br J Nutr 99:598–605

Kuzdraliński A, Solarska E, Mazurkiewicz J (2013) Mycotoxin content of organic and conventional oats from southeastern Poland. Food Control 33(1):68–72

Lombardi-Boccia G, Lucarini M, Lanzi S, Aguzzi A, Capelloni M (2004) Nutrients and antioxidant molecules in yellow pluma (*Prunus domestica* L.) from conventional and organic productions: a comparative study. J Agric Food Chem 52:90–94

Li M, Chen FS, Yang B, Lai SJ, Yang HS, Liu KL, Bu GH, Fu CL, Deng Y (2015) Preparation of organic tofu using organic compatible magnesium chloride incorporated with polysaccharide coagulants. Food Chem 167:168–174

Lu C, Toepel K, Irish R, Fenske RA, Barr DB, Bravo R (2006) Organic Diets Significantly Lower Children's Dietary Exposure to Organophosphorus Pesticides. Environ Health Perspect 114 (2):260–263

Lv JL, Huang HQ, Yu L, Whent M, Niu YG, Shi HM, Wang TTY, Luthria D, Charles D, Yu LL (2012) Phenolic composition and nutraceutical properties of organic and conventional cinnamon and peppermint. Food Chem 132(3):1442–1450

Maeder P, Hahn D, Dubois D, Gunst L, Alföldi T, Bergmann H, Oehme M, Amado R, Schneider H, Graf U, Velimirov A, Fließbach A, Niggli U (2007) Wheat quality in organic and conventional farming: results of a 21 year field experiment. J Sci Food Agric 87:1826–1835

Maffei DF, Silveira NFDA, Catanozi MDPLM (2013) Microbiological quality of organic and conventional vegetables sold in Brazil. Food Control 29:226–230

Magkos F, Arvaniti F, Zampelas Antonis (2006) Organic food: buying more safety or just peace of mind? a critical review of the literature. Crit Rev Food Sci Nutr 46:23–56

Malissiova E, Tsakalof A, Arvanitoyannis IS, Katsafliaka A, Katsioulis A, Tserkezou P, Koureas M, Govaris A, Hadjichristodoulou C (2013) Monitoring Aflatoxin M1 levels in ewe's and goat's milk in Thessaly, Greece; potential risk factors under organic and conventional production schemes. Food Control 34:241–248

Mazzoncini M, Antichi D, Silvestri N, Ciantelli G, Sgherri C (2015) Organically vs conventionally grown winter wheat: effects on grain yield, technological quality, and on phenolic composition and antioxidant properties of bran and refined flour. Food Chem 175:445–451

McMahon MAS, Wilson IG (2001) The occurrence of enteric pathogens and Aeromonas species in organic vegetables. Int J Food Microbiol 70:155–162

Mikkonen TP, Maatta KR, Hukkanen AT, Kokko HI, Torronen AR, Karenlampi SO, Karjalainen RO (2001) Flavonol content varies among black currant cultivars. J Agric Food Chem 49:3274–3277

Miranda JM, Guarddon M, Vázquez BI, Fente CA, Barros-Velázquez J, Cepeda A, Franco CM (2008) Antimicrobial resistance in Enterobacteriaceae strains isolated from organic chicken, conventional chicken and conventional turkey meat: a comparative survey. Food Control 19:412–416

Mukherjee A, Speh D, Dyck E, Diez-Gonzalez F (2004) Preharvest evaluation of coliforms, *Escherichia coli, Salmonella, and Escherichia coli* O157:H7 in organic and conventional produce grown by Minnesota farmers. J Food Prot 67:894–900

Nunez-Delicado E, Sanchez-Ferrer A, Garcia-Carmona FF, Lopez-Nicolas JM (2005) Effect of organic farming practices on the level of latent polyphenol oxidase in grapes. J Food Sci 70:74–78

Oliveira M, Usall J, Viñas I, Anguera M, Gatius F, Abadias M (2010) Microbiological quality of fresh lettuce from organic and conventional production. Food Microbiol 27:679–684

Olsson ME, Andersson CS, Oredsson S, Berglund RH, Gustavsson KE (2006) Antioxidant levels and inhibition of cancer cell proliferation in vitro by extracts from organically and conventionally cultivated strawberries. J Agric Food Chem 54:1248–1255

Organic Trade Association (OTA) (2015) Market analysis: US organic industry survey 2015. https://www.ota.com/what-ota-does/market-analysis

Piqué E, Vargas-Murga L, Gómez-Catalán J, Lapuente JD, Llobet JM (2013) Occurrence of patulin in organic and conventional apple-based food marketed in Catalonia and exposure assessment. Food Chem Toxicol 60:199–204

Pla M (2008) A comparison of the carcass traits and meat quality of conventionally and organically produced rabbits. Livestock Sci 115:1–12

Pussemier L, Larondelle Y, Peteghem CV, Huyghebaert A (2006) Chemical safety of conventionally and organically produced foodstuffs: a tentative comparison under Belgian conditions. Food Control 17:14–21

Rembiałkowska E, Kazimierczak RS′, Rednicka D, Bien′ ko K, Bielska M (2008) Different aspects of organic and conventional food consumers lifestyle. New Med 1:16–19

Ren H, Endo H, Hayashi T (2001) Antioxidative and antimutagenic activities and polyphenol content of pesticide-free and organically cultivated green vegetables using water-soluble chitosan as a soil modifier and leaf surface spray. J Sci Food Agric 81:1426–1432

Rist L, Mueller A, Barthel C, Snijders B, Jansen M, Simões-Wüst AP, Huber M, von Kummeling I, Mandach U, Steinhart H, Thijs C (2007) Influence of organic diet on the amount of conjugated linoleic acids in breast milk of lactating women in the Netherlands. Br J Nutr 97:735–743

Rosenquist H, Boysen L, Krogh AL, Jensen AN, Nauta M (2013) Campylobacter contamination and the relative risk of illness from organic broiler meat in comparison with conventional broiler meat. Int J Food Microbiol 162:226–230

Rossi F, Godani F, Bertuzzi T, Trevisan M, Ferrari F, Gatti S (2008) Health-promoting substances and heavy metal content in tomatoes grown with different farming techniques. Eur J Nutr 47:266–272

Samman S, Chow JWY, Foster MJ, Ahmad ZI, Phuyal JL, Petocz P (2008) Fatty acid composition of edible oils derived from certified organic and conventional agricultural methods. Food Chem 109:670–674

Sato K, Barlett PC, Kaneene JB, Downes FP (2004) Prevalence and antimicrobial susceptibilities of *Campylobacter* spp. Isolates from organic and conventional dairy hens in Wisconsin. Appl Environ Microbiol 70:1442–1447

Stracke BA, Rüfer CE, Bub A, Briviba K, Seifert S, Kunz C, Watzl B (2008) Bioavailability and nutritional effects of carotenoids from organically and conventionally produced carrots in healthy men. Br J Nutr 20:1–9

Trocino A, Xiccato G, Majolini D, Tazzoli M, Bertotto D, Pascoli F, Palazzi R (2012) Assessing the quality of organic and conventionally-farmed European sea bass (Dicentrarchus labrax). Food Chem 131:427–433

Uckoo RM, Jayaprakasha GK, Patil BS (2015) Phytochemical analysis of organic and conventionally cultivated Meyer lemons (*Citrus meyeri Tan.*) during refrigerated storage. J Food Compos Anal 42:63–70

Vallverdú-Queralt A, Medina-Remón A, Casals-Ribes I, Lamuela-Raventos RM (2012) Is there any difference between the phenolic content of organic and conventional tomato juices? Food Chem 130(1):222–227

Veberic R, Trobec M, Herbinger K, Hofer M, Grill D, Stampar F (2005) Phenolic compounds in some apple (Malus domestica Borkh) cultivars of organic and integrated production. J Sci Food Agric 85:1687–1694

Vrček IV, Bojić M, Žuntar I, Mendaš G, Medić-Šarić M (2011) Phenol content, antioxidant activity and metal composition of Croatian wines deriving from organically and conventionally grown grapes. Food Chem 124:354–361

Vrček IV, Čepo DV, Rašić D, Peraica M, Žuntar I, Bojić M, Mendaš G, Medić-Šarić M (2014) A comparison of the nutritional value and food safety of organically and conventionally produced wheat flours. Food Chem 143:522–529

Winter CK, Davis SF (2006) Organic foods. J Food Sci 71:122–124

Woese K, Lange D, Boess C, Bogl KW (1997) A comparison of organically and conventionally grown foods—results of a review of the relevant literature. J Sci Food Agric 74:281–293

Worthington V (2001) Nutritional quality of organic versus conventional fruits, vegetables, and grains. Journal of Alternative Complement Medicine 7:161–173

Websites

Avery DT (2002) Hidden angers in organic food. Hudson Institute. Available at http://www.cgfi.org/materials/articles/2002/jun_25_02.htm

PRNewswire—USNewswire (2014) American appetite for organic products breaks through $35 billion mark. PRNewswire. Available via Dialog. http://www.prenewswire.com/news-releases/american-appetite-for-organic-products-breaks-through-35-billion-mark-259327061.html. Accessed 15 May 2014

USDA http://www.ams.usda.gov/AMSv1.0/ams.fetchTemplateData.do?template=TemplateA&navID=NationalOrganicProgram&leftNav=NationalOrganicProgram&page=NOPUndersandingOrganicLabeling&description=Understanding%20Organic%20Labeling&acct=nopgeninfo

USDA- Ecomomic Research Service: http://www.ers.usda.gov/topics/natural-resources-environment/organic-agriculture/organic-market-overview.aspx

Chapter 5
Plant-Parasitic Nematode Problems in Organic Agriculture

Shabeg S. Briar, David Wichman and Gadi V.P. Reddy

Abstract Crop protection approaches differ widely among organic growers both globally and regionally, yet organic farming faces the same plant-parasitic nematode (PPN) issues as conventional farming. Due to the restrictions on use of synthetic chemical inputs and the limited number of options for nematode management in organic fields, organic producers are often at greater risk to nematode problems than their conventional counterparts. While worldwide estimates of crop losses of about 12 % annually of food and fiber due to nematode damage are reported in the literature, such information for organic farming systems is scarce. Comparative studies of organic and conventional farming systems and surveys conducted in organic farms in distinct regions show that the genera of nematodes attacking organic crops are similar to that in conventional fields, including species of root-knot (*Meloidogyne* spp.), cyst (*Heterodera* and *Globodera* spp.), and root lesion (*Pratylenchus* spp.) nematodes, among others. For PPN management, organic farmers employ practices such as crop rotation, use of cover crops or resistant crop cultivars, and soil amendments. In many instances, however, these methods may not be sufficient for PPN management. Although resistant cultivars of some crops are available for root-knot and cyst nematodes, they are resistant to only a few races or species of nematodes and new races develop over time. Biological control, using microbial pathogens, endophytes, or antagonists may help control PPNs in organic production of some crops but have had limited commercial success. In contrast, use of soil amendments has provided some level of suppression of PPNs under field conditions. Increased populations of predatory nematodes or other beneficial species grazing microbial films and stimulating soil nutrient mineralization have been observed in organic systems, indicating an improvement in the soil health. Further studies are needed to estimate yield losses caused by the

S.S. Briar (✉) · D. Wichman
Central Agricultural Research Center, Montana State University,
52583 US Highway 87, Moccasin, MT 59462-9512, USA
e-mail: shabeg.briar@montana.edu

G.V.P. Reddy
Western Triangle Agricultural Research Center, Montana State University,
9546 Old Shelby Rd, PO Box 656, Conrad, MT 59425, USA

D. Nandwani (ed.), *Organic Farming for Sustainable Agriculture*,
Sustainable Development and Biodiversity 9,
DOI 10.1007/978-3-319-26803-3_5

economically important PPN species in organic systems and to develop suitable nematode management strategies given for organic farming.

Keywords Nematodes · Organic agriculture · Implications · Management

5.1 Introduction

Nematodes are unsegmented, bilaterally symmetric roundworms, usually microscopic in size in the phylum Nematoda. They are one of the most important and abundant groups of animals and are essentially aquatic, living in water, moisture films, or host tissues. Plant-parasitic nematodes (PPNs) are obligate parasites which feed mainly on plant roots with common aboveground symptoms of stunting, yellowing, wilting, and yield losses and belowground root malformation due to direct feeding damage. In general, nematode bodies taper toward both head and tail, but females of some of species may be pear, lemon, or kidney-shaped. All major food crops are damaged by at least one species of nematodes, and the economic consequences of nematode infestations are many and varied, reducing crop quality and yield (Norris et al. 2003; Agrios 2005). Feeding of many PPNs creates entryways into plant roots for secondary pathogens, while feeding of some species directly transmits plant viruses (Rowe and Powelson 2002).

Organic farming has increased significantly worldwide over the last several years and is expected to grow further (Moyniham 2010). Briefly, organic farming is a set of plant and animal production practices that emphasize reliance on sustainable and renewable biological processes. Nutrients are supplied through the decomposition of cover crops of nitrogen-fixing legumes or animal manures or products. Pest management relies on an integrated approach of promoting plant health, vegetation management, and biological control (van Bruggen and Semenov 2000; McSorley 2011a). No synthetic inputs (such as broad spectrum fumigants) are allowed, leaving soil biological activities intact while the incorporation of plant and animal-derived organic materials enhances the soil food web. Soil food webs of organic farming systems are generally more diverse in terms of species richness and abundance compared to conventional systems (Ferris et al. 2001; Mäder et al. 2002; Aude et al. 2004). Organic farming practices also affect the abundance of PPNs (Hallmann et al. 2007; Chen et al. 2012; Adam et al. 2013).

Limited information is available on PPN densities and their damage in organic farming systems. Therefore, to illustrate the potential for nematode damage and opportunities or constraints for their management in organic agriculture, we discuss studies comparing conventional and organic farming systems and surveys of organic farms in different geographic locations. Although the majority of comparative studies showed reduction in the number of PPNs in organic farming compared to conventional systems (Griffiths et al. 1994; Ferris et al. 1996), results differed among nematode species within studies. For example, higher population densities of plant-parasitic root lesion nematodes (*Pratylenchus* spp.) were observed in organic plots than conventional fields (Clark et al. 1998; Neher and Olsen 1999; Berkelmans et al. 2003). In another study, root-knot nematode populations increased under organic tomato production conditions (van Bruggen and Termorshuizen 2003). Recent surveys have found increased populations of several genera of PPNs in fields under organic production (Hallmann et al. 2007; Chen et al. 2012; Adam et al. 2013). These studies demonstrate that organic farming has plant-parasitic nematode problems the same as conventional farms and that under certain circumstances, organic farms may experience even higher nematode pest pressure compared to conventional production systems. Higher densities of weeds, use of legumes to enhance nutrients, and continuous cropping to prevent nutrient leaching on organic farms provide perpetual hosts for several potentially damaging groups of nematodes.

5.2 Plant-Parasitic Nematode Problems in Organic Agriculture

For detailed information on the important PPN species, worldwide crop losses from nematodes, and options for their management for general crop production, see Chen et al. (2004), Luc et al. (2005), Perry and Moens (2006), Perry et al. (2009), and Stirling (2014a). Broadly, PPNs are categorized into two groups based on their feeding strategies. Ectoparasitic nematodes feed on roots without entering the root tissue, while endoparasites undergo at least one stage of development inside the plant host. Symptoms of feeding of PPNs on plant roots can be confused with symptoms of nutrient or water deficiency. Aboveground symptoms of nematode damage include stunting, yellowing, wilting (Fig. 5.1), and yield loss, while belowground indicators are malformations of roots (Fig. 5.2) such as galls, lesions, and distortions depending upon the type of nematode specie. Several genera of PPNs including both ectoparasites and endoparasites are observed in fields under organic production (Chen et al. 2012; Hallmann et al. 2007; Adam et al. 2013). However, little attention is being paid to the detrimental effects of PPNs can cause to the organic cropping systems, and it is now that farmers have started to realize nematodes as being pests in their organic fields (Hallmann et al. 2007).

Here, we briefly review PPNs, concentrating on the most economically important species reported from organic farming systems. Additional species, not discussed here, may also cause losses to some organic crops. Nematodes of greatest importance in organic crops appear to be sedentary endoparasites in the family *Heteroderidae* including the cyst nematodes (e.g., species of *Heterodera* and *Globodera*) and root-knot nematodes (*Meloidogyne* spp.), and migratory endoparasites of family *Pratylenchidae* (*Pratylenchus* spp.). Cyst nematode species including soybean cyst, potato cyst, and cereal cyst nematodes causes huge crop

Fig. 5.1 Above ground symptoms of nematode damage: patchy and stunted growth of barley crop due to cereal cyst nematode infestation (*Courtesy* Shabeg Briar, CARC, Montana State University)

Fig. 5.2 Below ground symptoms of root nematode damage. Distorted tomato root system due to knot formation (*Courtesy* Jack Kelly Clark, Univ. of California Statewide IPM Program)

losses (Wrather and Koenning 2006; Mai 1977). Cyst nematode-infected plants may develop bushy roots system, and individual roots may have knotted appearance with several females at each knot. Root-knot nematode (*Meloidogyne* spp.) are economically important polyphagous obligate plant parasites, distributed worldwide, and are known to parasitize nearly every plant species of higher plants (Moens et al. 2009). As a result of feeding of root-knot nematode, small to large galls or "knots" can form throughout the root system of infected plants. In contrast, root lesion nematodes (*Pratylenchus* spp.) are migratory endoparasites and feed within the root system. They are distributed worldwide and have wider host ranges, and several species are reported to cause economic injury to the crop plants (Thompson et al. 1999). As the name implies, root lesion nematodes produce characteristic necrotic lesions (darkened areas of dead tissue) on the surface and throughout the cortex of infected roots.

5.3 Nematode Management: Options and Constraints for Organic Agriculture

Basic methods of nematode management are exclusion, eradication, and protection (Norris et al. 2003; Agrios 2005). The concept of integrated pest management relies on a combination of approaches with minimum or judicious use of synthetic pesticides. Plant protection in organic agriculture, however, relies primarily on creating a favorable ecological equilibrium among soil biota without the use of synthetic pesticides (Delate et al. 2003). Letourneau and van Bruggen (2006) outlined three basic approaches for pest management in organic crops, including preventing pest colonization, regulating the abundance of pests or pathogens at low levels through biological processes, and employing curative materials permitted in organic farming, through which crop protection is generally achievable.

Nematodes can only move very short distances and are therefore unable to spread from one field to another on their own. They are usually transported to other locations by farming machinery, in plant material, on animals, or by water or wind. Routine practices such as sanitation of farm equipment and clean planting material can prevent their spread. Crop rotation with non-hosts and planting time adjustments can prevent the colonization and establishment of PPNs. All these steps to prevent nematode spread and establishment are practiced in both organic and conventional farming systems. However, measures to prevent nematode entry into fields under organic production are more critical because of restrictions on the use of curative synthetic chemicals such as fumigants. Similar to other soil-borne pathogens, PPNs once introduced and established into the field are difficult to eradicate and their ongoing management will then be necessary. Among methods for nematode management, we discuss application of organic amendments, cover crops, crop rotation, nematode trap crops, antagonistic crops, and biological control.

5.3.1 Soil Organic Amendments

Application of animal and plant by-products into the soil is best known for crop management especially where synthetic inputs are not permitted. Although the primary reason for using soil amendments is to enhance nutrient supplement, increase organic matter levels, and improve soil structure, numerous amendments have been assessed and recommended for the management of PPNs (Akhtar and Alam 1993; Akhtar and Malik 2000; Oka 2010; Rodríguez-Kábana and Ivey 1986; Trivedi and Barker 1986). Organic amendments can be divided into two broad categories: (a) amendments that are cultivated in situ and are incorporated into the soil such as green manure, cover crops, or trap crops and (b) the amendments transported from elsewhere into the field such as composted animal manure and composted yard material or animal waste.

Several mechanisms have been proposed for the probable cause/s for nematode suppression using organic amendments and have been reviewed in detail by McSorley (2011a). The effects of amendments in general are accepted as indirectly causing the nematode suppression through enhanced activity of naturally occurring antagonists (such as bacteria, fungi, and predatory nematodes) (Akhtar and Malik 2000; Oka 2010). Various antagonistic fungi and bacteria have been observed in compost including species of *Trichoderma*, *Penicillium*, *Aspergillus*, *Bacillus*, *Pseudomonas*, *Pantoea*, and *Actinomycetes* (Hoitink and Boehm 1999) which help control soil-borne pathogens and PPNs (Sharon et al. 2001; Kluepfel et al. 2002; Mekete et al. 2009). Additionally, plant residues and other organic amendments such as composted animal manure may release nematicidal compounds such as ammonia directly lethal to nematodes (Oka 2010; Thoden et al. 2011; Rodríguez-Kábana and Ivey 1986; Rodriguez-Kabana et al. 1987).

5.3.1.1 Cover Crops

Cover crops and green manure crops (intended for soil incorporation prior to maturity) are grown between cash crop cycles primarily to improve soil fertility and soil structure and prevent soil from erosion. Various grassy and legumes as cover crops appear to suppress nematodes in soil, including the following cover crop/nematode combinations: (1) sun hemp (*Crotalaria juncea* L.)/root-knot nematode (*Meloidogyne incognita*) (Wang et al. 2004), (2) velvet bean (*Mucuna pruriens*)/root-knot nematode (*Meloidogyne incognita*) (Quénéhervé et al. 1998), (3) sorghum or Sudan grass (*Sorghum bicolor*, *S. sudanense*)/root-knot nematodes (*Meloidogyne* spp.) (McSorley et al. 1994), and (4) pearl millet (*Pennisetum glaucum*)/root lesion nematodes (*Pratylenchus* spp.) (Bélair et al. 2005). Some of the green manure cover crops have also been identified for their antagonistic or allelopathic effects on PPNs. For example, root exudates of marigold (*Tagetes* spp.) possess nematicidal properties and helped in the suppression of several genera of PPNs (Siddiqui and Alam 1987).

The selection of a cover crop in the rotational sequence depends on the economics and its adaptability to a specific region (McSorley 1998, 2011b). The best choice, however, would be a crop that is poor host or non-host for the PPNs prevalent in the field. Therefore, care should be taken in selecting a cover crop in organic farming, as a crop resistant to one species of nematode may be a good host for other type of nematode (McSorley et al. 1994). More often, cover crops are mechanically incorporated into the upper layers of the soil using heavy tillage operations thereby leaving negative impacts on the soil food web and especially detrimental to the disturbance of sensitive predatory organisms (Briar et al. 2007). An alternate would be to apply the amendments on the soil surface as mulches which may be less detrimental to soil food web and also help in the suppression of PPNs (Wang et al. 2008).

5.3.1.2 Animal Manures

Composted animal manure is one of the most popular organic amendments for soils. Poultry or livestock manure has been tested for nematode management (Nahar et al. 2006; Akhtar and Alam 1993; Akhtar and Malik 2000; Rodríguez-Kábana and Ivey 1986; Trivedi and Barker 1986). Numerous studies have found positive correlations between the addition of compost and suppression of PPNs including the economically important species such as root-knot and root lesion nematodes (e.g., Marull et al. 1997; LaMondia et al. 1999; McSorley and Gallaher 1994, 1995, 1996; Everts et al. 2006; Kaplan et al. 1992). The degree of nematode suppression, however, is variable depending upon factors such as the type of manure, application rate, and natural microflora in it (McSorley 2011a).

5.3.2 Crop Rotation and Other Cultural Practices

As described previously in this chapter, nematodes do not move long distance on their own and by reducing their population below the damaging levels may result in increased crop yield. Planting non-host crop in the rotation would remove food source for the PPNs and consequently decline in their population below the damaging levels (Rodríguez-Kábana and Ivey 1986; LaMondia 1999). However, the effectiveness of rotation in suppressing PPN population depends upon the type of the nematode specie/s present in the field, host range, and the duration of time pest nematode can survive in the field in the absence of the host (Halbrendt and LaMondia 2004). In general, for specialized host-specific plant-parasitic nematode (such as root-knot and cyst nematodes species) selection of non-host crop is relatively less difficult as compared to the nematode with a wider host range (such as root lesion nematode) (LaMondia 1999). Nevertheless, accurate identification of plant-parasitic nematode/s prevalent in the field would help in selecting a non-host crop and planning a long-term rotation with a focus on nematode management in organic farming.

Other cultural practices to prevent colonization and establishment of plant-parasitic nematodes such as sanitation, nematode-free vegetative-propagating materials, adjustment of planting time, and removal of host weeds are recommended for both organic and conventional agriculture. However, they are even more important for organic farming, because curative measures such as synthetic fumigant nematicides applications are restricted (Letourneau and van Bruggen 2006). Prophylactic measure such as nematode-free planting material, cleaning equipment, and quarantine measures would help in minimizing the chances of nematode entry into the field and further spread. Adjustment of planting date (early or late) to coincide with the conditions when the temperature is too low or too high for nematode infection and development has been shown to be effective method for nematode management in vegetable cropping systems (Bridge 1996). Soil solarization using transparent polyethylene sheets to trap solar heat is usually considered effective method in hot and arid climates. For example, solarization experiments in Israel and tropical India were helpful in suppressing population of root-knot, cyst, and root lesion nematodes (Oka et al. 2007; Sharma and Nene 1990). Trapping solar energy for raising soil temperature to the level detrimental to nematodes appears to be effective, and the only disadvantage would be that it may not be cost-effective and feasible for large-scale farming systems.

5.3.3 Host Plant Resistance

Host resistance is often the most cost-effective tool for nematode management for organic farming systems. Resistant plants are defined as those that support lower levels of nematode reproduction compared to susceptible plants (Roberts 2002; Cook and Starr 2006), and the extent of nematode-resistant crop varieties has been reviewed (e.g., Roberts and Ulloa 2010; Williamson and Roberts 2009; Starr and Roberts 2004; Starr et al. 2010). Progress has been made in identifying genes for resistance to the economically important nematode species (Williamson and Kumar 2006). These include the *Hs1pro-1* gene that provides resistance to the sugar beet cyst nematode (*Heterodera schachtii*), the *Mi* gene that affects several species of root-knot nematodes in *Meloidogyne*, and the *Gpa2* gene that confers resistance against some isolates of the potato cyst nematode (*Globodera pallida*) (Williamson 1998, 1999). Nematode-resistant plant carrying resistant genes is either characterized by a rapid localized cell death that occurs near the anterior end of the nematode in the region of the root where feeding site initiation occurs or neither the feeding site nor the nematode is able to progress to the next developmental stage (Branch et al. 2004; Williamson 1999).

Ideally, resistance should be broad in nature, affecting many nematode species. For instance, the *Mi* gene confers resistance against four species of root-knot nematode (Huang et al. 2006). Most genes for resistance, however, provide effective suppression against only single specie or even just particular race of plant-parasitic nematode (Williamson and Roberts 2009). The continual emergence of new and more virulent races of PPNs sometimes leads to failure of resistant crop varieties planted over a longer period of time. In some cases, resistance is sensitive to temperature. Resistant crops developed in a colder region may be susceptible to the same pest nematodes in warmer regions of the world (Williamson and Roberts 2009). Another constraint in choosing resistant cultivars for organic farming is that some of the commercially available nematode-resistant cultivars also possess a modified gene for herbicide tolerance and are therefore not permitted in organic agriculture. Chen et al. (2012) observed higher numbers of soybean cyst nematode in fields under organic production as compared to conventional fields where genetically modified cultivars resistant to soybean cyst nematode were planted. A system of integrated control with a rotation of resistant and non-resistant crop varieties to slow selection for new virulent races is recommended for different cropping systems where limited numbers of resistant cultivars are available, and this approach seems to be effective for organic farming as well. In addition, optimizing plant health with adequate nutrition helps sustain high plant productivity, while suppressing PPNs and maintaining efficacy of resistant cultivars over a longer period of time (Williamson and Roberts 2009).

5.3.4 *Biological Control*

Because organic growers do not, by definition, choose to use synthetic nematicides, they must maximize beneficial organisms in soils to help manage PPNs together with cultural practices. Curative biological control can sometimes be accomplished through inundative releases of selected biological control agents obtained from commercial suppliers (Stirling 2014c). Microbial pathogens, endophytes, and antagonists are important in the regulation of PPNs, independent of farming system (Kerry 1990; Siddiqui and Mahmood 1996, 1999; Morton et al. 2004; Hallmann et al. 2009; Stirling 2014c). However, the introduction of beneficial organisms to the soil for nematode management via augmentative biological control has had limited success (Sikora et al. 2008). There are few commercial biological control products for nematode management are available in the market that might be considered for use in organic farming. Biological control agents along with their advantages and disadvantages are discussed and enlisted recently by Hallmann et al. (2009) and Stirling (2014c).

5.3.4.1 Bacterial Pathogens and Antagonists

Several types of saprophytic bacteria that occur in the soil, rhizosphere (in the root zone), or endorhiza (inside roots) have been shown to be antagonistic to nematodes, with their unique modes of action. The most widely studied group of beneficial bacteria resides on the plant rhizosphere, and its members are commonly considered as plant growth-promoting rhizobacteria (PGPR). Some rhizobacteria are able to enter the root system, colonize, and become endophytic (Hallmann et al. 2001). Among the dominant genera (*Bacillus* and *Pseudomonas*), there are several species such as *Bacillus subtilis*, *B. sphaericus*, and *Pseudomonas flourescens* producing metabolites toxic to PPNs (Sikora 1992; Hallmann et al. 1999; Kloepper et al. 1991). A number of different mechanism have been proposed for nematode control by rhizobacteria including direct antagonism through release of nitrogenous compounds toxic to nematodes, induced systemic resistance, interference with plant–nematode recognition, and plant growth promotion. These mechanisms are reviewed in detail by Tian et al. (2007).

The Gram-positive obligate endoparasitic bacteria of the genus *Pasteuria* are parasites of all the economically important genera of PPNs. Different species of *Pasteuria* have been reported, which differ in their host ranges and pathogenicity against PPNs (Trudgill et al. 2000; Timper 2009). Among them, the most widely studied is *Pasteuria penetrans,* parasitic on root-knot nematodes (*Meloidogyne* spp.) (Sayre and Starr 1985; Stirling 2014c). This bacterium has been successfully mass produced either in vivo (on nematode hosts) or in vitro in quantities only suitable to add to small-size microplots or in pots for nematode control (Stirling 1984; Trudgill et al. 2000; Hewlett et al. 2004). Since high application rates of pasteuria-based products are necessary for achieving effective nematode control, it may still not be cost-effective to apply at large-scale cropping areas (Stirling 2014b).

5.3.4.2 Fungal Pathogens and Antagonists

A wide range of fungi are known to parasitize PPNs and possess the potential for biological control of nematodes (Kerry 1990; Stirling and Smith 1998; Kerry 2000; Sayre and Walter 1991; Kerry and Hominick 2002; Sikora et al. 2008; Stirling 2014c). In particular, important for the purpose of controlling PPNs are nematophagous fungi, which may be either obligate or facultative parasites. Obligate fungal parasites species such as *Catenaria auxilaris* and *Nematophthora gynophila* use their spores to initiate infection either by adhering to the body of the nematodes or by being ingested and then penetrating the gastrointestinal tract. These fungal species have been reported to parasitize cyst nematode (Kerry and Crump 1980, 1998). Facultative parasites grow saphrophytically in the soil and parasitize nematodes by either way of specialized adhesive spores or trapping structures such as knobs, rings, or net structures that trap nematodes and kill them. Important fungal species in this group include *Dactylella* spp., *Dactylaria candida*, *Arthrobotrys botryospora*, *Paeciliomyces liliacinus*, *Verticillium chlamydosporium*, and *Hirsutella rhossiliensis* have been studied further in detail and possess the potential to be developed into biological control products (Hallmann et al. 2009; Stirling 2014c).

Although considerable progress has been made in the area of inundative application of fungal organisms for nematode control, the number of biotic and abiotic factors still limits their effectiveness in the field. For example, the biggest constraint in using nematophagus fungi for biological control is the difficulty in overcoming the competition from other resident soil organisms (Stirling 2014c). Abiotic factors such as soil texture, moisture, nutrients, organic matter, and pH also affect their survival and establishment directly and indirectly after their application (Chen and Dickson 2004).

5.3.4.3 Plant-Parasitic and Entomopathogenic Nematode Interactions

A number of entomopathogenic nematodes (EPNs) species (e.g., *Steinernema carpocapsae*, *S. feltiae*, *S. riobraveare*, *Heterorhabditis bacteriophora*, *Xenorhabdus nematophilus*) are better known than PPNs and are widely used to manage insect pests in agro-ecosystems (Grewal et al. 2005). However, their use for the management of PPNs is controversial, and the ecology of their interactions with PPNs is still not fully understood (Lewis and Grewal 2005). Only a few studies have observed EPNs to be antagonistic to PPNs and to consequently suppress them under field or greenhouse conditions (Bird and Bird 1986; Ishibashi and Kondo 1986; Grewal et al. 1999). Possible mechanism responsible for the suppression of PPNs seems to be either the allelochemicals produced by the EPNs itself (Grewal et al. 1999) or the antagonistic effects that are from the bacterial symbiont of the EPNs (Samaliev et al. 2000). Other studies have observed little or no effect on the suppression of PPNs (LaMondia and Cowless 2002; Nyczepir et al. 2004). Currently, there seems to be little realistic potential to use EPNs to manage PPNs cost-effectively, and growers would not consider relying on this method, especially if they are applying them with the sole goal of achieving PPN management under field conditions.

5.4 Conclusions

Conditions such as no application of synthetic inputs, higher levels of weed infestation, cultivation of leguminous crops in rotation for nutrient management, and lack of vegetation-free periods necessary to prevent nutrient leaching may be contributing to higher levels of nematode buildup over a longer period of time (Hallmann et al. 2007). For disease management, organic farming relies on cultural practices that may not be sufficient for nematode management under certain circumstances. Resistant cultivars of selective crops are usually resistant to only a few races or species of nematodes and may not last long due to the development of new races over time. Moreover, some of the commercially available resistant cultivars also possess genetically modified genes against herbicides which preclude their use in organic agriculture. Biological control measures involving the use of microbial pathogens, endophytes, and antagonists may play an important role in nematode management in organic crop production, but have shown limited success in their management at a commercial level due to higher cost, and their application may not be yet feasible in large-scale farming systems. Therefore, the emphasis should be on the conservation and enhancement of the existing pool of biological control agents through farming practices that support the survival and reproduction of natural enemies of PPNs in the soil (Stirling 2014b). For example, the addition of organic amendments to the soil has been shown to have some level of suppression of PPNs under field conditions.

A combined approach is needed for nematode management in organic farming systems. Measures such as appropriate crop rotation with less susceptible crop species and green manure crops, soil amendments with antagonistic crops, and consistent weed control are effective when used in together in concert (Hallmann et al. 2007). Identification of prevalent PPN specie/s through nematode diagnostic services should be considered for choosing non-hosts before planning long-term crop rotation for organic farming. Additional research is needed most in the development of resistant cultivars suitable for organic farming, nematode-free vegetative-propagating materials, and further steps in commercialization of biological control products permitted in organic farming systems.

References

Adam M, Heuer H, Ramadan EM, Hussein MA, Hallmann J (2013) Occurrence of plant-parasitic nematodes in organic farming. Int J Nematol 23:82–90

Agrios GN (ed) (2005) Plant pathology, 5th edn. Elsevier Academic Press, New York

Akhtar M, Alam MM (1993) Utilization of waste materials in nematode control: a review. Bioresour Technol 45:1–7

Akhtar M, Malik A (2000) Roles of organic soil amendments and soil organisms in the biological control of plant-parasitic nematodes: a review. Bioresour Technol 74:35–47

Aude E, Tybirk K, Michelsen A, Ejrnaes R, Hald AB, Mark S (2004) Conservation value of the herbaceous vegetation in hedgerows—does organic farming make a difference? Biol Conserv 118:467–478

Bélair G, Dauphinais N, Fournier Y, Dangi OP, Clément MF (2005) Effect of forage and grain pearl millet on *Pratylenchus penetrans* and potato yields in Quebec. J Nematol 37:78–82

Berkelmans R, Ferris H, Tenuta M, van Bruggen AHC (2003) Effects of long-term crop management on nematode trophic levels other than plant feeders disappear after 1 year of disruptive soil management. Appl Soil Ecol 23:223–235

Bird AF, Bird J (1986) Observations on the use of insect-parasitic nematodes as a means of biological control on root-knot nematodes. Int J Parasitol 16:511–516

Branch C, Hwang, CF, Navarre DA, and Williamson VM (2004) Salicylic acid in part of the *Mi*-1-mediated defense response to root-knot nematode in tomato. Mol Plant Microbe Interact 17:351–356

Briar SS, Grewal PS, Nethi Somasekhar, Stinner D, Miller SA (2007) Soil nematode community, organic matter, microbial biomass and nitrogen dynamics in field plots transitioning from conventional to organic management. Appl Soil Ecol 37:256–266

Bridge J (1996) Nematode management in sustainable and subsistence agriculture. Annu Rev Phytopathol 34:201–225

Chen S, Dickson DW (2004) Biological control of nematodes by fungal antagonists. In: Chen ZX, Chen SY, Dickson DW (eds) Nematology advances and perspectives, vol 2., Nematode Management and UtilizationCABI Publishing, CAB International, Wallingford, pp 979–1039

Chen ZX, Chen SY, Dickson DW (eds) (2004) Nematology advances and perspectives. CABI Publishing, CAB International, Wallingford

Chen SY, Sheaffer CC, Wyse DL, Nickel P, Kandel H (2012) Plant-parasitic nematode communities and their associations with soil factors in organically farmed fields in Minnesota. J Nematol 44:361–369

Clark MS, Ferris H, Klonsky K, Lanini WT, van Bruggen AHC, Zalom FG (1998) Agronomic, economic, and environmental comparison of pest management in conventional and alternative tomato and corn systems in northern California. Agric Ecosyst Environ 68:51–71

Cook R, Starr JL (2006) Resistant cultivars. In: Perry RN, Moens M (eds) Plant nematology. CABI Publishing, CAB International, London, pp 370–391

Delate K, Friedrich H, Lawson V (2003) Organic pepper production systems using compost and cover crops. Biol Agric Hortic 21:131–150

Everts KL, Sardanelli S, Kratochvil RJ, Armentrout DK, Gallagher LE (2006) Root-knot and lesion nematode suppression by cover crops, poultry litter, and poultry litter compost. Plant Dis 90:487–492

Ferris H, Venette RC, Lau SS (1996) Dynamics of nematode communities in tomatoes grown in conventional and organic farming systems and their impact on soil fertility. Appl Soil Ecol 3:161–175

Ferris H, Bongers T, de Geode RGM (2001) A framework for soil food web diagnostics: extension of the nematode faunal analysis concept. Appl Soil Ecol 18:13–29

Grewal PS, Lewis EE, Venkatachari S (1999) Allelopathy: a possible mechanism of suppression of plant-parasitic nematodes by entomopathogenic nematodes. Nematol 1:735–743

Grewal PS, Ehlers RU, Shapiro-Ilan DI (eds) (2005) Nematodes as biological control agents. CABI Publishing, CAB International, Wallingford

Griffiths BS, Ritz K, Wheatley RE (1994) Nematodes as indicators of enhanced microbiological activity in a Scottish organic farming system. Soil Use Manage 10:20–24

Halbrendt JM, LaMondia JA (2004) Crop rotation and other cultural practices. In: Chen ZX, Chen SY Dickson DW (eds) Nematology advances and perspectives, vol 2. CABI Publishing, CAB International, Wallingford, pp 909–930

Hallmann J, Rodriguez-Kabana R, Kloepper JW (1999) Chitin-mediated changes in bacterial communities of the soil, rhizosphere and within roots of cotton in relation to nematode control. Soil Biol Biochem 31:551–560

Hallmann J, Quadt-Hallmann A, Miller WG, Sikora RA, Lindow SE (2001) Endophytic colonization of plants by the biocontrol agent rhizobium etli G12 in relation to *Meloidogyne incognita* infection. Phytopathol 91:415–422

Hallmann JA, Frankenberg A, Paffrath A, Schmidt H (2007) Occurrence and importance of plant-parasitic nematodes in organic farming in Germany. Nematol 9:869–879

Hallmann J, Davies KG, Sikora R (2009) Biological control using microbial pathogens, endophytes and antagonists. In: Perry RN, Moens M, Starr J (eds) Root-Knot nematodes. CABI Publishing, CAB International, Wallingford, pp 380–411

Hewlett TE, Gerber JF, Smith KS (2004) In vitro culture of *Pasteuria penetrans*. Nematology Monogr Perspect 2:175–185

Hoitink HAJ, Boehm MJ (1999) Biocontrol within the context of soil microbial communities: a substrate-dependent phenomenon. Annu Rev Phytopathol 37:427–446

Huang GZ, Allen R, Davis EL, Baum TJ, Hussey RS (2006) Engineering broad root-knot resistance. transgenic plants by RNAi silencing of a conserved and essential root-knot nematode parasitism gene Proceedings of the national academy of science of the United States of America 103:14302–14306

Ishibashi N, Kondo E (1986) *Steinernema feltiae* (DD-136) and *S. glaseri*: persistence in soil and bark compost and their influence on native nematodes. J Nematol 18:310–316

Kaplan M, Noe JP, Hartel PG (1992) The role of microbes associated with chicken litter in the suppression of *Meloidogyne arenaria*. J Nematol 24:522–527

Kerry BR (1990) An assessment of progress towards microbial control of plant-parasitic nematodes. J Nematol 22:621–631

Kerry BR (2000) Rhizosphere interactions and exploitations of microbial agents for the biological control of plant-parasitic nematodes. Ann Rev Phytopathol 38:423–441

Kerry BR, Crump DH (1980) Two fungi parasitic on females of cyst nematodes (*Heterodera* spp.). Trans Br Mycol Soc 74:119–125

Kerry BR, Crump DH (1998) The dynamics of the decline of the cereal cyst nematode, *Heterodera avenae*, in four soils under intensive cereal production. Fund Appl Nematol 21:617–625

Kerry BR, Hominick WM (2002) Biological Control. In: Lee DL (ed) The biology of nematodes. Taylor and Francis, London, pp 483–509

Kloepper JW, Rodriguez-Kabana R, Mclnroy JA, Collins DJ (1991) Analysis of population and physiological characterization of microorganisms in rhizosphere of plant with antagonistic properties to phytopathogenic nematodes. Plant Soil 136:95–102

Kluepfel DA, Nyczepir AP, Lawrence JE, Wechter WP, Leverentz B (2002) Biological control of the phytoparasitic nematode *Mesocriconema xenoplax* on peach trees. J Nematol 34:120–123

LaMondia JA (1999) Influence of rotation crops on the strawberry pathogens *Pratylenchus penetrans*, *Meloidogyne hapla*, and *Rhizoctonia fragariae*. J Nematol 31:650–655

Lamondia JA, Cowles RS (2002) Effect of entomopathogenic nematodes and Trichoderma harzianum on the strawberry black root rot pathogens Pratylenchus penetrans and Rhizoctonia fragariae. J Nematol 34:351–357
LaMondia JA, Gent MPN, Ferrandino FJ, Elmer WH, Stoner KA (1999) Effect of compost amendment of straw mulch on potato early dying disease. Plant Dis 83:361–366
Letourneau D, van Bruggen A (2006) Crop protection in organic agriculture. In: Taji A, Kristianesen P, Reganold J (eds) Organic agriculture—a global perspective. CSIRO Publishing, Melbourne, pp 93–121
Lewis EE, Grewal PS (2005) Interactions with plant-parasitic nematodes. In: Grewal PS, Ehlers RU, Shapiro-Ilan DI (eds) Nematodes as biocontrol agents. CABI Publishing, CAB International, Oxon, pp 349–361
Luc M, Sikora RA, Bridge J (eds) (2005) Plant parasitic nematodes in subtropical and tropical agriculture, 2nd edn. CABI Publishing, CAB International, Wallingord
Mäder P, Fliessbach A, Dubois D, Gunst L, Fried P, Niggli U (2002) Soil fertility and biodiversity in organic farming. Science 296:1694–1697
Mai WF (1977) Worldwide distribution of potato-cyst nematodes and their importance in crop production. J Nematol 9:30–34
Marull J, Pinochet J, Rodrı′guez-Ka′bana R (1997) Agricultural and municipal compost residues for control of root-knot nematodes in tomato and pepper. Compost Sci Util 1:6–15
McSorley R (1998) Alternative practices for managing plant-parasitic nematodes. Am J Alterative Agric 13:98–104
McSorley R (2011a) Overview of organic amendments for management of plant-parasitic nematodes, with case studies from Florida. J Nematol 43:69–81
McSorley R (2011b) Assessment of rotation crops and cover crops for management of root-knot nematodes (*Meloidogyne* spp.) in the southeastern United States. Nematropica 41:200–214
McSorley R, Gallaher RN (1994) Effect of tillage and crop residue management on nematode densities on corn. J Nematol 26:669–674
McSorley R, Gallaher RN (1995) Effect of yard waste compost on plant-parasitic nematode densities in vegetable crops. J Nematol 27:545–549
McSorley R, Gallaher RN (1996) Effect of yard waste compost on nematode densities and maize yield. J Nematol 28:655–660
McSorley R, Dickson DW, de Brito JA, Hochmuth RC (1994) Tropical rotation crops influence nematode densities and vegetable yields. J Nematol 26:308–314
Mekete T, Hallmann J, Kiewnick S, Sikora R (2009) Endophytic bacteria from Ethiopian coffee plants and their potential to antagonize *Meloidogyne incognita*. Nematol 11:117–127
Moens M, Perry RN, Star JL (2009) *Meloidogyne* species- a diverse group of novel and important plant parasites. In: Perry RN, Moens M, Starr JL (eds) Root knot nematodes. CABI Publishing, CAB International, Wallingford, pp 1–17
Morton O, Kerry B, Hirsch P (2004) Infection of plant-parasitic nematodes by nematophagous fungi—a review of the application of molecular biology to understand infection processes and to improve biological control. Nematology 6:161–170
Moyniham M (2010) Status of organic agriculture in Minnesota: a report to the Minnesota Legislature 2010. Minnesota Department of Agriculture. http://www.mda.state.mn.us/~/media/Files/news/govrelations/organicstatusreport.ashx Files/news/govrelations/organicstatusreport.ashx. Accessed 26 June 2015
Nahar MS, Grewal PS, Miller SA, Stinner D, Stinner BR, Kleinhenz MD, Wszelaki A, Doohan D (2006) Differential effects of raw and composted manure on nematode community, and its indicative value for soil microbial, physical and chemical properties. Appl Soil Ecol 3:140–151
Neher DA, Olson RK (1999) Nematode communities in soils of four farm cropping management systems. Pedobiologia 43:430–438
Norris RF, Caswell-Chen EP, Kogan M (eds) (2003) Concepts in integrated pest management. Pearson Education Ltd., New Jersey

Nyczepir AP, Shapiro-Ilan DI, Lewis EE, Handoo ZA (2004) Effect of entomopathogenic nematodes on *Mesocriconema xenoplax* populations in peach and pecan. J Nematol 36:181–185
Oka Y (2010) Mechanisms of nematode suppression by organic soil amendments- a review. Appl Soil Ecol 44:101–115
Oka Y, Shapira N, Fine P (2007) Control of root-knot nematodes in organic farming system by organic amendments and soil solarization. Crop Prot 26:1556–1565
Perry RN, Moens M (eds) (2006) Plant nematology. CABI Publishing, CAB International, Wallingford
Perry RN, Moens M, Starr JL (2009) (eds) Root knot nematodes. CABI Publishing CAB International, Wallingford
Quénéhervé P, Topart P, Martiny B (1998) *Mucuna pruriens* and other rotational crops for control of *Meloidogyne incognita* and *Rotylenchulus reniformis* in vegetables in polytunnels in Martinique. Nematropica 28:19–30
Roberts PA (2002) Concepts and consequences of resistance. In: Starr JL, Cook R, Bridge J (eds) Plant resistance to parasitic nematodes. CABI Publishing, CAB International, New York, pp 23–41
Roberts PA, Ulloa M (2010) Introgression of root-knot nematode resistance into tetraploid cottons. Crop Sci 50:940–951
Rodríguez-Kábana R, Ivey H (1986) Crop rotation systems for the management of *Meloidogyne arenaria* in peanuts. Nematropica 16:53–63
Rodriguez-Kabana R, Morgan-Jones G, Chet I (1987) Biological control of nematodes: soil amendments and microbial antagonists. Plant and Soil 100:237–247
Rowe RC, Powelson M (2002) Potato early dying: Management challenges in a changing production environment. Plant Dis 86:1184–1193
Samaliev HY, Andreoglou FI, Elawad SA, Hague NGM, Gowen SR (2000) The nematicidal effects of the bacteria *Pseudomonas oryzihabitans* and *Xenorhabdus nematophilus* on the root-knot nematode *Meloidogyne javanica*. Nematol 2:507–514
Sayre RM, Starr MP (1985) *Pasteuria penetrans* (exThorne, 1940) nom. rev., comb. n., sp. n., a mycelial and endospore-forming bacterium parasitic in plant-parasitic nematodes. Precis Helminthol Soc Wash 52:149–165
Sayre RM, Walter DE (1991) Factors affecting the efficacy of natural enemies of nematodes. Annu Rev Phytopathol 29:149–166
Sharma SB, Nene YL (1990) Effects of soil solarization on nematodes parasitic on chickpea and pigeonpea. J Nematol 22:658–664
Sharon E, Bar-Eyal M, Chet I, Herrera-Estrella A, Kleifeld O, Spiegel Y (2001) Biological control of the root-knot nematode *Meloidogyne javanica* by *Tricoderma harzianum*. Phytopathology 91:687–693
Siddiqui MA, Alam MM (1987) Control of plant parasitic nematodes by intercropping with *Tagetes minuta*. Nematologia Mediterr 15:205–211
Siddiqui ZA, Mahmood I (1996) Biological control of plant parasitic nematodes by fungi: a review. Bioresour Technol 58:229–239
Siddiqui ZA, Mahmood I (1999) Role of bacteria in the management of plant parasitic nematodes: a review. Bioresour Technol 69:167–179
Sikora RA (1992) Management of the antagonistic potential in agricultural ecosystems for the biological control of plant parasitic nematodes. Annu Rev Phytopathol 30:245–270
Sikora RA, Pocasangre L, zum Felde A, Niere B, Vu TT, Dababat AA (2008) Mutualistic endophytic fungi and in-planta suppressiveness to plant parasitic nematodes. Bio Control 46:15–23
Starr JL, Roberts PA (2004) Resistance to plant parasitic nematodes. In: Chen ZX, Chen SY, Dickson DW (eds) Nematology advances and perspectives, vol 2., Nematode management and utilizationCABI Publishing, CAB International, Wallingford, pp 879–907

Starr JL, Moresco ER, Smith CW. Nichols RL, Roberts PA, Chee P (2010) Inheritance of resistance to *Meloidoygne incognita* in primitive cotton accessions from Mexico. J Nematol 42:352–358

Stirling GR (1984) Biological control of *Meloidogyne javanica* with *Bacillus penetrans*. Phytopathology 74:55–60

Stirling GR (ed) (2014a) Biological control of plant-parasitic nematodes. CABI Publishing, CAB International, Wallingford

Stirling GR (2014b) Integrated soil biology management. In: Stirlling G (ed) Biological control of plant-parasitic nematodes. CABI Publishing, CAB International, Wallingford, pp 304–341

Stirling GR (2014c) Biological products for nematode management. In: Stirlling G (ed) Biological control of plant-parasitic nematodes. CABI Publishing, CAB International, Wallingford, pp 342–389

Stirling GR, Smith LJ (1998) Field tests of formulated products containing either *Verticillium chlamydosporium* or *Arthrobotrys dactyloides* for biological control of root-knot nematodes. Biol Control 11:231–239

Thoden TC, Korthals GW, Termorshuizen AJ (2011) Organic amendments and their influence on plant-parasitic and free-living nematodes: a promising method for nematode management? Nematology 13:133–153

Thompson JP, Brennan PS, Clewett TG, Sheedy JG, Seymour NP (1999) Progress in breeding wheat for tolerance and resistance to root-lesion nematode (*Pratylenchus thornei*). Australas Plant Pathol 28:45–52

Tian B, Yang J, Zhang KQ (2007) Bacteria used in the biological control of plant-parasitic nematodes: populations, mechanisms of action, and future prospects. FEMS Microb Ecol 61:197–213

Timper P (2009) Population dynamics of *Meloidogyne arenaria* and *Pasteuria penetrans* in a long term crop rotation study. J Nematol 41:291–299

Trivedi PC, Barker KR (1986) Management of nematodes by cultural practices. Nematropica 16:213–236

Trudgill DL, Bala G, Block VC, Daudi A, Davies KG, Gowen SR, Fargette M, Madulu JD, Mateille T, Mwageni W, Netscher C, Phillips MS, Sawadogo A, Trivino GC, Vouyoukallou E (2000) The importance of tropical root-knot nematodes (*Meloidogyne* spp.) and factors affecting the utility of *Pasteuria penetrans* as a biocontrol agent. Nematology 8:823–845

van Bruggen AHC, Semenov AM (2000) In search of biological indicators for soil health and disease suppression. Appl Soil Ecol 15:13–24

van Bruggen AHC, Termorshuizen AJ (2003) Integrated approaches to root disease management in organic farming systems. Australas Plant Pathol 32:141–156

Wang KH, McSorley R, Gallaher RN (2004) Effect of *Crotalaria juncea* amendment on squash infected with *Meloidogyne incognita*. J Nematol 36:290–296

Wang KH, McSorley R, Gallaher RN, Kokalis-Burrelle N (2008) Cover crops and organic mulches for nematode, weed and plant health management. Nematology 10:231–242

Williamson VM (1998) Root-knot nematode resistance genes in tomato and their potential for future use. Annu Rev Plant Path 36:277–293

Williamson VM (1999) Plant nematode resistance genes. Curr Opin Pl Biol 2:327–331

Williamson VM, Kumar A (2006) Nematode resistance genes in plants: the battle underground. Trends Genet 22:396–403

Williamson VM, Roberts PA (2009) Mechanisms and genetics of resistance. In: Perry RN, Moens M, Starr J (eds) Root-knot nematodes. CABI Publishing, CAB International, Wallingford, pp 301–325

Wrather JA, Koenning SR (2006) Estimates of disease effects on soybean yields in the United States 2003 to 2005. J Nematol 38:173–180

Chapter 6
Breeding for Organic and Sustainable Production

George E. Boyhan and Suzanne P. Stone

Abstract Plant breeding has been with humankind since the beginning of civilization. Modern plant breeding, however, is a relatively recent development—just over 100 years. Recently with the increase in popularity of organic farming and farm sustainability, there is a growing interest in breeding for organic or low-input farming systems. These endeavors are in their earliest stages bringing to bear both traditional and modern plant breeding techniques to address the specific needs of organic and low-input farming. This chapter gives an overview of these early efforts, some of the techniques involved, as well as some of the social and philosophical concerns with breeding and crop improvement for low-input farming.

6.1 Introduction

Breeding for organic and sustainable production can be seen as occurring since the beginning of agriculture. Although not a formal process, cultivated plants underwent fundamental changes due to human selection. This process has been ongoing for approximately 12,000 years (National Geographic 2014). This type of selection over the millennia is referred to as evolutionary participatory breeding (EPB) (Murphy et al. 2005). Although the term EPB has been coined for this type of selection, it was not breeding in the modern sense, nor was there an understanding of the underlying mechanisms of inheritance.

Modern plant breeding began in earnest in the early twentieth century with the rediscovery of Mendel's work on inheritance (Kingsbury 2009). A great deal of early breeding work was done with corn. Corn, a wind-pollinated crop that generally

G.E. Boyhan (✉) · S.P. Stone
Department of Horticulture, University of Georgia, 1111 Miller Plant Science Bldg, Athens, GA 30602, USA
e-mail: gboyhan@uga.edu

S.P. Stone
e-mail: sstone2@uga.edu

D. Nandwani (ed.), *Organic Farming for Sustainable Agriculture*,
Sustainable Development and Biodiversity 9,
DOI 10.1007/978-3-319-26803-3_6

outcrosses, can be organized into individual homogenous lines through selfing over several generations. The characteristics of these inbreds can then be combined into hybrids that exhibited heterosis or hybrid vigor. The progeny or F_1 (filial) generation exhibits characteristics that are superior to either parent (Simmonds 1979). Seed saved, however, from the F_1 generation do not perform as well. The F_2 progeny segregate into a population with a myriad of characteristics without the consistent performance exhibited by the F_1 generation. This observation accelerated the development of modern breeding and selection methods. F_1 hybrids not only allow for rapid improvement of many crops, but also benefit seed companies' bottom line. As long as seed companies kept their inbred lines proprietary, growers have to return to the seed company every year to purchase the high performing F_1 hybrids. Thus, the advent of F_1 hybrids effectively took control of seed out of the hands of growers and placed it with seed companies. Although growers are limited in their access to such seed, there are benefits that growers enjoy from this arrangement. Seed companies can produce high-quality seed without foreign matter and with high germination rates. Growers reliably obtain consistent high-performing seed each year.

F_1 hybrids are not just limited to outcrossing crops such as corn. Crops such as tomatoes that predominately self-pollinize in nature can also benefit from this method of variety development, although heterosis is less likely to occur (Deppe 2000). Whether a crop is naturally outcrossing or inbred, there are two important hybrid seed crops must possess: There must be a way of making controlled crosses and sufficient seed must be produced from the crosses to make it economically feasible. In the case of corn, detasseling (removing male flowers) from one population makes controlled crosses via wind pollination possible. Tomatoes, although labor intensive, can be hand pollinated with sufficient seed produced from each cross to make it economically feasible. There are crop examples where F_1 hybrids are not possible, either because controlled crosses cannot be made or the amount of seed produced from controlled crosses is not sufficient. All beans and peas, for example, are open-pollinated varieties because each cross produces one pod of just a few seeds, thus making it uneconomic. Canola, Polish-type (*Brassica napus*), is an example where controlled crosses cannot be readily made, so F_1 hybrids are currently unavailable (Canola Council of Canada 2014).

Not all breeding efforts have focused on F_1 hybrids. Traditionally, public breeding programs have focused on developing open-pollinated varieties (Bassett 1986). This has included a broad range of vegetable crops across many land-grant institutions. For example, at Auburn University in the early 1980s, there were three plant breeders in the Horticulture Department breeding seven different crops. Today there are none. Popular older varieties, which were open pollinated, have consistently been available, although active breeding is not conducted with these varieties. Examples include "Hale's Best Jumbo" cantaloupe, "Crimson Sweet" watermelon, "Marketmore 76" cucumber, "California Wonder" pepper, "Waltham" butternut squash, and "Rutgers" tomatoes to name a few (Anonymous 2015).

The best breeding method that is chosen is based on the type of plant population and crop characteristics (Table 6.1). It is beyond the scope of this book to discuss in detail all of the approaches used in plant breeding; however, there are many

Table 6.1 Breeding strategies for different mating schemes and types of propagation

Mating scheme	Life cycle	Population type	Genetics and characteristics	Propagation
Self-pollinated	Annual	Inbred pure lines	Homozygous, homogeneous	Seed (open-pollinated)
Outcrossing	Annual, biennial, or perennial	Open pollinated	Heterozygous, heterogenous	Seed (open-pollinated)
Outcrossing	Annual, biennial, or perennial	Open pollinated	Heterozygous, homogenous	Seed (hybrids)
Outcrossing	Perennial	Open pollinated	Heterozygous, homogenous	Vegetative (clones)

excellent textbooks that can be consulted. Outbreeding seed propagated annuals or perennials form open-pollinated populations. Outbreeding seed propagated annuals or biennials can produce hybrids as discussed above. Inbreeding seed propagated annuals can form inbred pure lines. Finally, outbreeding perennials can be vegetatively propagated (Simmonds 1979). The techniques used with these four basic population types are generally the same in conventional plant breeding and in organic plant breeding; however, some goals and challenges are unique to organic systems.

6.2 Breeding Goals for Organic Agriculture

With the advent and growth of organic production, there has been an increasing interest in setting breeding goals that accommodate this type of production (Lammerts van Bueren and Myers 2012). For a more complete treatment of organic plant breeding for specific crops, Lammerts van Bueren and Myers (2012) *Organic Crop Breeding* gives a through presentation of the present state of this research.

What constitutes organic breeding? Variously, it has been defined by the goals that organic growers want (Table 6.2), as well as, the techniques that should be

Table 6.2 Goals of conventional breeders compared to organic plant breeders

Conventional values	Organic values
Maximize yield with high levels of inputs	Maximize yield with inputs from organic sources
Uniformity	Unique visual characteristics
Ease of mechanization	Emphasis on nutrition and taste
Processing and shipping efficiency	Variation in time to maturity
Specific disease resistances	Specific disease resistances
Abiotic stress resistance	Abiotic stress resistance
	Weed competitiveness
	Heirlooms and OPs

allowed in organic breeding (Arnchen and Thommen 2003; Dawson et al. 2011). Improving heritage or heirloom varieties with their unique characteristics by producing better yields and increasing disease resistance is one important goal of organic plant breeding. Nutrient use efficiency particularly for nitrogen is also an important goal of organic plant breeding (Baresel et al. 2008). Nutrient use efficiency not only involves the plant's physiology but also its root morphology (Melo 2003). Competitiveness with weeds is also an important characteristic (Hoad et al. 2008). Evaluation of current varieties for their performance under organic conditions is currently underway as an important first step in organic breeding.

An estimated 95 % of varieties grown on organic farms were actually bred for conventional, high-input systems (Lammerts van Bueren et al. 2011a, b). Growers rely on variety trial data from land-grant universities to choose the highest performing varieties for their region. However, the top performing varieties in organic systems are not always the same as the top performing varieties in conventional systems (Murphy and Jones 2007). Campion et al. (2014) found that performance of winter wheat varieties under high-input conditions are not good indicators of performance under organic conditions. These researchers did find that low-input (no fungicides/growth regulators, lower sowing rate, low nitrogen, and use of herbicides) production correlated well with organic management methods. Therefore, organic plant breeding should begin with a systematic evaluation of existing varieties in organic or low-input systems. The Northern Organic Vegetable Improvement Collaborative has been trialing a number of vegetables including broccoli, carrots, edible-podded peas, sweet corn, and winter squash (Myers et al. 2012). Characteristics that were evaluated included productivity of cool season crops during summer months, early season germination of sweet corn, weed competitiveness in overwintering carrots, and storability of butternut squash.

Not all evaluations reveal differences in rank among varieties in organic versus conventional systems. In experiments evaluating testcrosses of flint and dent corn over a range of environments in Germany, Burger et al. (2008) did not find any entries that showed a distinct advantage under organic production practices. There were, however, a fraction of hybrids that did perform well under both organic and conventional production. An organic trial of tomato varieties, including both modern commercial types and those popular with organic growers, were evaluated in Georgia (Boyhan et al. 2014). Modern F_1 varieties did better overall; however, many of the popular varieties among organic growers are indeterminate types not well suited to commercial staked tomato production. This study highlighted traits that organic tomato breeders should address.

Many varieties that are of particular interest to organic growers fall into the category of heirloom or heritage varieties. These older open-pollinated varieties had fallen out of favor with conventional growers for a variety of reasons including lack of disease resistance, unsuitability for shipping to market, and lack of uniformity. Organic growers favor many heirloom varieties because of unusual shapes, colors, and flavors they offer. The problems that plague these varieties, particularly the lack of disease resistance, continue to offer challenges in organic production. Improving disease resistance among such varieties while maintaining their unique characteristics would be a benefit to organic growers. In the Netherlands, research has been underway to reduce black spot disease in organic carrots and Fusarium wilt in spring wheat (Bueren et al. 2008). Other proposed work includes breeding for Anthracnose resistance in lupins, Ascochyta resistance in peas, late blight resistance in potatoes, and winter hardiness and drought tolerance in red clover (Vogt-Kaute 2002).

A great deal of work with organic breeding has been conducted on agronomic crops, particularly wheat. Evaluation of historic and modern wheat varieties showed no yield differences under organic growing conditions (Murphy and Jones 2007), probably because modern varieties have been bred in conventional, high-input conditions. Results of this evaluation suggest that selection under organic conditions would be advantageous in breeding wheat varieties for organic production. Several winter wheat cultivars have been released in Austria using a unique breeding method that initially screens germplasm using conventional low-input selection, which correlates with organic production (Löschenberger et al. 2008). The wild wheat species (*Triticum timopheevil)* has been investigated as a potential source of characteristics suited for organic production, particularly resistance to biotic and abiotic stresses. *T. timopheevil* has good disease resistance that is especially evident under organic conditions; however, crossing with traditional bread wheats and producing fertile offspring have been difficult but not impossible (Mikó et al. 2011). In another study, diallel crosses of wheat were evaluated as composite populations under organic conditions beginning with the F_2 generation (Kovács et al. 2010). Composite populations from tetraploids had lower yields than wheat composite populations, but better yield stability. Modern wheat varieties have also shown a reduction in many important nutrients such copper, iron, magnesium, manganese, phosphorus, selenium, and zinc compared to older varieties among soft white wheat, but not in hard red wheat, suggesting that selection in soft white wheat with increased nutrient content may be possible (Murphy et al. 2008). Mineral nutrient content is particularly important for organic and low-input agriculture that occurs in many parts of the world.

In evaluation of lentils under organic and conventional production, it was found that some varieties did well under both environments exhibiting broad adaptability, while others did better under organic production (Vlachostergios and Roupakias 2008). It is suggested that those entries that did better under organic production should be utilized in further breeding efforts. Finally, forages were evaluated and it was found that tetraploid Lolium did better than diploid Lolium and therefore may be better for sustainable/organic breeding options (Boller et al. 2008).

Beyond variety trials, breeders must determine which traits are ideal for organic farming. In a winter wheat study in Latvia, winter hardiness, resistance to snow mold, weed competitiveness, early maturity, prostrate growth, and large leaf size were all considered important traits for organic production (Kronberga 2008). Exploring traits necessary for weed competitiveness is an essential in organic agriculture because conventional pesticides are not permitted. Hoad et al. (2008) evaluated different wheat cultivars and one oat cultivar for their competitive ability against weeds. They devised a method of assessing weed suppression in which a value was calculated from the difference of weed growth in each plot compared to weed growth in adjacent unplanted areas. The ideal plant architecture for weed competitiveness is unique for each crop; therefore, a great deal of research is needed.

There are breeding goals that are unique to organic production systems. One such area is nutrient use efficiency. In conventional production, fertilizer is not a limiting factor; any shortcoming in a variety's ability to use fertilizer can be overcome by the addition of more fertilizer. One study investigated the root density in onions (Melo 2003). Onions generally have a poor root system that requires a great deal of fertility. Investigating differences in root morphology did not find much variation in onions. However, *Allium fistulosum*, a related species, does have a more robust root system. Improved root morphology was found in onion populations of the background (*A. cepa*) × (*A. roylei* × *A. fistulosum*), indicating the feasibility of breeding for this trait.

Some goals of organic plant breeding coincide with conventional breeding. For example, breeding for resistance to fruit cracking in tomatoes is a goal useful to both organic and conventional growers. Bender et al. (2005) evaluated several tomato varieties grown under organic conditions to evaluate them for cracking and found that "Maike" and "Valve" had the lowest fruit cracking. In the former case, the fruit were particularly small and in the latter, it had only two locules. An important general goal of breeding for organic production is performance stability over time or location under low-input conditions. This has been investigated with Durham wheat, tomatoes, and peas in Croatia (Lotti et al. 2008). Conversely, some organic breeders think breeding for local conditions would be a good investment particularly for local and regional markets (Rey et al. 2014).

6.3 Challenges in Developing Organic Seed Systems

Organic plant breeding is such a new area that issues associated with developing viable programs have arisen. Bozhanova and Dechev (2009) discuss issues unique to breeding for a whole farm system where biodiversity is to be maintained. In addition, they discuss issues related to identifying ideotypes suitable for organic production, as well as funding this type of research.

Maintaining biodiversity and a suitable gene pool that can be freely shared is very important in these breeding programs and many conventional, commercial breeding programs discourage this (Bueren and Osman 2001). The development of GMO varieties and the increased government regulation of these varieties have further reduced access to germplasm for breeding purposes. The consolidation of the seed industry, the widespread use of F_1 varieties, and the narrow control of genetic resources have been concerns of organic growers particularly as breeding programs that cater to their needs come into play (Navazio et al. 2012a). This consolidation of the seed industry and the effective elimination of grower rights to freely use seed is believed to be undermining biodiversity (Shiva 1997). This has been entrenched by treaty such as the International Convention for the Protection of new Varieties of Plants and the Trade Related Intellectual Property Rights treaty. This is particularly problematic in a country like India where 70 % of the seed supply is farmers' seed. The USDA as well as other organizations around the world do maintain germplasm collections that are available to plant breeders. Such collections have been an ongoing endeavor for many decades and can be an important resource for organic plant breeders.

Possible impediments to organic/sustainable plant breeding are the current rules governing organic production both in the USA and Europe. Although these rules require the use of organic seed, there is an exemption if organic seed of a specific variety is unavailable (Döring et al. 2012; USDA 2011). In France, however, there is quite a high adoption of organic seed use; 45–70 % for cereals and 75–100 % for vegetables suggesting that requiring organic seed when available is not much of an impediment to organic plant breeding (Rey et al. 2014). In Europe, standardized testing or value for cultivation and use (VCU testing) is used to assess cereal varieties. This means that wheat varieties available in the European Union must undergo testing under conventional farming practices and be registered for use by farmers. It was found that a separate performance review for organic production would be suboptimal and that combining performance from organic and non-organic trials is desirable (Przystalski et al. 2008). This program may restrict the availability of diverse types of varieties that may be more suitable for organic growers (Serpolay et al. 2011).

Organic growers have also complained about the discontinuation of many varieties, limited seed supply of older varieties, and the poor seed quality of these varieties (Navazio et al. 2012b). The limited supply of varieties suitable for organic production and the limited number of organic breeding programs is partially due to the size of this market. It is too small for many seed companies to invest the necessary resources. Conventional breeding programs may, however, be meeting many of the goals that organic growers are looking for. In a survey of Dutch onion breeders and organic growers, it was found that storage and bulb quality attributes were the same between these two groups. Breeders, however, did not give as much attention to field performance as organic growers would like. In addition, the use of cytoplasmic male sterility in producing hybrid onions was considered incompatible with organic principles (Osman et al. 2008).

There is an ongoing discussion of what techniques should be allowed in organic plant breeding. The use of genetically modified organisms (GMOs) is specifically prohibited by the National Organic Program (NOP) rules, while other techniques and approaches in plant breeding are not addressed (USDA 2011). A broad interpretation that encompasses most conventional breeding methods has been generally accepted (Leibinger and Reiners 2002). This has included modern techniques such as the use of DNA markers and the development of F_1 hybrids.

Some proponents of organic breeding contend that, along with the banning of GMO varieties, certain breeding techniques such as tissue culture should be avoided in favor of whole plant methods (Bueren et al. 2003). It has been argued that in vitro techniques be avoided so that breeding respects the whole living unit. This living unit can be defined either as the whole plant or a single cell. Depending on which definition these governing bodies choose, use of modern breeding techniques for organic systems may be restricted further. If living unit is defined as the whole plant, then tools such cytoplasmic male sterility and in vitro methods such as embryo culture or rescue, haploid culture, and protoplast fusion will be off-limits to plant breeders looking to improve crops for organic systems.

The International Federation of Organic Agriculture Movements has addressed the debate as to which practices should be allowed in organic plant breeding (Müller 2002). They have laid out a framework of acceptable breeding techniques that maintain fertility and diversity. They call for the exclusion of protoplast fusion techniques, artificial mutations particularly from radioactive sources, using cytoplasmic male sterility, and anther or microspore culture. Other breeding methods, although not banned, have been called into question relating to their "naturalness" (Haperen et al. 2010). These methods include cisgenetics and reverse breeding, or using doubled haploids to produce homozygous lines. Cisgenetics is disallowed because it involves the insertion of a gene directly into the plant genome, even though the gene is from the same species. Proponents of the whole plant approach wish to emphasize the intrinsic value of plants beyond human uses and incorporate this idea into the whole farm system of ecologically based farming.

6.4 Frontiers in Organic Plant Breeding

Although organic production in many ways looks to the past for more sustainable methods of production, this does not mean that modern techniques, such as marker-assisted selection and Quantitative trait loci (QTL) mapping, cannot be applied to breeding for this type of production. One unique challenge to organic plant breeding is the whole farm system of production. This can have an impact on the selection process but may be accelerated and improved with marker-assisted selection (Bueren et al. 2010). Understanding the genotypic characteristics of a species may be even more important for organic production because of the greater degree of genotype by environment interactions (Backes and Østergård 2008). QTL can be an important tool in estimating a trait's response to the environment. Evaluation of 188 barley entries indicated that alleles close to QTL for yield increased in frequency among lines that were selected for this trait in breeding programs (Pswarayi et al. 2008). In addition, this study indicated that increasing such allele frequency might be possible under low-input conditions as occur in organic production.

Redefining the genotype × environment interaction evaluates the genotype not only by specific plant characteristics such as yield, but also by farmer and consumers' needs (Desclaux et al. 2008). The environment portion of the interaction will also have a broader context since varieties developed for very local needs are to be emphasized over varieties adapted to a broad cross section of environments. The interaction in this context will be much more complex and the broader context will have to take into account other actors such as growers, regulators, society as a whole, and ultimately end users. The widely different environments that can occur within organic production necessitate the need to make selections in a relatively narrow environment for local production (Wolfe et al. 2008). This narrow environment can be affected by the whole farm system such as what rotation is being used and where the particular crop fits in that rotation. Beyond yield, such characteristics as nutrient uptake efficiency, disease resistance, and competitiveness with weeds need to be taken into account.

In conventional plant breeding, growers have been cooperators, but rarely if ever collaborators with breeders. On-farm variety trials and demonstrations have been an important part of conventional breeding efforts. These have, in many cases, been as much about advertising as variety evaluation. Often such plantings are not harvested, but are just evaluated visually with local growers invited to inspect new offerings. Organic plant breeders on the other hand are interested in a greater role for growers with participatory plant breeding (PPB). Organic plant breeders are more interested in identifying ideotypes suited to this type of production with greater input from growers, ecologically more sustainable, and adhering to ethical standards of organic production (Lammerts van Bueren 2003). Current varieties of cabbage and cauliflower used in Brittany, France, are judged not suitable for organic production because they were F_1 hybrids that had been widely used in the region beginning in the 1980s. Prior to these varieties were open pollinated and locally produced. A program of improvement involving these crops was initiated that utilized PPB (Chable et al. 2008). Growers are participating with mass selection of individuals that researchers then utilize in further selection and improvement. A new rice variety has been developed for the Kerala state of India that was bred using pedigree methods under organic production practices with grower participation (Vanaja et al. 2013).

Organic plant breeders may be interested in developing varieties for very local conditions since organic production systems are more complex with many more factors to consider (Dawson et al. 2011). Alternatively selecting a heterogeneous population for a wider geographic region that can be subject to local selection pressure is another approach. This type of breeding calls for greater grower participation particularly in the selection process. Although many governments and organizations are interested in conserving germplasm resources, PPB offers another avenue for germplasm conservation with growers working with PPB and actively saving and conserving seed (Dawson 2014). The Organic Seed Alliance is a non-profit organization that advocates for maintaining seed resources in the hands of farmers and those interested in sharing such resources as widely as possible as a means of conserving this resource (Organic Seed Alliance 2003). Conventional agriculture has been associated with a reduction in biodiversity while organic growers are interested in maintaining and increasing it. The socioeconomic nature of the problem has been studied and shown that most current plant breeding is in collaboration with large agribusiness institutions that tend to limit biodiversity (Mendum and Glenna 2010). It is believed that greater PPB can help solve the reduction in biodiversity trend in breeding efforts. PPB may be another resource in germplasm conservation that augments current efforts by the USDA and other organizations that conserve such resources.

Organic plant breeding is attracting a new generation of breeders that hope to fill this new niche in breeding. The annual Student Organic Seed Symposium brings together about 30 graduate students each year to showcase ongoing organic breeding work and to network with plant breeders and seed advocates from public, private, and nonprofit organizations (Luby et al. 2013). The mission of this group is to support new research in public breeding programs and facilitate collaboration among breeders and industry leaders to advance the organic seed movement.

In conclusion, organic plant breeding is in its infancy, but it does have a broad base of knowledge thanks to thousands of years of traditional breeding plus more than one hundred years of modern plant breeding. First, organic growers need to determine which commercially available varieties perform well in organic systems. Then, breeders will need to identify and select traits that are ideal for organic farming and low-input systems. How these tasks are to be carried out will be determined by the collaboration of farmers, breeders, policy-makers, and consumers alike. There are many problems that could be explored and possibly solved through breeding specifically for organic or low-input production (Fig. 6.1).

Fig. 6.1 Pumpkin breeding for disease resistance is a goal of both organic and conventional plant breeders

References

Anonymous (2015) High mowing organic seeds catalog. Wolcott, VT, pp 102

Arnchen C, Thommen A (2003) Organic plant breeding—a discussion of breeding strategies in organic farming. / Biologische Pflanzenzüchtung—Beitrag zur Diskussion der Züchtungsstrategien im Ökolandbau, Bericht über die Arbeitstagung 2002 der Vereinigung der Pflanzenzüchter und Saatgutkaufleute Österreichs gehalten vom 26. bis 28 November 2002 in Gumpenstein

Backes G, Østergård H (2008) Molecular markers to exploit genotype-environment interactions of relevance in organic growing systems. Euphytica 163:523–531

Baresel JP, Zimmermann G, Reents HJ (2008) Effects of genotype and environment on N uptake and N partition in organically grown winter wheat (*Triticum aestivum* L.) in Germany. Euphytica 163:347–354

Bassett MJ (ed) (1986) Breeding vegetable crops. AVI Publishing Co., Inc., Westport

Bender I, Vabrit S, Raudseping M (2005) Cracking resistance of tomato varieties at Jõgeva Plant Breeding Institute organic trial in 2001–2003. / Tomatisortide lõhenemiskindlus Jõgeva Sordiaretuse Instituudi mahekatses aastatel 2001-2003. Trans Est Agric Univ Agron 220:120–122

Boller B, Tanner P, Schubiger FX (2008) Breeding forage grasses for organic conditions. Euphytica 163:459–467

Boyhan GE, Tate S, McNeill R, McConnaughey J (2014) Evaluation of tomato varieties under organic production practices in Georgia. Horttechnology 24:252–258

Bozhanova V, Dechev D (2009) Plant breeding orientied to the needs of the organic agriculture-reasons, problems and perspectives. Selskostopanska Nauka (Agric Sci) 42:14–22

Bueren EL, Osman A (2001) Stimulating GMO-free breeding for organic agriculture: a view from Europe. LEISA 17:12–14

Bueren ETL, Struik PC, Tiemens-Hulscher M, Jacobsen E (2003) Concepts of intrinsic value and integrity of plants in organic plant breeding and propagation. Crop Sci 43:1922–1929

Bueren ETL, Goldringer I, Dawson JC, Rey F, Vettoretti A, Chable V, Lammerts van Bueren E, Finckh M, S. Barot S (2010) Future breeding for organic and low-input agriculture: integrating values and modern breeding tools for improving robustness. http://search.ebscohost.com/login.aspx?direct=true&db=lah&AN=20123096435&site=ehost-live. Accessed 19 Nov 2000

Bueren ETLV, Osman AM, Tiemens-Hulscher M, Timmermans B, Scholten O, Voorrips R, Burger K, Steenhuis G, Zijlstra S, Galvan G (2008) Examples of current breeding research for organic farming systems in the Netherlands. In: Proceedings of the 5th organic seed growers conference, Salem, Oregon, USA, 14–15 Feb 2008. Port Townsend, USA, pp 80–85

Burger H, Schloen M, Schmidt W, Geiger HH (2008) Quantitative genetic studies on breeding maize for adaptation to organic farming. Euphytica 163:501–510

Campion Al, Oury FX, Morlais JY, Walczak P, Bataillon P, Gardet O, Gilles S, Pichard A, Rolland B (2014) Is low-input management system a good selection environment to screen winter wheat genotypes adapted to organic farming? Euphytica 199:41–56

Canola Council of Canada (2014) Canola varieties. http://www.canolacouncil.org/crop-production/canola-grower's-manual-contents/chapter-2-canola-varieties/canola-varieties. 14 Feb 2015

Chable V, Conseil M, Serpolay E, Lagadec Fl (2008) Organic varieties for cauliflowers and cabbages in Brittany: from genetic resources to participatory plant breeding. Euphytica 164:521–529

Dawson J (2014) On-farm conservation of genetic diversity and organic plant breeding. In: Proceedings of the 7th organic seed growers conference, Corvallis, Oregon, USA, 30 Jan–1 Feb 2014. Port Townsend, USA, pp 156–158

Dawson JC, Rivière P, Berthellot JF, Mercier F, Kochko PD, Galic N, Pin S, Serpolay E, Thomas M, Giuliano S, Goldringer I (2011) Collaborative plant breeding for organic agricultural systems in developed countries. Sustainability 3:1206–1223

Deppe C (2000) Breed your own vegetable varieties the gardener's and farmer's guide to plant breeding and seed saving. Chelsea Green Publishing, White River Junction

Desclaux D, Nolot JM, Chiffoleau Y, Gozé E, Leclerc C (2008) Changes in the concept of genotype × environment interactions to fit agriculture diversification and decentralized participatory plant breeding: pluridisciplinary point of view. Euphytica 163:533–546

Döring TF, Bocci R, Hitchings R, Howlett S, van Bueren ETL, Pautasso M, Raaijmakers M, Rey F, Stubsgaard A, Weinhappel M, Wilbois KP, Winkler LR, Wolfe MS (2012) The organic seed regulations framework in Europe-current status and recommendations for future development. Org Agric 2:173–183

Haperen PF, Gremmen B, Jacobs J (2010) Shifting schemes of naturalness. Global food security: ethical and legal challenges. Wageningen, Netherlands, pp 488–493. 16–18 Sept 2010

Hoad S, Topp C, Davies K (2008) Selection of cereals for weed suppression in organic agriculture: a method based on cultivar sensitivity to weed growth. Euphytica 163:355–366

Kingsbury N (2009) Hybrid the history and science of plant breeding. University of Chicago Press, Chicago

Kovács G, Mikó P, Megyeri M (2010) Evolutionary breeding of cereals under organic conditions. Breeding for resilience: a strategy for organic and low-input farming systems? In: EUCARPIA 2nd conference of the organic and low-input agriculture section. Paris, France. 1–3 Dec 2010

Kronberga A (2008) Formation of triticale crop ideotype for organic farming. Options Méditerranéennes. Série A Séminaires Méditerranéens 81:391–393

Lammerts van Bueren ET (2003) Challenging new concepts and strategies for organic plant breeding and propagation. EUCARPIA leafy vegetables 2003. In: Proceedings of the EUCARPIA meeting on leafy vegetables genetics and breeding. Wageningen, Netherlands. 19–21 Mar 2003

Lammerts van Bueren ET, Myers JR (2011a) Pest and disease management in organic farming: implications and inspirations for plant breeding. http://search.ebscohost.com/login.aspx?direct=true&db=lah&AN=20133403483&site=ehost-live, http://onlinelibrary.wiley.com/doi/10.1002/9781119945932.ch3/summary

Lammerts van Bueren ET, Jones SS, Tamm L, Murphy KM, Myers JR, Leifert C, Messmer MM (2011b) The need to breed crop varieties suitable for organic farming, using wheat, tomato and broccoli as examples: a review. NJAS Wageningen J Life Sci 58:193–205

Lammerts van Bueren ET, Myers JR (eds) (2012) Organic crop breeding. Wiley-Blackwell, West Sussex

Leibinger T, Reiners E (2002) Demands of sector bodies for organic plant breeding. / Anliegen der Öko-Verbände. Beiträge zur Züchtungsforschung - Bundesanstalt für Züchtungsforschung an Kulturpflanzen Conference. Quedlinburg, Germany. 22–23 Nov 2001. 8:19–23

Löschenberger F, Fleck A, Grausgruber H, Hetzendorfer H, Hof G, Lafferty J, Marn M, Neumayer A, Pfaffinger G, Birschitzky J (2008) Breeding for organic agriculture: the example of winter wheat in Austria. Euphytica 163:469–480

Lotti C, Resta P, Pavan S, Marzano CF, Marcotrigiano AR, Zonno V, Ricciardi L (2008) Usefulness of genetic resources in organic farming: case studies relative to the 'SIAB' project. Agronomski Glas 70:355–363

Luby CH, Lyon AH, Shelton AC (2013) A new generation of plant breeders discovers fertile ground in organic agriculture. Sustainability 5:2722–2726

Melo PE (2003) The root systems of onion and Allium fistulosum in the context of organic farming: a breeding approach. Wageningen Universiteit (Wageningen University), Wageningen, Netherlands. Thesis

Mendum R, Glenna LL (2010) Socioeconomic obstacles to establishing a participatory plant breeding program for organic growers in the United States. Sustainability 2:73–91

Mikó P, Megyeri M, Kovács G (2011) Characterization of *Triticum timopheevii* gene bank accessions to gain useful materials for organic wheat breeding. Climate change: challenges and opportunities in agriculture. In: Proceedings of AGRISAFE Final Conference. Budapest, Hungary. 21–23 Mar 2011

Müller KJ (2002) Aims and methods of organic plant breeding—current situation of the international discussion. / Wege und Ziele einer ökologischen Pflanzenzüchtung—aktueller Stand der internationalen Diskussion. Beitr zur Züchtungsforschung—Bundesanst für Züchtungsforschung an Kulturpflanzen 8:24–26

Murphy K, Jones SS (2005) Genetic assessment of the role of breeding wheat for organic systems, wheat production in stressed environments. In: Proceedings of the 7th international wheat conference, Mar del Plata, Argentina, 27 Nov–2 Dec 2005

Murphy K, Lammer D, Lyon S, Carter B, Jones SS (2005) Breeding for organic and low-input farming systems: an evolutionary-participatory breeding method for inbred cereal grains. Renew Agr Food Syst 20:48–55

Murphy KM, Campbell KG, Lyon SR, Jones SS (2007) Evidence of varietal adaptation to organic farming systems. Field Crops Research. 102:172–177

Murphy KM, Reeves PG, Jones SS (2008) Relationship between yield and mineral nutrient concentrations in historical and modern spring wheat cultivars. Euphytica 163:381–390

Myers J, McKenzie L, Mazourek M, Tracy W, Shelton A, Navazio J (2012) Breeding peas, sweet corn, broccoli, winter squash, and carrots as part of the Northern organic vegetable improvement collaborative (NOVIC), strengthening community seed systems. In: Proceedings of the 6th organic seed growers conference, Port Townsend, Washington, USA, 19–21 Jan 2012, pp 44–45

National Geographic (2014) The development of agriculture. 3 Sept 2014. https://genographic.nationalgeographic.com/development-of-agriculture/

Navazio JP, Colley MR, Huber N, Campbell B, Tracy WF (2012a) Hybrids versus open-pollinated varieties: how do they fit into an evolutionary seed system? Strengthening community seed systems. In: Proceedings of the 6th organic seed growers conference, Port Townsend, Washington, USA, 19–21 Jan 2012. Port Townsend, USA, pp 52–58

Navazio JP, Zystro JP, Shelton AC (2012b) Introduction to organic on-farm plant breeding. Strengthening community seed systems. In: Proceedings of the 6th organic seed growers conference, Port Townsend, Washington, USA, 19–21 Jan 2012. Port Townsend; USA, pp 61–67

Organic Seed Alliance (2003) Organic seed alliance advancing the ethical development and stewardship of the genetic resources of agricultural seed. http://seedalliance.org/about_us. 12 Feb 2015

Osman AM, Almekinders CJM, Struik PC, Bueren ETLv (2008) Can conventional breeding programmes provide onion varieties that are suitable for organic farming in the Netherlands? Euphytica 163:511–522

Przystalski M, Osman A, Thiemt EM, Rolland B, Ericson L, Østergård H, Levy L, Wolfe M, Büchse A, Piepho HP, Krajewski P (2008) Comparing the performance of cereal varieties in organic and non-organic cropping systems in different European countries. Euphytica 163:417–433

Pswarayi A, Eeuwijk FAv, Ceccarelli S, Grando S, Comadran J, Russell JR, Pecchioni N, Tondelli A, Akar T, Al-Yassin A, Benbelkacem A, Ouabbou H, Thomas WTB, Romagosa I (2008) Changes in allele frequencies in landraces, old and modern barley cultivars of marker loci close to QTL for grain yield under high and low input conditions. Euphytica 163:435–447

Rey F, Sinoir N, Mazollier C, Chable V (2014) Organic seeds and plant breeding: stakeholders' uses and expectations—French inputs on vegetables. Acta Hortic 1041:133–139

Serpolay E, Dawson JC, Chable V, Bueren ELv, Osman A, Pino S, Silveri D, Goldringer I (2011) Diversity of different farmer and modern wheat varieties cultivated in contrasting organic farming conditions in western Europe and implications for European seed and variety legislation. Org Agric 1:127–145

Shiva V (1997) Biodiversity totalitarianism: IPRs as seed monopolies. Econ Polit Wkly 32:2582–2585

Simmonds NW (1979) Principles of crop improvement. Longman Group Limited, London

USDA (2011) Agricultural marketing service—national organic program. 19 July 2011. http://www.ams.usda.gov/AMSv1.0/nop

Vanaja T, Mammootty KP, Govindan M (2013) Development of organic indica rice cultivar (*Oryza sativa* L.) for the wetlands of Kerala, India through new concepts and strategies of crop improvement. J Org Syst 8:18–28

Vlachostergios DN, Roupakias DG (2008) Response to conventional and organic environment of thirty-six lentil (Lens culinaris Medik.) varieties. Euphytica 163:449–457

Vogt-Kaute W (2002) Plant breeding for organic farming (using the example of important field crops). / Pflanzenzüchtung für den ökologischen Landbau (am Beispiel wichtiger ackerbaulicher Kulturen). Beitr zur Züchtungsforschung—Bundesans für Züchtungsforschung an Kulturpflanzen 8:106–109

Wolfe MS, Baresel JP, Desclaux D, Goldringer I, Hoad S, Kovacs G, Löschenberger F, Miedaner T, Østergård H, Van Bueren ETL (2008) Developments in breeding cereals for organic agriculture. Euphytica 163:323–346

Chapter 7
Organic Agriculture: A Viable Option for Food Security and Livelihood Sustainability in Nepal

Khem Raj Dahal, Krishna Prasad Sharma, Dila Ram Bhandari, Basu Dev Regmi and Dilip Nandwani

Abstract Agriculture is the principal economic activity of Nepal contributing about 35 % to the national GDP and engaging about 66 % of the population. Existing low productivity of agriculture has resulted in food insecure and food deficit nation. Degradation of resources, mainly land, water, agrobiodiversity, and forest, is believed to be the immediate cause for the low productivity. Therefore, agricultural practices those conserve and promote productivity level while regenerating the degraded natural resource are of paramount importance in Nepal. Organic farming has been proven as one of such practices as it promotes and maintains soil and human health, manages and enhances biodiversity, and offers better nutrient cycling and mineralization with favorable microclimatic regimes, and thereby less risk to farmers. Till the recent past, agriculture in Nepal was organic with self-sustained method of production relying on integration of local biodiversity using traditional knowledge and wisdom. However, for the last three decades, use of high yielding exotic crops/varieties and agrochemicals, introduced under the banner of Green Revolution Agriculture (GRA), has become pervasive. Although, GRA served its short-term propose to some extent to increase the

K.R. Dahal (✉)
Institute of Agriculture and Animal Science, Tribhuvan University, Rampur, Chitwan
e-mail: d.khemraj@ymail.com

D. Nandwani
Department of Agricultural and Environmental Sciences, College of Agriculture, Human and Natural Sciences, Tennessee State University, Nashville, TN 37209, USA

K.P. Sharma · D.R. Bhandari
Department of Agriculture, Ministry of Agricultural Development, Hariharbhawan, Lalitpur, Nepal
e-mail: kp.sharma01@gmail.com

D.R. Bhandari
e-mail: dilarambhandari@yahoo.com

B.D. Regmi
Green Grand Australia, 2/19 Donovan St, Osborne Park, WA, Australia
e-mail: basu.regmi@gmail.com

D. Nandwani (ed.), *Organic Farming for Sustainable Agriculture*, Sustainable Development and Biodiversity 9,
DOI 10.1007/978-3-319-26803-3_7

production of major food crops, over the time, the indiscriminate use of external non-organic inputs resulted in soil degradation, loss of biodiversity, food poisoning, environmental pollution, and threat to sustainability and food security. This devastating scenario compelled the stakeholders to think of alternative resilient practices as a long-term solution to conserving the resource base and salvaging the environment. Organic agriculture, as advocated and promoted by International Federation of Organic Agriculture Movement (IFOAM) based on worldwide research results, can be instrumental to address the current as well as long-term problems of agriculture in Nepal. The practice of organic agriculture is not new for Nepalese farmers because it is a traditional mainstream food production system from the time immemorial. However, as a movement, organic agriculture has emerged as a new intervention in farming in recent years. Development of addictive sense toward chemical-based farming made difficult to convince farmers about immediate and long-term advantage of organic agriculture. Therefore, non-government organizations are involved to advocate, promote, and popularize organic farming with policy support from the government. Growing health and environmental consciousness against chemical farming among consumers has helped the movement advance faster. At present, many conscious farmers, entrepreneurs, and academic and development institutions are focusing their efforts to promote organic farming in the country. Government has also formulated some policies favoring the shift from chemical-based farming to organic farming. This chapter focuses on general features on the past and present of Nepalese agriculture; its resource base; declining productivity and sustainability; and the role of organic farming as a viable option for food security and livelihood sustainability in Nepal.

Keywords Nepalese agrobiodiversity · Traditional knowledge · Organic farming · Viable option · Food security

7.1 Agro-Environment and Agriculture of Nepal

Nepal is a small landlocked mountainous country situated in the southern lap of Hindu Kush Himalaya in South Asia covering an area of 147,181 km^2 (Fig. 7.1) and is the world's 93rd largest country by land mass and the 41st most populous country inhabited by about 2.8 million people mostly living in rural areas.

7.1.1 Physiography and Climate

The country has five distinct physiographic regions divided along south–north direction: the *Tarai*, *Siwaliks*, Middle Mountain, High Mountain, and High Himalayas (Fig. 7.1). The *Tarai* region has suitable climate and good accessibility

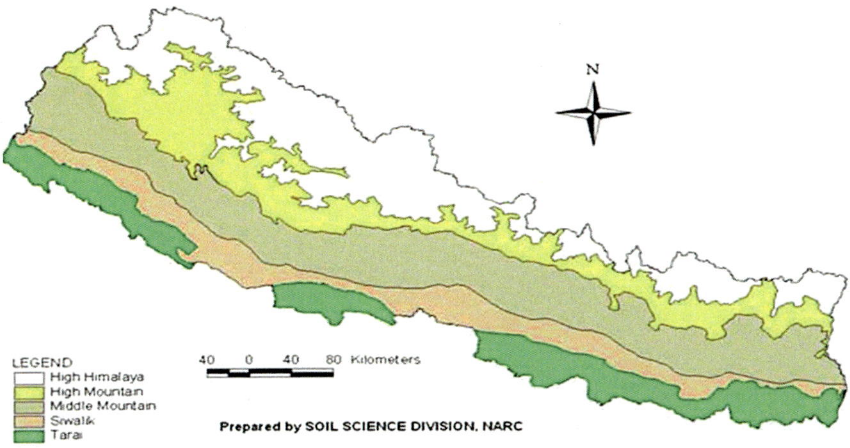

Fig. 7.1 Five physiographic regions of Nepal

to market, whereas High Mountain region has cold climate and limited accessibility.

Though small in size, Nepal is bestowed with almost all types of climate of the world. Summer and winter are two marked seasons determined by the thermal regime, although spring and autumn are also observed. Summer is hot and dry followed by intensive rainy months, and winter is generally dry and cold with occasional rain.

7.1.2 Agriculture and Its Changing Trend

Agriculture is the major occupation of most of the people and the mainstay of economic development of the country. Livelihood of about 66 % of the population depends directly on agriculture, and it contributes about 35 % of the national gross domestic products (GDPs) (ABPSD 2013). Although the contribution of agriculture to GDP has decreased by around 11 % since the 1990s (MoAC 2010), it still remains the crucial life-supporting system of the people. Increasing population and increased food demand have been putting pressure on producing more by increasing production per unit area and also by bringing more area under cultivation of food crops. This has resulted in the use of high yielding exotic varieties with chemical inputs and production package. The efforts of bringing more area under production have resulted in deforestation, use of fragile ecosystem causing land erosion and environmental degradation. The cumulative effect has caused the deterioration of soil fertility and production capacity threatening the sustainable future. This scenario has diverted the attention of everyone concerned to change the present practices of food production system. The concept of hunger, food, nutrition, and healthy eating is slowly emerging and spreading out among the consumers bringing change in eating habits.

7.1.3 Agriculture and Its Resource Base

Nepalese agriculture is based on five basic resources: forest, water, land, animal, and people. Land is the most precious resource for agriculture and livelihood. About 21 % of the total land area of the country is used for cultivation, and the principal crops are rice (45 %), maize (20 %), wheat (18 %), millet (5 %), and potatoes (3 %), followed by sugarcane, jute, cotton, tea, barley, legumes, vegetables, and fruits (ABTRACO 2008). Farmers, with local knowledge and wisdom about the soils and land resource, cultivate various crops based on their suitability and productive capacity. Unfortunately, over 28 % of the land in Nepal is estimated to be in degraded condition (MoEST 2008). The major causes of land degradation are attributed to fragile geological structure, deforestation, shifting cultivation, farming in steep slopes, flooding in the plain areas, overgrazing, forest fire, avalanches, and excessive use of chemical fertilizers (Acharya and Kafle 2009; Regmi 1999). Much of the hill and mountain areas are very fragile and vulnerable to landslides, whereas Tarai lands are regularly threatened by flooding and sedimentation. Urbanization and unplanned construction of roads and buildings have taken most of the fertile land, resulting in decreased area and total agricultural production.

Forest is an integral component of Nepalese farming systems and rural livelihood, especially in the hills, as it provides fuel, animal fodder, construction materials, manure to farmland, and contribution to household monetary and non-monetary incomes (Bhattari 2011). About 87 % of the total energy and about 42 % of the fodder requirements in Nepalese farming systems come from the forest (WECS 2010).

Water is vital for successful farming and healthy livelihood. Nepal is the second richest country in water resource in the world with 6000 rivers, streams, rivulets, and brooks (Shrestha 2012). But until present, only few of them are utilized as the source of irrigation and only 40 % of cultivated land is irrigated. The rests are the temporary source of water and are mostly the cause of landslide and erosion. Despite being the water-rich country, chronic shortages of water at various places of the country are a common problem (Deshar 2013). Therefore, the fate of total annual production of crops in Nepal still is determined by the occurrence and distribution of rainfall.

Livestock constitutes an integral part of Nepalese agriculture as the source of important human nutrition, soil/plant nutrition, organic matter, and household income for financial viability. However, shrinking grazing lands and labor shortage are causing the decline in the numbers of animals in traditional farming system. Improved breeds and feeding technology, in many cases, have compensated the total production. Loss of traditional hardy breeds, large stock of unproductive herds, declining transhumance systems in mountains, high feed requirement of exotic breeds, prone to diseases, and infertility problems are making this sector unsafe. Many success stories, however, with goat, pig, and poultry farming in specific areas of the country are appearing to contribute to sustainable livelihoods of farming families, and supporting the industrial sector as well (Bhandary 2013; Aryal 2013; Ghimire 2012).

Human resource in the form of farm labor, manager, and intellectual is important in all aspects of agriculture. However, due to seasonal nature of farming, there is rampant under as well as unemployment (Deshar 2013). At the same time, agriculture has mostly been supported by old generation and has not been a dignified profession. Increased annual out-migration of youth in search of job, for example, 249,000 in 2007/08 to 450,834 in 2012/13, has been a disappointing factor while considering the manpower requirement to advance agriculture and national development (Khatiwada 2012).

7.2 Patterns of Agricultural Practices in Nepal

Nepal encompasses high diversity regarding climate and ethnicity that form diversified agroecological pockets making heterogeneous farming systems and practices possible. Mountain farming systems are characterized by pristine setup, niche specificity, and diversity. Animal husbandry, mostly of transhumance nature, with the cultivation of few hardy crops such as wheat, barley, buckwheat, potato along with the temperate fruits such as apple and walnut is common in mountain regions. Transhumant pastoral systems, mainly with yaks and their crossbreds, sheep, and mountain goats, dominate the livelihood in the rural areas which exclusively depend on the utilization of common natural resources such as forests and alpine rangelands (Barsila 2011). Crop production is limited by low temperature and marginal lands with fragile resource base resulting in food shortage. Maize–millet-based farming system in upland terraces and rice–wheat- or potato-based system in river valleys can be found everywhere in the middle hills. Shifting cultivation in the slopping lands is still in practice in many rainfed marginal lands. Soil erosion is pervasive weakening overall productivity of the system. However, adoption of sustainable soil management practices (SSMPs), promotion of community forestry, rehabilitation of degraded lands through leasehold forestry, use of slopping agricultural land technology, and community practices on biodiversity conservation are some of the promising practices that are helping the farming systems in hills to sustain.

The river valleys in mid hills are the productive agricultural pockets in the country (Fig. 7.2). However, water-induced disasters such as floods, river bank erosion, inundation, and sedimentation are the recurrent problems in these areas. Southern plains are considered to be the grain basket of the country. Rice–wheat- and/or legume-based farming systems in lowland and maize–mustard- and/or legume-based farming systems in uplands are dominant. Unfortunately, the time of planting and crop performance is precipitation dependent. Therefore, irrigation is one of the major constraints to the successful farming in these areas. The problems caused by sedimentation, flooding, and water logging cannot be undermined.

Fig. 7.2 Terraced rice crop in foothills supported by forest and water resources

7.2.1 Primitive and Pristine

In the past, Nepalese agriculture used to be self-sustained based on a unique functional amalgamation of natural landscape and people. People managed location-specific production systems based on inherited knowledge and wisdoms accumulated from the generations-long experience that developed through coevolution and coexistence. The system is still functioning well in remote areas where so-called modern facilities have not reached yet. If most of us recall our childhood and the agriculture of that time, the ingenuity of our grandparents and the villagers in integrating the components of the farming system temporally and spatially still makes us mesmerized. No land was remained fallow in the season, and no off-season cultivation was in practice. During off-season, the soil used to take rest and thereby restore its fertility naturally, and the free-roaming animals also used to graze on the crop residues and grasses, trample them, and contribute to the fertility through their excreta. The next crop used to be bumper, almost as much or even higher than that we now get from increased cropping intensity. The farm basically relied on diversity of biotic resources fitted to the locality, from both the farmlands and the wild, as the source of nutrition for human as well as for the crops. However, increased population and food demand invited intervention of GRA slowly displacing the pristine food production system, the vital approach for human health and survival.

7.2.2 *Green Revolution Agriculture (GRA)*

Though GRA is intended to offer an efficient use of resources through scientific innovations in agricultural production, it relies heavily on external artificial inputs, energy based on fossil fuels and timely irrigation to realize yield potential. Moreover, the system is not well suited to rainfed, risk-prone marginal hinterlands (Reijntjes et al. 1992), which dominate our agricultural system. Modern inputs, improved varieties and fertilizers, are either far from the reach of general farmers or are not cost-effective. The fast disappearance of heirloom varieties and replacement of local land races by modern varieties have caused genetic erosion. Pesticide-use scenario is rather worse. Most of the sellers and the farmers using the pesticides do not have required knowledge for safe use of pesticides (Sharma et al. 2012). However, GRA still remains as the main mode of agricultural research, development, and technology recommendations in Nepal. The impact of GRA on Nepalese agriculture is shown in Table 7.1. Negative impact has drawn the attention of stakeholders to think about alternative option for correction and support to agricultural sustainability.

Table 7.1 Traditional agriculture and green revolution agriculture in Nepalese context

Particulars	Traditional agriculture	Conventional agriculture (CA)/GRA	Effect of CA on farming system	
			Immediate	Long term
Varieties	Tall and late maturing	Dwarf and early	Monoculture, increased cropping intensity	Land races erosion and decreased yield
Fertilizers	Farmyard manures, compost, residue	Imported chemical fertilizers	Increased production	Ill health of soil, costly and yield decline
Pesticides	Biological, botanical, and manual control	Imported synthetic pesticides control	Pest control and increased production	Pest resurgence, food and environment poisoning
Mechanics	Local tools, draft, and manual power	Tractors and combine harvesters with fossil fuels	Increased resource-use efficiency and time saving	Soil compaction, increased cost and decreased resource efficiency
Irrigation	Mostly rainfed and community-based irrigation system	Big irrigation projects	Assured irrigation and more lands brought into rice–wheat cultivation	Expensive maintenance, field siltation and nutrient imbalance
Capital	Low capital but labor intensive	Low labor but heavy capital investment	Better management of crops and increased yield	Problems with financers and loan payback for marginal farmers

Table 7.2 Area and production of cash crops in Nepal (2012/13)

S. no.	Crop	Area (Ha)	Production (Mt)	S. no.	Crop	Area (Ha)	Production (Mt)
1	Vegetables	246,392	33,011,684	7	Cotton	175	150
2	Oilseeds	215,600	179,000	8	Tea	19,036	20,588
3	Potato	197,234	2,690,421	9	Coffee	1750	366 (green bean)
5	Sugarcane	64,483	2,930,000	11	Cardamom	Na	5753
6	Jute	11,300	15,500	12	Ginger	Na	235,033

Approximately 27,000 farmers are involved in coffee farming
Source ABPS (2013)

7.2.3 Agriculture Production Systems in Transition

Majority of the farming systems in Nepal are of subsistence nature. However, rapid urbanization, changing life style and food habit, communication and market facilities, and increased need for cash at hand have encouraged farmers to go for cash crop farming (Table 7.2). At present, plethora of evidences shows that many farmers are crossing the traditional boundary of subsistent farming and trying to opt one or another agro-enterprise. In such endeavor, most of the entrepreneurs use external inputs which may put the sustainability system in question. At the same time, many people have started organic farming as potential agrobusiness based on demand from health conscious consumers, star hotels, and international markets. In recent years, farmers are practicing organic apple production and are getting good price in some pockets in mountains such as Jumla and Mustang. Similarly, animal product such as cheese, organic by default, is another valuable commodity coming in the market (Lucksom 2013).

7.3 Issues of Agricultural Sustainability and Food Security in Nepal

7.3.1 Soil Health and Agriculture Productivity

Agricultural productions and productivity are largely determined by the soil health, which, in turn, depends on the management practices. Farmers in Nepal are using techniques to manage the land and various sources of plant nutrients to improve and maintain soil fertility in their farming systems (Carson 1992). Managing local resources through close linkages between the forest, livestock, and crop production is the traditional way to manage soil fertility. Earlier to the introduction of mineral fertilizer into Nepal in 1952, crop production mainly depended on farmyard manure (FYM) (Pandey and Joshy 2000). In addition, soil fertility was largely maintained by the application of compost, crop residues, forest litters, and so on, and many farmers do so even today. But, in recent years, labor shortage, controlled or no access to forest resources, and shrinking grazing lands are making the linkage between forests, livestock, and cropping systems weak. At the same time, double and triple annual cropping are more nutrient demanding resulting in the application of increased amount of chemical fertilizers. Therefore, chemical fertilizer is becoming gradually a major source of crop nutrients in Nepal. Heavy use of chemical fertilizers especially in high cropping intensity areas is deteriorating the health of soils (Deshar 2013). Declining soil health and productivity are the direct threat to agricultural sustainability and food security.

7.3.2 Agroecology and Agrobiodiversity

Nepal has three ecological belts: *Tarai*, hill, and mountain along a south-to-north transect, whereas in transverse, there are three distinguished segments dissected by major river systems: Koshi, Gandaki, and Karnali. These rivers have shaped landscape making numerous ecological pockets and niches to emerge, giving home to high level of biocultural diversity. Nepalese farming communities have been managing a high level of on-farm biodiversity to meet their specific needs. They have been using 3000 or more plant species for food and sustaining livelihood, through cultivating and trading (Koirala and Thapa 1997). Crop species in Nepal owe their variability to the presence of about 120 wild relatives of the commonly cultivated food plants. Integration of crops and livestock accommodates high level of diversity in home gardens, farm fields, agroforestry parcels, and forest gardens (Fig. 7.3)—a strong foundation of livelihood sustainability.

Unfortunately, majority of the land races of major crops are lost, and many of them are under the threat. Agrobiodiversity in high hill is relatively better maintained due to specificity of landraces, undisturbed forests, and remoteness, whereas *Tarai* area has experienced maximum genetic erosion especially of the landraces of major crops mainly due to modern agriculture. Degrading agroecology and declining agrobiodiversity as a result of so-called modern agriculture based on synthetic inputs and exotic crop varieties have emerged as serious issue in sustainability of desirable level of agroecology and agricultural sustainability.

7.3.3 Food Security, Nutrition, and Human Health

At present, two-fifths of 3.4 million landholdings in Nepal produce enough food only for less than six months. The productivity of major cereal crops except maize and millet and horticultural crops is far below regional average compared to the neighboring countries. Food production in the mountains remains short of the requirement by 34–45 %. In the hills, the deficit is between 15 and 30 %. The current scenario of requirement and production of cereal in the country shows a

Fig. 7.3 Crop diversification and agroforestry practices in hill farming systems of Nepal

little surplus of 400,000 ton (Table 7.3). That too is not at the reach of needy people due to distribution problem, among others. This makes the country to import agricultural products worth of millions of rupees every year, and food import was amounted to worth Rs. 144 billion in the period of 2010–2012 (KD 2013).

Almost 50 % of Nepal's population is undernourished. Forty-one percent of children under five are stunted, 29 % are underweight, and 11 % is wasted (acutely malnurished). More than 2.4 billion people get key nutrients from rice, wheat, maize, soybeans, field peas, and sorghum (Leachy 2014). Present-day chemical-intensive agriculture has produced nutrient poor hollow food that lacks vital micronutrients (Myers et al. 2014). Micronutrient deficiencies in Nepal are widespread; in particular, 46 % of children aged 6–59 months, 35 % women of reproductive age, and 48 % pregnant women are anemic (Uprety et al. 2014).

Table 7.3 Dynamics of edible cereal production and requirements (ton) in recent decades

Indicator	Time in 10-year interval					
	1964/65	1974/75	1984/85	1994/95	2004/05	2012/13
Production	2,212,000	2,410,000	2,752,000	3,397,760	4,942,553	5,648,265
Requirement	1,919,000	1,871,000	2,579,000	3,882,917	4,779,710	5,239,823
Balance	293,000	539,000	173,000	−485,157	162,843	408,442

Source Koirala and Thapa (1997) and ABPSD (2013)

Fig. 7.4 Chemical-intensive agriculture—rice transplanting in Nepal (Leachy 2014)

"Hollow food production method" addicted to the use of chemicals in order to produce more has resulted in nutrient poor food in Nepal attributing to poor human health (Fig. 7.4). Realizing the problem of poor nutrition, government has planned and developed appropriate policies to address the deficiency of food and nutrition on short-term basis. The government has made food security a national priority and has achieved some progress in combating the hunger. However, sustainable production of quality and quantity of enough and nutritious food requires sincere efforts of all concerned. Hollow food-producing practices have to be stopped by promoting nutrient-rich production practices.

7.3.4 Agricultural Sustainability

Nepal is witnessing three basic types of farming systems in last fifty years. (1) traditional farming systems where no external inputs are used. The system is self-reliant and completely organic; (2) farming system where low or medium level of external inputs agriculture (LEIA) is used. Most of Nepalese farming systems fall under this category; (3) high external input agriculture (HEIA) which, handful though, is common in market-oriented agricultural pockets and is on the rise. Evidences show that none of these systems are sustainable in the long run. Due to population pressure and degradation of resources, the first one is no more capable of surviving and growing as such. Increased reliance of the second system on external inputs, but with little knowledge about them in general and about agrochemicals in particular, raises the question to its sustainability. Soaring cost of inputs, unsure market, and environmental concerns are making the third system unsustainable. Pretty (1995) suggests with evidences that low-input agriculture can be made sustainable with the adoption of resource-conserving technologies and processes even in diverse, risk-prone conditions. But Serchan and Karki (2005) have reported disappointing results of the efforts of sustainable approach in Nepal. However, Deshar (2013), based on studies, suggests that farming with no or low use of agrochemicals would be a strategic destination toward achieving sustainable development of Nepalese agribusinesses. Based on the evidences, past and current policies and implementation procedures have not been proven capable of providing sustainability to agriculture development to protect soil and environment. Organic agriculture, slowly rising in popularity to address such issues of sustainable livelihood, may work as guiding light for Nepal in the long run.

7.4 Organic Farming—Viable Option

7.4.1 Definition

International alliance of sustainable agriculture (1990) defines organic farming as "a system of agriculture that encourages healthy soils and crops through such practices as nutrient recycling of organic matter (such as compost and crop residue), crop rotations, proper tillage and the avoidance of synthetic fertilizers and pesticides" (Reijntjes et al. 1992). This definition focuses on the technical aspects of agricultural production to be followed focusing on the input use as option. IFOAM (2008) has defined organic agriculture as "a production system that sustains the health of soils, ecosystems and people. It relies on ecological processes, biodiversity and cycles adapted to local conditions, rather than the use of inputs with adverse effects. Organic agriculture combines tradition, innovation and science to benefit the shared environment, and promotes fair relationships and a good quality of life for all involved" (IFOAM 2009). The focus of this definition is on all aspects of agriculture as a life-supporting system and as a part of the global ecosystem comprising landscape and biodiversity including people and their interrelationships that foster and ensure their coexistence. The definition well catches the notion of the shared role of many species including human being to make the ecosystem function and sustain as Mollison (1990) states, for examples, that recycling of nutrients and energy in nature is a function of many species. Our own survival demands that we preserve all existing species and allow them a place to live and promote. This definition also recognizes the shared importance of traditions, science, and innovation. However, when we turn our attention toward marketing of the organic products, the definition remains rather silent and manifests in the form of standards or rules set by the recipient nation, company, or organization to be involved in or followed while producing and processing the particular product in deal.

7.4.2 Evolution of the Concept and Its Development

World agriculture was purely organic till the beginning of the twentieth century when synthetic chemical fertilizers were introduced. Use of herbicides, pesticides, and bovine hormone increased slowly as additional input for boosting up the agricultural production. These inputs now have become an indivisible part of agriculture worldwide. In concurrence with the development of chemical-based agriculture, the present organic agricultural movement roots to the 1940s when insecticidal property of dichloro diphenyl trichloroethane (DDT) was discovered by Paul Muller and was pervasively used as a powerful weapon to control insect pests in agriculture. From that time itself, debates on whether the use of DDT in agriculture was good for the environment were begun. Naturalists strongly advocated against pesticides and seeded message of ill effects of synthetic chemicals on health and environment. However, introduction of GRA in the 1960s made the use of agrochemicals ever pervasive. All that changed with the publication of Rachel Carson's landmark book, *Silent Spring* (1962), in which in clear and powerful language, she explained the danger of DDT and pesticides in general (Gips 1987). The book created profound public awareness against agrochemicals. As a result, organic farming surged in the late 1960s and 1970s as a parallel lobby against the growth of chemical-based farming. By then, the use of agrochemicals and its misuse were so wide spread that the quest for the search of safer and healthier agricultural alternatives got momentum. The pursuit triggered a new thinking of organized movement against chemical reliant agriculture. This gave birth to a new organization, International Federation of Organic Agriculture Movements (IFOAM), in 1972 in France.

Long before the birth of IFOAM, in the late 1930s and early 1940s, Sir Albert Howard, widely recognized as the "father of organic farming," developed the practical approach to organic agriculture: application of scientific thoughts and principles to the traditional and natural methods in nourishing the soil and agriculture (Howard 1943). The major concepts and practices that he promulgated in farming were the living connections between soil fertility and plant as well as animal health, and method of composting about which he has described in detail in his book, "*An Agricultural Testament*" (1943). These concept and practices are now central to organic farming. He strongly advocated the recycling of all organic waste materials including sewage sludge back to farmland following the "The Law of Return," as he called it (Conford 2001). However, Walter Northbourne was apparently the first to apply the word "organic" in farming in his influential book *Look to the Land* (1940) in which he elaborated the idea of the farm as an "organic whole" (Heckman 2007) referring to organic as an entity "having a complex but necessary interrelationship of parts, similar to that in living things."

In the 1940s, J.I. Rodale, pioneer of organic farming in USA, founded a working organic farm for trials and experimentation, and established the Rodale Institute and the Rodale Press to teach, and advocate organic farming. Christopher Chapman (Canada) and Lady Eve Balfour (UK), and many others across the world contributed to the promotion of organic agriculture in the years to follow. In 1974, Bill Mollison (Australia) promulgated the Permaculture Principles and Practices authenticating that permanent agriculture with judicious integration of landscape and people is well possible without the use of chemicals (Mollison 1990). The first International Conference by IFOAM "Towards Sustainable Agriculture" (1977) and a report by World Commission on Environment and Development (WCED), "Our Common Future" (1987), became highly instrumental in promoting safer, healthier, and more eco-friendly agricultural production system (Gips 1987; UN 1987).

In addition, it is noteworthy that the practice of organic farming roots to the oriental countries as farmers in these countries heavily relied on organic way of fertilizing the field from the time immemorial. Canal mud; all kinds of human and animal manure; green manure; composts; and ash were the common fertilizers in Japan, China, and Korea (King 1911) till the beginning of twentieth century. Collection of night soils for fertilizing the land especially in Bhaktapur area of Kathmandu valley was common till the recent past in Nepal as well. Hence, the practice of organic agriculture per se is not new for Nepali farmers, for it used to be main stream of agriculture production system until the 1950s. However, development of organic agriculture as a movement with new perspective was started in 1986 with the establishment of Institute of Sustainable Agriculture Nepal (INSAN) (Dahal 2012).

At present, organic farming is well-adopted world wide as one of the viable approaches toward sustaining agriculture, enhancing ecosystem health and maintaining livelihood of the farming communities. In the first decade of this century, organic production and certification spread in major developed nations all over the world with various regulations and certification standards. Developing nations with big question of food security have understood organic farming as possible alternative for livelihood sustainability rather recently. At present, more and more nations, provinces, and states are declaring themselves as total organic to safe guard food safety/security, human health, and environment.

Organic agriculture is developing rapidly with the share of agricultural land (0.98 % in 2013 as compared to 0.8 % in 2009) covering 78 million ha managed organically by more than 2 million producers, including smallholders in 170 countries with 82 countries having organic regulations (IFOAM 2015). The countries with the highest numbers of producers are India, Uganda, and Mexico. About a quarter of the worlds organically managed land, and more than 80 % producers are in developing countries with emerging market. More than 35 million hectares are under organic wild collection areas for beekeeping and the majority of which is in developing countries. Since the early 1990s, the retail market for organic farm produce has grown about 20 % annually due to increasing consumer's demand. According to current statistics, worldwide annual sales of organic products are over US$72 billion. Revenues have increased almost fivefolds since 1999 (Willer and Lernoud 2015).

7.4.3 Organic Farming for Safer Food and Environment

Organic agriculture has been promoted as a farming system that is specifically aimed at producing food in a more environmentally friendly way. It has attracted increasing attention over the last decades because it is perceived to offer some solutions to the problems currently besetting the agricultural sector (Charyulu and Biswas 2010). Organic agriculture has long been appreciated for its ecological soundness, environmental health, productivity, and sustainability. A review of organic farming systems in Latin America encompassing 14 farmers' groups in 6 countries covering more than 5000 farmers managing 9000 ha showed number of environmental benefits of organic farming in terms of erosion control and soil fertility, among others (IFOAM 2006). It has the potential to improve soil fertility and help build both nutrient and carbon stocks (IFOAM 2008). Organic farming is far superior to conventional systems when it comes to building, maintaining, and replenishing the health of the soil (RI 2011). IFAD (2003) reports that the major advantages to small farmers shifting to organic production are as follows: the enhancement of soil fertility; the closeness to traditional and existing systems; reversal of soil erosion; and the low cost of the technology which enhances self-reliance.

Biodiversity conservation is one of the well-established contributions of organic agriculture (FAO 2011). A meta-analysis of 766 scientific papers published in Europe showed a higher degree of biodiversity in organic farms than in conventional farms (Rahmann 2011). A 21-year-long field trial in Switzerland comparing organic and non-organic farming systems showed dramatic differences in microorganisms responsible for soil fertility, and delivering nutrients to the roots of crops; up to 85 % is higher in the organically managed field than that non-organically managed (Fließbach et al. 2000). The management and enhancement of biodiversity in organic farming offer more beneficial interactions among the components, higher resource-use efficiency, higher associational resistance, higher nutrient cycling and mineralization, better microclimatic regimes, and less risk to farmers. In addition, most of the world's biodiversity is located in developing countries with marginal rainfed lands and fragile environment. Therefore, if organic agriculture was more widely adopted, the higher yields obtained in these highly biodiverse areas would allow for preservation of more wild land in regions where it matters most. Organic agriculture can help maintain the fertility of these fragile lands, thereby contributing to both maintaining levels of agricultural productivity on agricultural lands and avoiding the loss of biodiversity (IFAOM 2008).

Organic farming offers safe, healthy, nutritious, and mineral-rich tasty food (Worthington 2001). It has been demonstrated that organically produced foods have lower levels of veterinary drug residues and, in many cases, lower nitrate content (FAO 2000). Evidences have shown that organic plant-based food products generally contain higher amounts of antioxidants, vitamins, minerals, and other beneficial substances (IFOAM 2008; Woese et al. 1997; Worthington 2001). The concentrations of a range of antioxidants such as polyphenolics were found to be substantially higher in organic crops/crop-based foods, with higher percentage of phenolic acids, flavanones, stilbenes, flavonols, and anthocyanins. Dietary intervention and epidemiological studies (Barański et al. 2014) have shown to be linked to reduced risk of chronic diseases, including cardiovascular and neurodegenerative diseases, and certain cancers. In addition, organically processed products do not contain hydrogenated fats and other additives whose negative health impacts are widely acknowledged (IFOAM 2008). Clear health benefits from consuming organic dairy products have been demonstrated in regard to allergic dermatitis (Crinnion 2010). Improvement in taste and nutritional content of the products produced by the farmers converted into organic system has also been reported by Parrott and Marsden (2002). Similarly, a survey research in western *Tarai* Nepal reported people's preference to organic food because of its good taste (Aryal and Dahal 2010).

It is generally perceived that organic farming yields much lower than its conventional counterpart. No use of synthetic fertilizers is believed to result in lower yields (Muller 2009). Studies conducted in different countries have shown, in general, that there is decline in production when conventional farms are converted into organic farms, with gradual increases after conversion (Ricker 1997). However, the reduction of yield depends on the region, resource endowment, level of management, and crop in question. Agricultural diversification practices such as multi-cropping and crop rotations substantially reduce the yield gap (Ponisio et al. 2014). Studies also have shown that organic farms can be almost as productive as conventional farms. Over the 30 years of the trial in USA, organic corn and soybean yields were equivalent to conventional farming (RI 2011). IFOAM (2008) states that yield in organic agriculture may be around 20 % less than in conventional agriculture in developed countries, but, in general, are higher than in conventional agriculture in developing countries where most of the farmers practice their farming under drought prone conditions and in rainfed areas with low level of external inputs. For example, in Karnataka state (India), rice farmers using high yielding varieties and chemical fertilizers saw their crops reduced by more than 50 % during the 2001–2002 droughts, whereas the region's organic farmers lost less than 20 %. Similarly, sugarcane losses were 58 and 1 %, respectively. These developments drew the attention of the other farmers who began to adopt organic methods and convert the following year (IFAD 2005). Similar was the result obtained in USA where organic corn yields were 31 % higher than conventional in years of drought (RI 2011).

The widespread assumption that converting to organic means a decline in yields is no more a truth (Parrott and Marsden 2002). Green manures and cover crops have increased yields of maize by between 20 and 250 % in Brazil; composted plots produced between three and five times higher than those treated only with chemicals in Ethiopia; agroecological practices increased yield by 175 % in Nepal (ISIS undated). Countries where most of the farmers rely on integrated farming and organic production systems achieve equal or even higher yields, as compared to the current conventional practices, which translate into a potentially important option for food security and sustainable livelihoods for the rural poor (Scialabba and Müller-Lindenlauf 2010). If an estimation of the value of natural resource depletion were included, organic farming may seem even more profitable (Ricker 1997). IFAD (2005), in line with this, states that the transition from "traditional" agriculture in rainfed areas to organic farming very frequently leads to increased yields. Given the assumption of higher relative yields in most organic crops compared with existing low-input agriculture (Pretty and Hine 2001; IFAD 2003), there is potential for improving local food security in South and Southeast Asia (SSA) if non-certified organic farming is supported by capacity building and research. A study using the food policy, IMPACT, model showed that conversion of 50 % of agricultural area in SSA results in increased self-sufficiency and decreased net food import to the region (Halberg et al. 2006). Similarly, organic agriculture has major potential for reducing agricultural greenhouse gas emissions and enabling ecosystems to better adjust to the effects of climate change (FAO 2007). This is due to the ability of organic agriculture to be both a significant carbon sink, better buffer to adjust environmental stress and to be less dependent on fossil fuel-based inputs, all desirable attributes for sustainability.

7.4.4 Organic Farming in Nepalese Context

Soil health and productivity are vital to sustain the livelihood of Nepalese people. Inherent fertility of soil determines the fate of agricultural production, in addition to other natural inputs and processes, because external inputs are still far from the reach of general farmers. In recent years, agricultural intensification has led to an increased demand of nutrients in the soil. Locally available sources, mainly FYM, compost, inherent soil nutrients pool, and biologically fixed nitrogen are not sufficient to meet that demand. This, along with the government policy (subsidy) in favor of chemical fertilizers, has promoted their use. The heavy use of chemical fertilizers, mainly in suitable and remunerative farming pockets, with other pollutant technologies has resulted in the degradation of farmlands (Adhikari 2012; Deshar 2013). Use of urea- and ammonium-based fertilizers has increased soil acidity and created an imbalance in soil–plant nutrients system. Intensive farming with high chemical inputs, monoculture, and other interventions such as excessive tillage practices leads to a reduction in soil organic carbon (Dahal 2010).

Soil Management Directorate (SMD), Department of Agriculture, reports that Nepal is facing a serious problem of soil quality decline as a result of recent changes in agricultural practices and increasing resource constraints. Recent analysis of the soil samples from all over the country by SMD revealed that about 53, 13.45, and 33.51 % samples were acidic, alkaline neutral (Table 7.4), respectively.

Table 7.4 Soil analysis and fertility categorization

Fertility parameters	Low	Medium	High
Total nitrogen (n = 17,000)	56.25	29.5	14.25
Available phosphorus (n = 17,000)	41.76	29.54	30.70
Available potassium (n = 17,000)	49.99	26.09	23.92
Organic matter (n = 5718)	44.47	40.89	14.64

Note Nitrogen: low: <0.1 %; medium: 0.1–0.2 %; and high: >0.2 %; Phosphorus: low: <26 kg/ha; medium: 26–55 kg/ha; and high: >55 kg/ha; and Potash: low: <110 kg/ha; medium: 110–180 kg/ha; high: >280 kg/ha
Source SMD (2014)

Majority of the soil samples fall under either low or medium category regarding the major nutrient content confirming the declining status of soil fertility in Nepal. It is important to note that the percentage of sample falling in higher range is declining year by year (SMD 2014).

Generally, low and imbalance contents of micronutrients are common in Nepalese soil. In some places, zinc has decreased, and at other places, sulfur has increased and the soils have become hard and heavy due to the use of chemicals (Sharma 1990). Boron deficiency is universal affecting 80 to 90 % of agricultural soil. The impoverishment is mainly due to the increased crop intensity and associated application of the synthetic fertilizers containing only one or two major elements, and decreased soil organic amendments. The availability of compost and animal manure has declined due to a fall in animal husbandry and labor shortage (Adhikari 2012). Depletion of the organic matter content has been the center of the overall soil fertility decline.

Although traditional resources such as FYM are important to enrich soil fertility, not adequate attention has been given to their preparation, storage, and application. Research has shown that saving FYM pit from rain and sun significantly increases the nitrogen content (2.28 vs. 3.41) of the manure. Similarly, incorporation of manure immediately after application saves significant amount of nitrogen (SSMP 2010). However, the general practice of scattering FYM in the open field long before the field preparation results into heavy loss of nitrogen from the manure (Fig. 7.5).

Nowadays, consciousness has aroused among the researchers, farmers, and development activists on the importance of soil health. Therefore, different approaches are being tested and promoted in order to prevent degradation and improve soil health. There is scope of improving nutrient-use efficiency through the careful management of organic manures such as FYM, compost, green manuring, and vermicompost through the adoption of organic farming.

Studies revealed that there is significant role of the improved soil management techniques in maintaining and improving soil organic matter, an important indicator of soil health. Major soil indicators (OM, N, P, K, and pH) were also appeared to be heavily influenced by the increased use of organic manures (Regmi et al. 2006).

Fig. 7.5 General practice of FYM application in most farmers field in Nepal

Table 7.5 Results of benchmark sites after intervention of organic-based inputs (n = 236)

	pH	N (%)	P kg/ha	K kg/ha	OM (%)
Year 1	5.9 (0.08)	0.17 (0.08)	31 (32)	477 (201)	3.2 (1.4)
Year 3	6.0 (0.90)	0.19 (0.08)	36 (34)	462 (345)	3.6 (1.6)
Difference	+0.1	+0.02	+ 5	−15	+0.4

Figure in parentheses indicates STDEV
Source Regmi et al. (2006)

This indicates that SSMP can maintain the soil health, particularly through the conservation of the top-soils effectively and efficiently (Table 7.5).

In this context, it is noteworthy that organic farming is gaining momentum gradually among the farming communities especially where NGOs with "sustainable community development" mission have reached and farmers are aware of the negative impacts of agrochemicals. Soil nutrient management with locally available resources and managing pests without synthetic pesticides are the two thrusts on the way to shift to organic system. Depending upon the location, farmers opt two groups of approaches for soil, crop, and pest management: **traditional and non-traditional**. Traditional approach in managing the soil encompasses crop rotation with inclusion of legumes; use of crop residues and kitchen wastes; soil cakes; farmyard manures; penning animals in the field; transhumant keeping of herds; different composts; biogas slurry; oilcakes; mulching; and in situ as well as ex situ green manure.

Non-traditional measures of managing soil nutrient include vermicompost, *Rhizobia* culture, cattle urine-based liquid manure (*Gitimal*), and *Bokashi* and industrial biofertilizers. In recent years, many preparations (industrial biofertilizers), produced in the country as well as imported from outside, are floated in the market and are being advertised daily as successful aids to organic agriculture. Some of the bio-organic fertilizers such as *Jaibik* Superphosphate; *jaibik dhulo* and plant tonic; *jhol mal*, HB 101, Bonsoon Super *Prangarik Mal*, Green Gold Super *Prangarik Mal*, *Nasabike Mal*, etc., are produced in Nepal. Bio-organic fertilizer, Chao Nang granules, Super green plus, Super green plant, super green mix, and Quine Thang are imported from Thailand; and Primimum Azosp, Primimum phospofix, and Primimum Azotoplus are imported from India. Among all these biofertilizers, vermicompost is getting popularity rather quickly as farmers prepare this biofertilizer in many places in the country at household levels (Bhattari 2014) as well as at industrial levels. The number and the import of biofertilizers aiming at promoting organic agriculture in Nepal are in increasing trend. However, their quality and consequent effect on soil and ecosystem have not been well tested so far in our condition. Nepal has just initiated the regulatory mechanism to quality control of production, import, and sells of such products. In the absence of quality control at present, farmers are relying only on the advertisements and are using the products (Dahal 2013). Fertilizing crop fields, especially kitchen gardens for vegetables, with human urine is also becoming popular among the farmers in certain locality such as Darechock (Chitwan) and Sotang (Solukhumbu) by building "ecosan" toilets (Mallapati 2012).

Plant protection presents itself rather difficult challenge for the farmers who want to for organic farming. Traditionally, farmers were managing pests through manual methods; hand picking and killing, cultural methods; crop rotation and mixed cropping; use of genetic resistance, escaping (adjustment of planting time), spraying of diluted animal urine, ash, different oils and use of local plant materials with pesticidal property. It is estimated that of the 2400 plants with pesticidal value worldwide, 425 plants are found in Nepal and farmers are widely using most of them for the purpose. The most common among them are (*Azedirachta indica*), garlic (*Allium sativum*), pudina (*Mentha arvensis*), ginger (*Zingiber officinalis*), turmeric (*Curcuma domestica*), tite pati (*Artemesia alatum*), marygold (*Tagetes patula*), timur (*Xanthoxylum alatum*), asuro (*Adhatoda visica*), tulasi (*Ocimum sactum*), bakaino (*Melia aderachata*), papaya (*Carica papaya*), sisnu (*Urtica dioica*), tobacco (*Nicotianum tobacum*), pire ghas (*Polygonum hydropeper*), sarifa (*Annonaa squamosa*), sital chini (*Moringha oleifera*), onion (*Allium cepa*), siundi (*Euphorbia royaleana*), sajiwan (*Jatropha curcus*), and simali (*Vitex nigundo*). Gradual loss of traditional knowledge and wisdom has limited the use of these plants for organic pest management. Therefore, these practices are no more common among most of the present-day crop producers. However, in recent days farmers in many villages are trained to prepare local biopesticides by various NGOs working on sustainable agriculture and are reviving the old tradition with new flavor.

The concept of sustainable agriculture for better human health and environment has introduced **Integrated Pest Management (IPM)** practice in the country since 1995. Although its impact seems weak until now, IPM has scope and is considered as the bridge between chemical pest control and organic pest management system. Farmers using IPM are familiar with techniques and safer means from hand picking to the use of botanical and fungal preparations to manage the pests. Some of the marketed products, such as various pheromone traps (Spodolure, Helilure, DBM lure, etc.), are being common in farms especially in cucurbits. Various preparations based on fungi, such as *Trichoderma* spp, *Beuveria, and Metarhizium anisopliae*, are also available in the market, and few farmers are using them. Nuclear polyhedrosis virus (NPV) is becoming popular among the framers in controlling pod borer as it can be prepared in farmers' fields. Zibatu, a special preparation consisting of a group of microbial population developed in Nepal, and EM (effective microorganism, a Japanese formula prepared in Nepal) for both pest and nutrient management are also available in the market and are common in some localities. However, the use of these means is site specific and needs scaling up.

Organic farming as a movement: Organized effort to promote organic farming in Nepal was started in 1986 with the establishment of INSAN followed by Judith Chase, US citizen, who started an organic farm in Gamchcha, Bhaktapur, in 1987 to promote organic vegetable production and make local people aware of organic agriculture. At present, many NGOs, few cooperatives, entrepreneurs, hoteliers, and conscious farmers are involved in the development and promotion of organic farming in the country. The organic way of life is slowly becoming less of a passing trend and more of lifestyle, for an increasing number of people (Bisht 2011) especially in cities and among health conscious circles. The feeling of need to revitalize soil by organic approach in villages and communities is slowly emerging and growing with popularity. Conscious and progressive farmers are getting organized in cooperatives to promote organic farming in various parts of the country. Jumla, a remote far western district, has declared itself as "organic district" by the 14th session of the District Development Council in 2007. Although there is no official data on the status of organic agriculture in Nepal, information available is showing sixteen cooperatives and many individual farmers producing organic products, and twenty-five private companies merchandizing them in national as well international markets. An estimate suggests both local and export transaction of organic products exceeding seven million US dollars annually. At local level in Kathmandu valley, five organic outlets, four weekly farmers' markets, nine supermarkets, and about 35 hotels are engaged in organic business. Opening of such outlets in other parts of the country is in process. The products of trading are: orthodox and leaf tea and coffee; honey; high land beans; buckwheat; root and leafy vegetables; bread and pastas; essential oils and herbs; soap and raw materials for cosmetics anddetergent; wild fruit syrups; and fiber for textile. The major export commodities include tea (green, leaf and herbal), coffee (raw beans, roasted beans and powder), beans (pinto and adzuki), buckwheat, spices (ginger, turmeric, coriander seed, and super hot chili), essential oils, herbs (wild and cultivated), textile, and raw materials for cosmetics.

There are 13 industries producing about 19,000 tons of biofertilizers, about 1000 tons is imported, and the number of industries desiring to produce such inputs is increasing. There is no official data on the area and production under organic farming in Nepal; however, about 26 % of agriculture is estimated to be still organic by default. IFOAM reports based on 2013 data indicated that 9361 ha (0.2 % of total agricultural land) cultivated by 687 producers are organic in Nepal (IFOAM 2015). Most of the coffee producing 418 tons of green beans in 1760 ha is either certified organic or in the process of citification (NTCDB 2014). About 3150 tons (9 %) of orthodox tea produced in Nepal and exported is certified organic. Farmers who once were very strong proponents of chemical-based agriculture are slowly converting into organic farmers. It is noteworthy that nonresident Nepalese (NRN) are interested to invest in organic agriculture and young people in different parts of the country, especially returnees from foreign labor market, are interested in and attracted toward organic farming where they have seen their future.

Government initiatives: Organic agriculture was first appeared in the 10th Five-year Plan of the Government of Nepal in 2003 (NPC 2003) and has been mentioned in various plans and policies thereafter, such as National Agriculture Policy (NAP) 2004; Agribusiness Promotion Policy (ABP) 2007; Agricultural Biodiversity Policy 2007; Three-Year Interim Plan 2007–2010; National Adaptation Plan of Action (NAPA) to climate change 2010; Agricultural Development Strategy (2013); the Thirteenth Plan (FY 2013/14–2015/16); and National Seed Vision (2013–2025). Agriculture Policy, 2004, has provision to support organic certification. The newly formulated Agricultural Development Strategy (ADS), a policy-level guiding document of the country for agricultural development for coming 20 years implemented recently, also mentions about organic farming as a viable option for sustainable agriculture (ADS 2013). In line with this, government also have some related acts and regulations enacted such as Pesticides Act 1991; Pesticide Regulation 1994; Environment Protection Act and Environmental Protection Regulation 1997; and Seed Regulations 2013. First National Organic Farming Workshop organized in 2006 helped raise interest of national scientists toward organic farming. National Coordination Committee on Organic Agriculture Production & Processing System (NCCOAPPS), a high-level body, and National Organic Agriculture Accreditation Body (NOAAB) have been formulated. Ministry of Agricultural Development has developed policy documents such as National Technical Standard for Organic Agricultural Production and Processing System Directive, 2007 (Revision 2008); Incentives for Establishment of Organic Fertilizer Production Industry work procedure 2009; National Recognition for Organic Agriculture Related Agencies, Work Procedure, 2012; Participatory Quality Guaranty System for Organic Agriculture Production directive 2012; Collective Certification on Organic Agricultural Production for Internal Control System Directive, 2012; and Subsidy on Certification Fee for Organic Agricultural Production Export, Work Procedure, 2012. Based on these policies, government has taken some initiatives and has endorsed and implemented some programs such as the setup of Nepal Organic Technical Committee and effective implementation of subsidy on organic certification for export. Recently, the government has provisioned 50 % subsidy for the production of organic fertilizer including vermicompost to the farmer group cooperatives. The government has also mentioned in the budget speech of Fiscal Year 2014/15 that the Village Development Committees (VDCs) that are involved in organic farming without use of pesticides and chemical fertilizers will be provided additional 25 % grant on top of regular grant (MoF 2015). Training Directorate, Department of Agriculture, has incorporated organic agriculture in its regular training program. Government organizes National Organic Agricultural Fair each year, starting from 2007, with the aim to popularize organic production system among the stakeholders, and 8th such fair was held recently in Pokhara where 134 organic farmers participated with their products (WRADO 2015).

Role of NGO: Organic agricultural movement in Nepal was initiated by INSAN, a non-government organization. Since then, the role of NGO remained always critical for the promotion of organic farming among the farmers. This is because of most of the NGO works with deprived communities who have nothing except a small parcel of lands, few animals, and piece of forest nearby. In such situation, food production relies solely on the resources that they have, and organic farming is quite adoptable and appropriate option in that circumstances. Nepal Permaculture Group, a national network established in 1992, working in the field of sustainable agriculture, has been instrumental in promoting organic agriculture in Nepal through advocacy, lobbying, training, organizing workshops, and seminars for all stakeholders including policy makers. Other national NGOs such as Utilitarian Service (USC) Nepal, Ecological Service Centre, HASERA, Local Initiatives for Biodiversity Research and Development (LIBiRD), Sustainable Agricultural Development Program Nepal (SADP) Nepal, Namsaling Community Development Center (NCDC), Nepal Community Support Group (NECOS), Jajarkot Permaculture Program (JPP), Lotus Land Agriculture Farm (LLAF), Community Welfare and Development Society (CWDS), Bansun Agro-Organics, Organic Nepal Co-operatives have worked and are working effectively at the field level to promote organic farming directly or indirectly. International humanitarian foundations and research and development institutions such as Helvtas, Switzerland; GIZ and EED, Germany; Agriculture Institute, Canada; and SNV, the Netherlands have been working to take organic agriculture forward through various programs and projects in Nepal. Vedic Agriculture Foundation Nepal has long-term plan with an ambitious mission to convert Nepal into the first organic country in the world (NMVF 2010). Nepal as a member of WTO has an enormous potential to show the presence in international market with fresh organic products.

Certification: Certification is one of the critical issues in organic farming in Nepal. Helping farmers in certification process by NGO such as Helvetas (coffee) and SNV (tea) is in progress, and private certifying agencies have started certification for export commodities. Internal control system (ICS), Participatory

Guarantee System (PGS), and third-party (standard) certification process by the authorized certifiers are becoming gradually a common practice among the producers desiring to take their products to international market. National certifying company, Organic Certification Nepal (OCN), and many international agencies such as NASAA (Australia), ECOCERT (Belgium), Control Union (The Netherlands), IMO (Switzerland), and OneCert (USA) are working on the certification of various organic products in Nepal. Cost for certification, scattered farmlands; farmers' unfamiliarity with the process; subsidy on certification only for export market; and volume of production for export are some of the major problems at present which need proper attention. Given such circumstances, third-party certification to comply with the standard of organic product may not be appropriate in general in case of Nepal. But shifting to organic production practices is very vital for long-term survival of sustainable food production system. Therefore, flexible policies to adjust all different options from soft to hard certification standard should be developed and adopted. In this connection, it is imperative to develop local market where producers and consumers meet frequently and know each other. Consumers are ready to pay premium price, provided that they are sure that the products are pure organic (Bhatta et al. 2008). If they know the producers, it is easier to build the trust between them. It avoids the need for going through the lengthy process of certification and transportation of the products far. At the same time, there is chance of strengthening local economy and salvaging local ecology.

7.5 Organic Agriculture for a Viable Future of Nepal

Everything in this planet earth "our home" is in constant motion, and we are not exception. We are changing every fraction of second stepping toward our destination. The inner and outer environment is constantly changing, and nature has provided the instinct to adapt well to the change. Our ability and capacity to adapt well to changing environment depend on how well we use our wisdom hidden within each of us. Nature has programmed each living being with strong sense of self-interest for survival, joy, peace, and comfort without any discrimination. Human beings are given extra sense of responsibility to understand the interest of nature and maintain balance for sustainable survival. Charles Darwin understood the operation of natural selection and wrote "On the Origin of Species" and based on his doctrine Herbert Spencer used the term "Survival of the Fittest." No one can deny the doctrine and still survive. Nature has set reproductive boundaries, but technological innovation has crossed the boundaries in genetic engineering and genetic modification of organism. It is the absolute truth, and we now need to wake up and correct our wrong deeds against nature and make home safe for the generations to come. One of the wake-up calls now is transforming our agriculture production system from unfriendly to nature friendly where our wellness and viability of livelihood exist.

7.5.1 Need of Organic Agriculture

Agriculture serves as the basis of livelihood and economic prosperity of Nepal. Despite its importance, the agricultural sector faces several challenges and limitations when it comes to meeting the demands of its growing population. Agrochemicals, introduced aiming at increasing food production, did not brought the expected result rather the country is becoming more and more dependent for food from outside. At present, the country spends about NRs. 113,7361,7000 in chemical fertilizers and about NRs. 380,000,000 for pesticide annually (MoAD 2014). Despite the huge amount of money spent in agrochemicals, the productivity of food crops is either stagnant or increasing insignificantly.

Unscientific use of agrochemicals, particularly the pesticides, not only produces nutritionally poor food products but also results into health hazard to human and the environment. Haphazard use of pesticides has not only killed the beneficial fauna and flora above as well as underground indiscriminately but also has offered resistance among the pests demanding ever stronger pesticides threatening the present and future human health. It is estimated that about 85 % of the pesticides are used in vegetable farming which are generally consumed fresh, meaning direct threats to the public health. A study suggests that the health of 1.5 million farmers has been affected by pesticides. According to the data from seven major hospitals in the country in a little over the last one decade, there has been a twofold increase in the number of patients admitted to hospital for cancer. There were 3251 in year 2000 and 7212 in 2012—as reported by National Cancer Registry Program (NCRP). The number may be higher because of the paucity of exact national data and poor reporting mechanism in hospitals. NCRP estimates suggest that 30,000–40,000 cancer cases are diagnosed annually (TKP 2015). Further, farmers are becoming ever more dependent on external inputs, mainly for agrochemicals. In addition, even in rural areas, water quality is deteriorating and environmental health problems are in rise. In this context, organic farming that enhances soil productivity by keeping soil biologically alive is the only alternative for food security and sustainability. Given the rich natural wealth such as varied numerous agroecological zones, distinct seasonal variations, biodiversity, and indigenous knowledge, Nepal has the potential to manage agricultural production organically. In addition, organic farming offers such noble benefits as self-reliance and resilience; conservation of resources; support for local economy through jobs creation; help family farming and maintain social cohesion; provides tasty and natural foods; preservation of the culture of agriculture and mitigation and adaptation to climate change for future viability.

7.5.2 Critical Issues for Organic Agriculture

Majority of the farmers view chemicals as an integral part of their production system and think impossible to produce crops without chemicals. This is partly because chemicals are easily available, and partly because our national agricultural research and development system is still engulfed into the promotion of technological packages based on high yielding varieties and associated chemicals and is not in position to offer the alternatives to chemicals. Production of crop without synthetic inputs and maintaining crop quality requires skills, knowledge, and patience. Lack of sound knowledge and standard alternative technology on organic inputs, especially for nutrient management and plant protection, and knowledge on their preparation has limited the expansion of organic farming. Majority of the present-day farmers, especially the newer generation, have lost the feeling of ownership on what they have: traditional knowledge and wisdom or that level of knowledge may not be sufficient in the context of growing food demand. The knowledge needs scientific perspectives to refine it as per the time has demanded. Organic has become a popular and fashion cosmetic word today which is often misused in unethical market where both consumers and genuine producers are being cheated. There are very few outlets for organic products in the country, and in common market, there is no price difference between the products grown organically or by using chemicals.

There are difficulties to export organic products because of complications in certification process, volume of production, transportation cost, and international standards. Different countries have their own standards to which producers should comply and the process of qualifying for these is costly and time consuming. Most of producers are small holders, and volume of production is very low. Continuous supply of products with sufficient quality and quantity at competitive price is quite difficult for them. Organic seeds are lacking, seed production is mostly at the hand of big companies, and the farmers have to depend on unreliable and high-cost seed. More importantly, the country has no suitable and effective government intervention for organic agriculture, and pro-organic policies are being developed recently but slowly lagging behind the time. Organic farming has not been entered into the mainstream of national agricultural research and development system yet. Therefore, a national strategy to support, advocate, and adopt organic agriculture, as the only alternative for sustainable food system, is the need of time for Nepal dominated by diversified agroecology and the farmers with small farm holdings.

7.5.3 Opportunities for Successful Organic Agriculture

Fortunately, about 26 % of Nepalese farmers still follow traditional production practices that can correct many negatives of chemical farming. Shifting to organic is not that difficult for Nepalese farmers, provided that they are educated with newer perspective, incentive, and technical know-how and easy access to needed inputs. INGOs and NGOs working toward sustainable development and food security in Nepal are focusing their activities to promote organic farming. The comparative multi-dimensional benefits of organic farming and its products are slowly spreading among consumers creating awareness about the demerits of chemically produced food and the merits of the organic one. Health conscious consumers are ready to pay even higher price for organic products, provided that the quality is assured. Youth and new generations have seen the increasing prosperity in organic agriculture, and as a result large numbers of organic farms from different parts of the country are already bringing the products in the market. Entrepreneurs are also willing to invest in organic agriculture. Government is subsidizing the establishment cost of inputs factories producing biofertilizers and certification cost for export commodities. Production of organic manures and fertilizers in and outside farms in sufficient quantity to supply needed amount of plant nutrients may slowly reduce and replace the use of chemical fertilizers, and build farmers confidence to adopt standard organic production practices. Small-holding and integrated farming system of Nepal makes organic farming easier to practice with high success. Fortified with physical and biological heterogeneity in a short vertical distance nourishing a high level of biodiversity, Nepal can offer myriads of fresh and unique organic products for domestic as well as international markets. Only with these products, the country can insure food security and secured livelihood of the people and compete in international market for which sound policy and smart management from both public and private sector is sought most.

7.5.4 Policy Need for the Promotion of Organic Agriculture in Nepal

Government needs to formulate concrete agricultural policies favoring a shift from present farming system to a viable organic farming in order to promote sustainable agricultural production and livelihood system. National policy on organic agriculture still requires a clear, holistic, and coordinated approach to direct agricultural practices toward organic with the following points in focus: (a) define clearly organic farming to be promoted among the common farmers; (b) total ban of synthetic pesticides; (c) develop appropriate, effective, and plausible national organic standards; (d) develop easy mechanism for certification; (e) develop reliable market; (f) provide the premium price for organic products; (g) delineate critical zones and hot spots that need intervention favoring organic system; (h) embed organic agriculture into national agricultural research and development system; (i) start research on technology development in problematic area/s and or crop/s; (j) outline policy to shift present subsidy on chemical fertilizers to in situ preparation of organic manures; (k) make a national declaration that organic production system is the only alternative way for sustainable healthy living and food security;

(l) incorporate organic agriculture in school curricula; (m) establish a separate shell in the Ministry of Agricultural Development to look after organic agriculture; and (n) mainstreaming of organic agriculture in academic institutions teaching agriculture and environmental sciences.

7.6 Conclusion

Nepal is basically an agrarian country with heterogeneous biophysical and socioeconomic conditions. Despite the agrarian nature, agriculture is not being able to provide the people with sufficient food and nutrition because of its poor performance in recent decades which is mainly due to the degradation of resources. Since the 1960s, government has adopted the policy for scientific research and development of agriculture. Advancement of science and technology and assistance from developed countries, and conventional high-input technology was introduced as an effort to increase production per unit of land and feed the increasing population. Use of high yielding exotic varieties of various crops and associated agrochemicals fulfilled the immediate purpose to address the issues of food shortage and food security but failed to consider the long-term impact on production sustainability and food security. Use of high yielding varieties and agrochemicals has resulted in the loss of local biodiversity (land races and heirloom varieties), soil structure and quality deterioration, fertility decline, food poisoning, environmental pollution, ecosystem damage, and natural resource degradation. Agrochemical-based food production system has been proven unsustainable and unhealthy from both nutritional and environmental point of view. Therefore, a safer and resilient method of food production system needs to be identified, promoted, and protected for livelihood sustainability in Nepal.

Scientific research and innovation of new technology to improve agriculture through use of better high yielding genes and supporting agrochemicals slowly displaced natural organic input-based agriculture production. Some genius activist and naturalist advocated the organic concept and got organized internationally as a strong body against conventional or GRA. There have been tremendous amount of research evidences supporting the organic agriculture for correcting present ill effects brought out against humanity and environment. Organic production system is a default of agriculture in Nepal which has been proven now as right and safe way of long-term livelihood sustainability. Nepal is also influenced by international organic movement and local NGO in collaboration with INGOs, and donor agencies have been successful in promoting the traditional wisdom as new intervention in agriculture production system. Integrated farming of various components such as different species and varieties of food crops (grain, fruit, vegetables, tubers, root), cash crops (flowers, tea, coffee, fiber, narcotic and sugar), aromatic and medicinal herbs, and livestock such as buffaloes, cows, goat, sheep, pigs, fisheries, poultry, and bees raising is common even in small farm. A model-integrated diversified crop components-based production system can support in a symbiotic manner providing buffer and crop security. Organic agriculture which has been proven to increase yield by enriching soil health with respect to fertility and adaptive capacity against moisture stress can be the viable option for Nepal. Public advocacy and increasing popularity of organic agriculture through genuine subsidy support of the government are good news. However, academic institutions and research organizations involved in research and development of agriculture should revise the curricula and research mandate focusing toward organic agriculture through technology, innovation and development. Organic standard development and certification should be based on practical capacity focusing toward slowly converting agrochemical-based agriculture to organic agriculture without losing household supporting level of yield. Organic products meeting third-party certification standard could be produced at organized cooperative farms and/or government farms aiming to demonstrate the profitability of organic farming during the first few years.

References

ABPSD (2013) Statistical information on Nepalese agriculture 2012/2013. Agri- Business Promotion and Statistics Division, Ministry of Agriculture and Co- operatives, Government of Nepal, Singha Durbar, Kathmandu, Nepal

ABTRACO (2008) Country report on the state of the Nepal's plant genetic resources for food and agriculture. Agri-Business and Trade promotion Multi-cooperative Limited, Kathmandu 92p

Acharya AK, Kafle N (2009) Land degradation issues in Nepal and its management through agroforestry. J Agri Environ 10:115–123

Adhikari J (2012) Farmers in lurch. The Kathmandu post, July 8, 2012. Web site http://www.ekantipur.com/kantipur/

ADS (2013) Agricultural development strategy (ADS) final report prepared for Government of Nepal, Kathmandu, Nepal, 274 p

Aryal M (2013) Yuwa farkera bakhra palan ma (Youth returnees in goat farming). Kantipur Daily, 6th Shrawan, 2070 (May 20, 2013). Web site http://www.ekantipur.com/kantipur/

Aryal KP, Dahal KR (2010) Report on the status of organic agriculture in Tarai, Nepal, submitted to Nepal permaulture group. Babarmahal, Kathmandu, 43 p

Barański M et al (2014) Higher antioxidant and lower cadmium concentrations and lower incidence of pesticide residues in organically grown crops: a systematic literature review and meta-analyses. Br J Nutr 26. doi: 10.1017/S0007114514001366

Barsila SR (2011) Transhumance livestock herding in the eastern mountains of Nepal. Colloquium "Selected aspects of sustainable development", 13 May and 18 May 2011, North-South Centre research for Development. Web site http://www.northsouth.ethz.ch/colloquium/past_colloquia/collo_2011/Barsila_abstract

Bhandari GP (2013) Bakhra Palera Chhorilai Boarding (Sending daughter to boarding by goat farming) Kantipur Daily, 4th Shrawan, 2070 (July 19, 2013). Web site http://www.ekantipur.com/kantipur/

Bhatta G, Doppler W, KC KB (2008) Problems and potentials of organic agriculture development in Nepal. Paper presented in conference on International research on food security, natural resource management and rural development, Tropentag 2008 University of Hohenheim, 7–9 Oct 2008

Bhattari RC (2011) Economic impact of community forestry in Nepal: a case of mid-hill districts of Nepal. Econ J Develop Issues 13 and 14(1–2):75–96

Bhattari RC (2014) Jitpure ko Argnic Yatra: Ghar Ghar ma Gadyaule mal (Organic Journey of Jitpure) Annapurna Post (25 May 2014). Web site http://www.annapurnapost.com/

Bisht K (2011) An organic revolution. ECS Nepal magazine (April 2011). ECS Media Pvt. Ltd., Lalitpur, pp 46–54

Carson B (1992) The land, the farmer, and the future. A soil fertility management communities. In: Proceedings of the National Academy of Sciences (USA), pp 1–6

Charyulu DK, Biswas S (2010) Economics and efficiency of organic farming vis-à-vis conventional farming in India. Indian Institute of Management, Ahmedabad. W. P. No. 2010-04-03, 26 p

Conford P (2001) The origins of the organic movement. Floris Books, Glasgow

Crinnion WJ (2010) Organic foods contain higher levels of certain nutrients, lower levels of pesticides, and may provide health benefits for the consumer. Altern Med Rev 15(1):4–12

Dahal KR (2010) Ecological agriculture in Nepalese context. In: Jha PK, Karmacharya SB, Balla MK, Chettri MK, Shrestha BB (eds) Sustainable use of biological resources in Nepal, pp 52–61

Dahal KR (2012) Education on organic agriculture in Nepal. In: Poudyal KP, Khatiwada MP, Khatri BB, Subba N (eds) Proceedings of the 7th National Horticulture Seminar, 12–14 June, 2011, Lalitpur, Horticulture Research Division, NARC and Nepal Horticultural Society, Kathmandu, pp 229–235

Dahal KR (2013) Technologies for organic crop nutrient management system (OCNMS) in Nepal: practices and promises. In: Proceedings of the fourth SAS-N convention (4–6 April 2012), Lalitpur. Nepal Agricultural Research Council (NARC) and Society of Agricultural Scientists-Nepal (SAS-N), pp 383–391

Deshar BD (2013) An overview of agricultural degradation in Nepal and its impact on economy and environment. Glob J Econ Social Develop 3(1):1–20

FAO (2000) Food safety and quality as affected by organic farming, Report of the 22nd regional conference for Europe, Portugal, 24–28 July 2000. Food and Agriculture Organization, Rome

FAO (2007) Report on International conference on organic agriculture and food security, food and agriculture organization, Rome

FAO (2011) Organic agriculture and climate change mitigation. A report of the round table on organic agriculture and climate change, food and agriculture organization, Rome

Fließbach A, Mäder P, Dubois D, Gunst L (2000) Results from a 21 year old field trial: organic farming enhances soil fertility and biodiversity, FiBL-Dossier No. 1, 16 p

Ghimire LN (2012) Gai palera ekkaish barshamai safal (A successful cow farmer already in 21 years). Kantipur Daily, 21 Ashoj 2069 (7 Oct 2012)

Gips T (1987) Breaking the pesticides habit: alternatives to 12 hazardous pesticides. International Organization of Consumers Unions, Penang p 348

Halberg N et al (2006) The impact of organic farming on food security in a regional and global perspective. In: Halberg N, Alroe HF, Knudsen MT, Kristensen ES (eds) Global development of organic agriculture: challenges and prospects. CABI Publishing, Chapter 10, pp 277–322

Heckmen J (2007) History of organic farming: transitions from Sir Albert Howard's war in the soil to the USDA national organic program. Health topics—Farm and Ranch. Web site http://www.westonaprice.org on 29 Mar 2014

Howard A (1943) An agricultural testament. Oxford University Press, New York and London 228 p

IASA (1990) Planting the future, a resourceguideto sustainable agriculture in the third world. Minneapolis:IASA

IFAD (2003) The adoption of organic agriculture among small farmers in Latin America and the Carribean—thematic evaluation, International Fund for Agricultural Development (IFAD). Report No. 1337

IFAD (2005) Organic agriculture and poverty reduction in Asia: China and India focus thematic evaluation. Report 1664, office of evaluation, IFAD, Rome

IFOAM (2006) Organic agriculture and food security. International Federation of Organic Agricultural Movement (IFOAM), Germany

IFOAM (2008) Criticisms and frequent misconceptions about organic agriculture: the counter-arguments, p 71. Web site www.ifoam.org

IFOAM (2009) The world of organic agriculture: statistics & emerging trends 2009, International Federation of Organic Agricultural Movement/BioFach, Bonn and FiBL, Frick

IFOAM (2015) The World of organic agriculture, statistics and emerging trend 2015. FiBl and International Federation of Organic Agricultural Movement, Frick Bonn

ISIS (undated) Organic agriculture fights back. ISIS report, The Institute of Science in Society, Web site http://www.i-sis.org.uk/OrganicAgriculture.php

KD (2013) Agricultural products worth Rs 144 billion imported in three years, Karobar Dainik, 6th July 2013. Web site http://np.karobardaily.com/

Khatiwada Y (2012) Kam Khojdai Jhandai Char Lakh Yuba Bidesie (Outmigration of about four hundred thousand youths in search of job), Kantipur Daily, 16 July, 2012, Kathmandu. Available at http://www.ekantipur.com/kantipur/

King F (1911) Framers of forty centuries or permanent agriculture in China, Korea and Japan. Website http://www.permaculturenews.org/files/farmers_of_forty_centuries.pdf

Koirala GP, Thapa GB (1997) Food security challenges: where does Nepal stand?. HMG/N, MOA/Winrock International, Kathmandu 41p

Leachy S (2014) CO_2 producing hollow food. UXBRIDGE, Canada, 7 May 2014 (IPS). Website http://www.ipsnews.net/2014/05/co2-producing-hollow-food/

Lucksom L (2013) The history of Nepali cheese. In: ECS Nepal magazine (August, 2013). ECS Media Pvt. Ltd. Lalitpur, pp 46–54. Web site http://ecs.com.np/features/the-history-of-nepali-cheese

Mallapati S (2012) Liquid gold: farmers in Nepal find resourceful way to fertilize crops, Environmental news. Web site http://www.environmentalhealthnews.org/ehs/news/2012/urine-as-fertilizer

MoAC (2010) National Agriculture Sector Development Priority (NASDP). Ministry of Agricultural Development, Nepal

MoAD (2014) Annual Progress Report, Department of Agriculture, Ministry of Agriculture Development, Government of Nepal, Kathmandu

MoEST (2008) State of the Environment (Agriculture, Forest and Biodiversity), Ministry of Science, Technology and Environment, Government of Nepal Singh Durbar, Kathmandu

MoF (2015) Budget speech of fiscal year 2014/15. Ministry of Finance, Government of Nepal, Kathmandu

Mollison B (1990) Permaculture: a practical guide for a sustainable future. Island Press, USA 579 p

Muller A (2009) Benefits of organic agriculture as a climate change adaptation and mitigation strategy for developing countries. Environment for Development (EfD), Swedish International Development Cooperation Agency (Sida)

Myers S et al (2014) Increasing CO_2 threatens human nutrition. J Nature 510:139–142

NMVF (2010) Nepal to become the first organic nation, Nepal Maharishi Vedic Foundation, Good news from around the world. Web site http://globalgoodnews.com/environmental-news-a.html?art=128726087317162355 on 12 March 2011

NPC (2003) Tenth Five Year Plan, His Majesty's Government of Nepal, National Planning Commission, HMGN, Kathmandu

NTCDB (2014) Coffee data base in Nepal. National Tea and Coffee Development Board, Nepal 66 p

Pandey SP, Joshy D (2000) Fertilizer consumption and food grain production in Nepal: (1965/66–1995/96), Soil Science Division, Nepal Agricultural Research Council (NARC), ISBN: 99933-557-0-4, Lalitpur

Parrott N, Marsden T (2002) The real green revolution. Greenpeace Environmental Trust London, United Kingdom 151 p

Ponisio LC et al (2014) Diversification practices reduce organic to conventional yield gap. doi:10.1098/rspb.2014.1396. Web site http://rspb.royalsocietypublishing.org/content/282/1799/20141396

Pretty JN (1995) Regenerating agriculture. Earthscan Publication Limited 320 p

Pretty JN, Hine R (2001) Reducing food poverty with sustainable agriculture: a summary of new evidence, University of Essex

Rahmann G (2011) Biodiversity and organic farming: what do we know? Agri For Res 3 (61):189–208

Regmi PP (1999) Agricultural development through eco-restructuring in different ecological zones across Nepal. PhD Dissertation AIT AC 99-2. Asian Institution of Technology, Bangkok, 419p

Regmi BD, Paudel CL, Rajbhandari NP, Hada N (2006) Combating desertification process in hills of the Himalayan region in Nepal through sustainable soil management practices. J Arid Land Stud 15(4):227–230

Reijntjes C, Haverkort B, Waters-Bayer A (1992) Farming for the future, an introduction to low-external input and sustainable agriculture. Macmillan Press Limited, London 250 p

RI (2011) The farming system trials celebrating 30 years. Rodale Institute, USA. Web site http://rodaleinstitute.org/our-work/farming-systems-trial/farming-systems-trial-30-year-report/

Ricker HS (1997) Review (untitled). Am J Agric Econ 79(1):280–282

Scialabba N, Müller-Lindenlauf M (2010) Organic agriculture and climate change. Renew Agric Food Syst 25(2):158–169

Sharma CK (1990) Chemical pollution of the soil and groundwater in the Kingdom of Nepal, Groundwater Monitoring and Management (Proceedings of the Dresden Symposium, March 1987). IAHS Publ. no. 173

Sharma DR, Manandhar HK, Thapa RB, Pradhan SB (2012) Use of pesticides in Nepal and impacts on human health and environment. J Agri Enviorn 13:67–74

Sherchan DP, Karki KB (2005) Plant nutrient management for improvingcrop productivity in Nepal paper number 3. Web site http://www.fao.org/docrep/010/ag120e/AG120E10.htm

Shrestha BN (2012) Rivers of Nepal, Posted on June 6, 2012 by Border Nepal. Web site https://bordernepal.wordpress.com/2012/06/06/rivers-of-nepal/ retrieved on 10 Jan 2015

SMD (2014) Soil fertility status of Nepal (FY 2013/14): a report by soil management directorate, Department of Agriculture, Ministry of Agriculture Development, Lalitpur, 13 p

SSMP (2010) Gothemal thatha Compost Byabastapan Talim Pustika in Nepali (Training Manual on FYM and Compost Management), Sustainale Soil management Program and Soil Management Division, DoA, MoAC, Nepal 97 p

TKP (2015) The Kathmandu post daily newspaper, 3 Feb 2015. Website e.kantipur.com

UN (1987) Our common future. Report of the World Commission on Environment and Development, United Nations, 247 p

Upreti BR, Sharma SR, Paudel SB (eds) (2014) Food security in post conflict Nepal: challenges and opportunities. Kathmandu University and Nepal Centre for Contemporary Research (NCCR), Kathmandu, 172 p

WECS (2010) Energy sector synopsis report, Water and Energy Commission Secretariat (WECS), Governemnt of Nepal, Singa Durbar Kathmandu

Willer H, Lernoud J (2015) The world of organic agriculture: summery. In: Helga W, Lernoud J (eds) The world of organic agriculture. Statistics and Emerging trend 2015. FiBL and IFOAM Report. Research Institute of Organic agriculture (Frick) Switzerland and IFOAM Organic International, Bonn

Woese K, Lange D, Boess C, Bogl KW (1997) A comparison of organically and conventionally grown foods: results of a review of the relevant literature. J Sci Food Agri 74:281–293

Worthington V (2001) Nutritional quality of organic versus conventional fruits, vegetables, and grains. J Altern Complement Med 7(2):161–173

WRADO (2015) Proceedings of the 8th National agricultural fair 2071, Western Agricultural Development Office, Pokhara, 33 p

Chapter 8
The Potential of Silvopastoral Systems for Milk and Meat Organic Production in the Tropics

F.J. Solorio, S.K. Basu, L. Sarabia, A. Ayala, L. Ramírez, C. Aguilar, J.A. Erales, J.C. Ku and J. Wright

Abstract The demand for livestock products is rising rapidly in tropical areas as a consequence of increased human population. As demand for food increases, deforestation and land degradation occur. Though varying by country and region, the conversion of forest into cattle pastures has been one of the main driving forces of this degradation. In various Latin American countries, the creation of livestock farms, with government support, has been the single most important source of deforestation. This expansion of cattle ranching is also one of the principle causes of the increase in greenhouse gas emissions. Agriculture releases significant amounts of CO_2, CH_4, and N_2O into the atmosphere. For example, CO_2 is released largely from microbial decay or the burning of plant litter and soil organic matter produced during agricultural processes. Recently, silvopastoral systems (SPSs) have been advocated as promising alternatives to current practices by reconciling conservation and development needs. SPS is the production of livestock on land in a system which combines multipurpose leguminous shrubs at high densities together with grasses to improve both the yield and quality of fodder, resulting in milk and meat products with a high potential to attract an organic premium. This SPS plays an important role in healthy milk and meat production. Recent research advances have proven that *Leucaena* grass pastures are the most productive, profitable, and sustainable pasture-fed option for agroecological cattle production. Because the levels of input have traditionally been relatively low in the production of meat and milk from extensive grassland systems, they are among the easiest to convert to organic production. However, the long-term prospects for organic systems are not clear. There is continued pressure to

F.J. Solorio (✉) · L. Sarabia · A. Ayala · L. Ramírez · C. Aguilar
J.A. Erales · J.C. Ku
Campus de Ciencias Biológicas Y Agropecuarias, Universidad Autónoma de Yucatán, Carretera Mérida-Xmatkuil Km. 15.5. C.P., Mérida 97100, Yucatán, Mexico
e-mail: ssolorio@uady.mx

S.K. Basu
Department of Biological Sciences, University of Lethbridge, AB T1K 3M4, Lethbridge, Canada

J. Wright
Centre for Agroecology, Water and Resilience, Coventry University, Coventry, UK

D. Nandwani (ed.), *Organic Farming for Sustainable Agriculture*, Sustainable Development and Biodiversity 9,
DOI 10.1007/978-3-319-26803-3_8

ensure that all livestock systems and agriculture in general develop in a way that has minimal environmental impact. The differences between organic and conventional systems may diminish over time, and the pressure in demand for organic products may slow. Compared with other sources of fodder for meat and milk production, SPSs can provide a cheap source of feed. The SPS produces double the amount of milk and meat compared to pastures in monocrops with the minimum use of external inputs. The objective of this chapter is to describe the potential of SPS for organic milk and meat production.

Keywords Silvopastoral systems • Organic productions • Milk • Meat • Livestock • Agriculture • Fodder

Abbreviations

AU	Animal units
CP	Crude protein
DM	Dry matter
FAO	Food and Agriculture Organization
GHG	Greenhouse gases
LME	Liquid milk equivalent
ME	Metabolizable energy
Mmt	Million metric tons
NDF	Neutral detergent fiber
SPSs	Silvopastoral systems
UNEP	United Nations Environmental Program
WHO	World Health Organization

8.1 Introduction

Increases in population, rising salaries, and urbanization place an enormous pressure on the demand for high-quality animal protein products such as milk and meat. This increase in demand originated the Livestock Revolution, reported by Delgado (2003); however, this demand-led transformation is also happening in the context of global warming and climate change. Livestock industry needs to increase production while decreasing the emission of greenhouse gases (GHG) derived from the entire production cycle, and this needs to be done on an already set amount of land (Gerbens-Leenes and Nonhebel 2002, 2005). Many reports describe the sources and amounts of GHG emissions through livestock production, raising public concern regarding the production system they are buying products from.

The expansion of food production has depleted land cover and biodiversity, with diverse negative consequences for human well-being and health; major nutrient cycles are being disrupted (McMichael et al. 2007). At the same time, an estimated 1 billion poor depend on livestock for food and income (FAO 2014).

Nitrogen is the most important nutrient that limits food production. Half of the nitrogen demand is met through fossil fuel–driven fixation of nitrogen, while leguminous fixation of nitrogen provides the other half. Fertilizer use has vastly increased the concentration of bioactive nitrogen compounds in the atmosphere. With the current focus on reducing emissions of GHG while simultaneously increasing biomass production for food, fiber, feed, and fuel, there is a strong case for improving the use efficiency of the leguminous nitrogen (Hogh Jensen 2011).

Consumers are increasingly interested in the provenance of the products they buy. In terms of animal protein sources, global warming and climate change as well as animal well-being are some of the major concerns for the consumer (UNEP 2005). There is increasing uncertainty around temperature and precipitation regimes, making agriculture more uncertain for the millions of livelihoods who depend on the dryland crop production.

Organic products are perceived by the consumer as a better option for health and nutrition, and demand has continuously increased over the past decades, as more evidence shows that industrial food refining, marketing, and overconsumption increase the risks of some non-communicable diseases (Salman et al. 2008; McMichael et al. 2007). Prohibiting the use of chemical fertilizers, pesticides, hormone growth promoters, and antimicrobials provides a product free from contaminants and also assures the consumer that the environment is not receiving such pollutants (UNEP 2005; Finch 2014). However, the vast majority of farms in the tropics are small, and most of these are mixed crop and livestock farms. A common characteristic of these farms is their strong dependence on the use of native and/or introduced pastures. Despite the important role of pastures in livestock production, more than 60 % of the pasture land in some Latin American countries shows symptoms of degradation.

Silvopastoral systems, the introduction of multipurpose trees and shrubs into grasslands, can provide food, fodder, energy, and increased cash income, as well as contribute to the retention of soil moisture and improvement of the quality of land. Silvopastoral management can contribute to increase good-quality milk and meat production, and maintain soil productivity and the ecosystem services necessary in tropical areas. SPSs meet several of the most important criteria for the transition from conventional livestock production to organic livestock production, as they seek to decrease reliance on external nutrient sources and to produce the animals outdoors in pasture-based systems.

8.2 Importance of Grasslands for Tropical Livestock Production

Extensive grazing by cattle is the major land use in tropical grasslands. Livestock can use grasses and other forage plants more efficiently than humans as they can convert indigestible plants into food for human consumption. Tropical grasses are used for extensive livestock production, particularly for meat and milk production.

Livestock production based on pastures is one of the best ways to reduce costs related to animal production. Natural grasslands are the world's largest multi-functional terrestrial ecosystems, covering about 40 % of the global land surface (Suttie et al. 2005). Tropical forages are very important as livestock feed in both commercial and traditional systems. Animal performance (the rate of live weight gain or milk production level) depends on the nutritive value of forage which is also related to the year, season, and the stage of maturity of the swards being grazed (this is the digestibility of the feed). One way to maintain or increase milk or meat production levels is to supply animals with concentrates rich in nutrients. In this sense, levels of milk yield or animal live weight gain are basically determined by the daily dry matter (DM) intake of net energy. However, the use of maize, sorghum, or soya for the production of concentrates is not recommendable as these food items are more efficiently used in feeding humans directly.

8.3 Fluctuation of Fodder Quality and Production Levels in Tropical Regions

Tropical regions have a great potential for animal production, since the abundance of resources necessary for growth (water, light, and temperature) can be used efficiently by tropical grasses. Nevertheless, the fluctuation of forage availability limits animal productivity over the year. Water may also be limited in some regions and seems irrigation can help to minimize this fluctuation where irrigation costs are low.

Animal production in tropical regions is based on forage production, mainly obtained from grasses, which produce large amounts of biomass, in general of low quality, particularly during the rainy season (Jank et al. 2005). According to FAO (2002), almost half of the world's beef comes from tropical and subtropical countries. One of the greatest advantages of the tropics and subtropics is the fact that temperatures remain high during most of the year, which enable plants (especially C_4 plants, such as tropical grasses) to grow all year around. Nevertheless, tropical regions vary in their potential to produce biomass due to the different edaphoclimatic conditions and pasture management. For instance, in the humid tropics, where annual rainfall may be higher than 2000 mm, water availability does not limit plant growth, but other climatic factors do, such as temperature or photoperiod, whereas in the seasonal tropics, where rainfall lasts six months or less, plants are limited by the lack of available water during the dry season. Biomass production fluctuates throughout the year, and this brings the challenge of how best to match animal requirements with forage availability in a production system during the year. Usually, producers compensate this by letting animals either lose weight or overgraze their paddocks, with detrimental consequence to the production system. Only in few cases, do farmers use other sources of feed than grass, such as maize straw or grain, molasses, grass hay, and maize silage,

Fig. 8.1 Extensive grassland for beef or dairy production under tropical conditions

to minimize the effects of biomass production fluctuation, since these trade actions will bring more costs to the production operation (Fig. 8.1).

In addition to the fluctuation in biomass production over the year, tropical grass also varies enormously in terms of quality, especially protein and fiber content, and digestibility. These fluctuations limit animal productivity. Ideally, animals should have diets with consistent quality in such a way that the rumen environment remains constant and animals render more milk and meat. Unfortunately, this seldom happens and animal productivity is limited by the fluctuation in dietary quality. Reduction in energy availability in the diet, as a consequence of a reduction in forage digestibility, limits rumen microbial activity, and consequently, animals do not fill their requirements and they need to mobilize their reserves to compensate for this limitation.

Tropical grasses are known for their great capacity to produce forage, since they follow the C_4 photosynthesis pathway. They are better able to utilize the abundant growth resources in the tropics, light, and water, than other types of plants. One of the characteristics of these grasses is, therefore, their high forage yield compared with legumes and temperate grasses. Nonetheless, this high yield depends on several factors such as the availability of growth resources and the genetic makeup of the grass. Water availability throughout out the year is critical for plant growth, productivity, and reproduction; timing and quality of water determine the efficiency of its use for forage production. Unfortunately, as a likely consequence of climate change, current rainfall patterns are changing resulting in flooding and dry periods, both of them limiting grass yield and contributing to increasingly degraded pastures. In some cases, water quality, especially Ca and Na contents, may also limit plant growth.

Irrigation has been a key to achieve food security in many parts of the world (Rosegrant et al. 2005), but non-agricultural uses compete for water availability. The increasing costs of water may limit its use for food production. Rosegrant et al. (2005) stated that the main effective way to deal with water scarcity is to increase water use efficiency.

8.4 Milk and Meat Demand

It is becoming clear that meat and dairy products are the foods carrying the greatest environmental burden, accounting for approximately half of food-generated GHG emissions and 18 % of global GHG emissions (FAO 2006). However, global consumption of livestock products is growing. Demand for meat and milk is set to double (FAO 2006) by 2050 (Garnett 2009). Garnett mentions that there is a large and growing literature on the GHG emissions associated with livestock rearing. The findings broadly conclude that livestock products are GHG intensive compared with other food groups and that the vast majority of impacts occur at the farm stage, with subsequent processing, retailing, and transport playing relatively minor roles.

As human population increases, the demand for food increases as well; in particular, new generations living in countries with expanding economies have a tendency to buy increasing amounts of food of animal origin such as meat and milk, replacing other commodities. Urbanization is a major driving force influencing the global demand for livestock products. Urbanization stimulates improvements in infrastructure, including refrigerated storage and transport, which permit trade in perishable goods and thus an increase in the availability of food. Annual meat production is projected to increase from 218 million tons in 1997–1999 to 376 million tons by 2030 (WHO 2014; Fukase and Martin 2014).

There is a strong positive relationship between the level of per capita income and the consumption of animal protein (Fukase and Martin 2014). Developing countries are embarking on higher meat consumption at much lower levels of gross domestic product than industrialized countries did some 20–30 years ago.

Table 8.1 shows trends in per capita consumption of livestock products in different regions of the world. There has been a remarkable increase in the consumption of animal products in regions such as developing countries (Brazil) and East and Southeast Asia (China), although the levels are in general below the levels of consumption in North American and most other industrialized countries (WHO 2014).

As diets become richer and more diverse, the high-value proteins that livestock products offer improve the nutrition of the vast majority of people of the world. Not only livestock products can provide high-value proteins, but they are also important sources of a wide range of essential micronutrients, in particular minerals such as iron and zinc, and vitamins such as vitamin A. For the large majority of people in the world, particularly in developing countries, livestock products remain a desired food for nutritional value and taste. Excessive consumption of animal products in some countries and social classes can, however, lead to excessive intakes of protein and fat as well as of antibiotics, growth hormones, and other pharmaceuticals used in livestock production. Some authors propose education in high-income countries for a reduction in consumption, which would benefit human health mainly by reducing the risk of ischemic heart disease (especially related to saturated fat in domesticated animal products), obesity, colorectal cancer, and, perhaps, some other cancers (Chaudhri and Timmer 2003). However, there is increasing evidence that it

Table 8.1 Historic and projected consumption of livestock products per capita and per year per country or per region

Region	Meat (kg per year)				Milk (kg per year)			
	1964–1966	1997–1999	2007	2030	1964–1966	1997–1999	2007	2030
World	24.2	36.4	285.7	45.3	73.9	78.1	671.3	89.5
Developing countries	10.2	25.5	175.4	36.7	28.0	44.6	313.5	65.8
Near East and North Africa	11.9	21.2	9.6	35.0	68.6	72.3	36.4	89.9
Sub-Saharan Africa	9.9	9.4	9.3	13.4	28.5	29.1	24.3	33.8
Latin America and Caribbean	31.7	53.8	40.3	76.6	80.1	110.2	68.7	139.8
East and Southeast Asia	12.6	43.0	106.3	70.2	40.6	77.5	42.9	124.7
Industrialized countries	61.5	88.2	110.2	100.1	185.5	212.2	357.7	221.0
Transition countries	42.5	46.2	24.7	60.7	156.6	159.1	–	178.7

Adapted from FAO (2002, 2012)

is carbohydrates and especially in the form of processed sugars and grains that are the real cause of such diseases (Lustig 2012). Nevertheless, the meat-intensive diets enjoyed by many in the industrialized world lead to a greater degree of inequality in the use of environmental services than is apparent from the examination of the distribution of food consumption across countries (White 2000).

Still, there is a strong case that an increase in the consumption of animal products in low-intake populations, toward the proposed global mean figure, should benefit human health and development.

8.5 Meat and Milk Consumption

People in developing countries are increasing their consumption from the very low levels of the past, and they have a long way to go before coming near developed country averages. In developing countries, people consumed an annual average of 28 kg/capita meat and 32 kg/capita milk between 2002 and 2015; one-third the meat and one-fifth the milk per capita comes from the developed countries.

Per capita consumption is rising fastest in regions where urbanization and rapid income growth result in people adding variety to their diets. Across countries, per capita consumption is significantly determined by average capita income. Aggregate consumption grows fastest where rapid population growth augments income and urban growth. Since the early 1980s, total meat and milk consumption

Table 8.2 Annual per capita human food consumption (kg) and percent of calories from meat and milk livestock

Consumption (kg)	Developed countries			Developing countries		
	2002	2015	2030	2002	2015	2030
Annual meat per capita	78	83	89	28	32	38
Total	102	112	121	137	184	252
Annual milk per capita	202	203	209	44	55	67
Total	265	273	284	222	323	452

Source Calculated from data in the United Nations Food and FAO database, 2006, and projections reported by Thornton (2010)

Table 8.3 Meat and Milk consumption per capita (kg) by region and some countries in 1990–2010

Region	Meat			Milk		
	1990	2000	2010	1990	2000	2010
World	33	37	42	77	78	89
Developing countries	9.5	11	14	27	29	39
USA	113	122	120	257	257	251
China	24	44	57	126	159	221
India	3.9	4.1	4.4	52	62	80
Africa	15	16	18	36	37	46
Asia	17	25	31	32	41	57
Brazil	216	357	411	152	189	242
Latin America	36	48	59	90	95	101

Source FAOSTAT (2015) (http://faostat3.fao.org/download/FB/CL/E)

grew at 6 and 4 % per year, respectively, throughout the developing world. In East and Southeast Asia, during late 1980's and 1998, the per year growth for population was 2–3 % the income growth was 4–8 %, the meat consumption grew between 4–8 % per year (Table 8.2).

The Livestock Revolution has been most evident in East Asia, as illustrated by the per capita figures for China. China and Brazil play a dominant role in the meat part of the Livestock Revolution. However, the near doubling of aggregate milk consumption as food in India between the early 1990s and the late 2000s suggests that the Livestock Revolution goes beyond just meat and beyond China and Brazil. At 60 million metric tons (mmt) of liquid milk equivalent (LME) considered low by many Indian dairy analysts, Indian milk consumption amounted to 13 % of the world's total and 31 % of milk consumption in all developing countries (Table 8.3).

The medium-to-high milk consumption of Latin America in 1990–2000, at 90–95 kg/capita, is halfway between the developing world as a whole (28 kg/capita) and the USA (257 kg/capita), because of the very high level (75 %) of urbanization in Latin America.

FAO (FAOSTAT 2015) suggests that during 2010, developing country aggregate consumption growth rates of meat and milk are separately to be 3.0 and 2.9 %/year, respectively, compared to 0.8 and 0.6 %, respectively, in the developed countries.

Aggregate meat consumption in developing countries is projected to grow by 37 (kg/person/year) projected by 2050 (FAO 2012), whereas the corresponding figure for developed countries is 90 kg/person/year. FAO (2012) suggested that similarly additional milk consumption in the developed countries of 220 kg/person/year of LME will be dwarfed by the additional milk consumption in developing countries of 75 kg/person/year.

The principal conclusion of the most recent projections is to confirm the view that the Livestock Revolution in developing countries will continue at least to 2020 and will increasingly drive world markets for meat, milk, and feed grains.

The main trade impact is that developing countries as a whole will increase their already large net imports of cereals to an annual amount in 2020 of about the same magnitude as the annual US corn crop (193 mmt). About half (92 mmt) of these net imports will be maize and cereals other than rice and wheat; most of the coarse grains will probably go to feeding. Meat and milk production increases in developing countries will largely match the big consumption increases, and meat exports from Latin America to Asia will soar.

Meat prices will fall in the range of 3 %, whereas the milk price is projected to fall 8 %. These falls would be substantially higher without the Livestock Revolution. On the positive side, increased consumption of meat and milk can improve the incomes of poor farmers and food processors in developing countries.

The rapidly growing demand for livestock products is a rare opportunity for smallholder farmers to benefit from a rapidly growing market and for their families to have a viable source of much-needed micronutrients and dense calories.

In developed countries, organic markets are better developed than those in less developed countries; however, in both cases, the most influencing force in organic food demand is price; consumers with reduced budgets will opt for less expensive products, such as skimmed milk, or for larger containers (Finch 2014). In the case of organic meat and milk, the main challenge is to be able to feed the animals with organic feeds, which is not easy at times of a shortage of forage and where organic regulations demand for grazing for at least 120 days per year and to comprise at least 30 % of food intake, as well as the need of organic grains (Kahn 2014).

8.6 Food Quality and Animal Welfare

The silvopastoral systems (SPSs) offer a diversity of tree and shrub species in association with grasses that allows cattle a variety of options with a better dietary quality, compared to a monoculture of pastures. Increased diet quality is attributable to low fiber, high protein, and digestibility of the foliage of trees and shrubs (Ibrahim et al. 2005), in addition to a marked increase in the concentration of crude protein (CP) of the associated grasses. Reports of Barros-Rodríguez et al. (2013)

Table 8.4 Estimates (% Ndfa) and N content (kg ha^{-1}) in different plant components of *L. leucocephala* under two pruning regimes (35 y 50 days)

Pruning frequency	Fodder content						
	DM (Kg ha)	N %	N (Kg ha)	FBN %	CP %	FDA %	FDN %
35	2042.1	4.65	26.4	84.39	29.0	24.1	40.3
50	2369.2	4.18	33.1	88.19	26.1	24.9	42.3

and Mayo Eusebio (2014) reported 11.4 and 10.7 % CP in *Panicum maximum* grass associated with *Leucaena leucocephala* leguminous shrub in SPSs for sheep and cattle, respectively. The higher values of CP in grasses in SPSs can be explained by the fixation of atmospheric nitrogen (Table 8.4), the natural enrichment of the soil by the legume (Sarabia-Salgado 2013), and the deposition of manure and urine from the livestock grazing the mixed pasture (Murgueitio and Ibrahim 2008).

In SPSs, positive interactions among trees associated with grasses may maximize above- and belowground resource utilization for fodder quality. Intercropping legume shrubs with grasses increases the opportunity for complementary N-use and improves nutrient cycling. Additionally, some grass species benefit from the shade of trees, and it has been reported increases in CP of *P. maximum* cultivar Tanzania from 9.6 to 12.9 % with 54 % shade (Cruz 1997). The accumulation of fallen leaves under trees is also an important source of organic matter and minerals for the associated pastures because the deeper roots of trees pump nutrients to the soil surface where grass roots benefit.

From an animal welfare perspective, trees and shrubs in pastures provide shade and comfort to the animals; this issue is particularly important in the tropics, where ambient temperatures often exceed 40 °C. A recent study in the dry tropic region of Michoacán, Mexico, found that the presence of trees in a SPS contributed to reduce the environmental temperature by 6 °C, which in turn reduced external body temperature by 2 °C, number of breaths per minute by 20 times (from 60 to 40), and the number of water drinks during the day from 17 to 12 times, favoring in overall a better animal welfare and productive performance in growing cattle (Utrilla 2013). Galindo et al. (2013) found a reduction of 4 °C in skin temperature of the cattle in a SPS compared to the cattle in a monoculture system. The tree components in the SPS are also important as barriers to the wind and represent shelter for cattle during rainfall. The trees also allow the animals to hide from other perceived dangers (Broom and Fraser 2007). The SPS contributes to the biodiversity by increasing insects, and the number of birds, which are natural predators of ticks and other ectoparasites of cattle. Significant decreases in the incidence of tick-related diseases such as anaplasmosis were reported in SPS in Colombia (Murgueitio and Giraldo 2009). On the other hand, the constant management of livestock in the SPS makes animals more docile and manageable, decreasing nervousness and stress (Blokhuis et al. 2003; Ocampo et al. 2011); this benefits animal productivity and thus farmer income (Fig. 8.2).

(a) (b)

Fig. 8.2 Silvopastoral systems provide good-quality forages and animal comfort under tropical conditions

8.7 Potential of Silvopastoral Systems for Meat and Milk Production in the Tropics

Feeding of cattle (beef, dairy) in tropical regions is based on the grazing of native and introduced pasture species. During the dry season, dry pasture is available which contains a low concentration of CP, high concentration of neutral detergent fiber (NDF), low apparent digestibility, and thus low metabolizable energy (ME) concentration. Under such conditions, DM consumption of ruminants is reduced due to the lack of rumen fermentable nitrogen; thus, the ME requirement for maintenance and production cannot be covered, leading to negative energy balance and weight losses, delaying the time (in months) for growing cattle to reach slaughter weight (470 kg). The implementation of agroforestal practices, such as silvopastoralism, allows the integration of trees and shrubs with animal production. With this model, a more rational approach to production can be developed with a lower impact on the ecological balance that can also improve animal performance (live weight gain, milk yield), as well as the quality of the products of animal origin and profitability (Nahed-Toral et al. 2013; Ferguson et al. 2013). Barros et al. (2012) reported that *L. leucocephala* foliage contains 29 % of CP and that its biomass yield is constant throughout the year. The incorporation of tree legumes such as *L. leucocephala* in silvopastoral systems is an alternative for increasing meat and milk production of ruminants, since they supply nutrient-rich forages which are essential for the growth of animals. Barros et al. (2012) obtained moderate live weight gains in Pelibuey sheep grazing 35,000 and 55,000 leucaena plants: 106 and 81 g/head/day, respectively. In the valley of Tepalcatepec, Michoacán, Mayo Eusebio (2014) in studies with cattle grazing a silvopastoral system associated with *P. maximum* var. Tanzania and 30,000 plants of *L. leucocephala* per hectare registered 765 g/head/day live weight gain and a stocking rate ranging from 1.4 to 3.5 animal units (AU) per hectare, and this weight gain is comparable to that reported by Shelton and Dalzell (2007) in Australia, with

Table 8.5 Milk production and composition at different stocking rates of lactating dual-purpose cows grazing intensive silvopastoral systems in the dry tropical region of Michoacán, México

Farm name	Stocking (SR)	Land (ha)	SR (AU/ha)	LW (kg/ha)	Milk composition		
					Protein	Fat	Lactose
Vivero	High	4.4	4	1800	3.1	4.6	4.2
Uricho	Medium	11	3	1350	2.9	3.6	4.1
Semillero	Low	4.5	2	900	3.1	3.5	4.5

AU animal unit = 450 kg LW

similar production systems. Studies carried out in Colombia by Mahecha et al. (2012), when comparing SPSs with degraded pastures, found that the amount of meat produced per ha increased from 74 to 1060 kg per year.

Ruiz-González (2013) found a milk yield of 7.7 kg/day in crossbred cows fed a ration consisting of 45 % leucaena foliage and 55 % chopped *P. purpureum* grass (DM basis) and supplemented with 2 kg/day of ground maize, while Arjona-Alcocer (pers. comm.) found no effect of the type of carbohydrate (sorghum, citrus by-product, cane molasses, or rice polishings) on milk yield (4–5 kg/day) of crossbred cows fed a ration of 45 % leucaena foliage and 55 % chopped *P. purpureum* grass (DM basis), although the cows were in mid-lactation. Bottini-Luzardo (pers. Comm.) found that DM intake of crossbred cows grazing a silvopastoral system (*Leucaena* + *Cynodon nlemfuensis*) and producing 10 kg milk per day was 11. 9 kg per cow per day, with an intake of *Leucaena* (DM) being one-third of that amount. Under silvopastoral systems, animals are able to eat huge amounts of a good-quality fodder and consequently produce more milk or reach better live weight gain. Leucaena in association with grasses increases feed intake, and the inclusion of up to 40 % *Leucaena* in the diet increases forage intake by 15–25 %. Table 8.5 shows the results from different stocking rates on the milk production (kg LW/ha), and the high stocking rate (4 unit animal/ha) increased the milk by 900 kg on average per lactating period (Fig. 8.3).

(a) (b)

Fig. 8.3 Dual-purpose cows grazing silvopastoral systems and hand-milking process

8.8 Conclusions

Beef and dairy livestock production from tropical pasture-based systems is of low productivity. The protein content of tropical pastures decreases rapidly as growth progresses, whereas the protein in shrub/tree species is maintained. The deficiency of CP in pasture can be improved by the use of tropical N-fixing legumes and other tree species that improve both production and nutrient cycling and eliminate the need for chemical N fertilizer. Silvopastoral systems provide several advantages for animal production in comparison with systems where feed comprises cereals and other grains rich in protein and energy. SPSs have a positive effect on the environment, but the most important aspect from the livestock production point of view is the direct benefit on animal performance under tropical conditions. SPSs are able to increase yield and fodder quality with the minimum use of external inputs. Animals can forage all day, and dairy or bull calves can be fed with more than 80 % of the forage grown naturally. Reconversion of tropical monocrop grasses to silvopastoral systems may be highly profitable, especially due to the high-quality forages produced at low cost, in an environmentally friendly manner.

Acknowledgments This publication was partially supported by the Fundacion Produce Michoacan (Grant number 2013-0012 and 2012-0003) and the SAGARPA-CONACYT (Grant Number 2011-14-175169).

References

Barros-Rodríguez M, Solorio-Sánchez J, Ku-Vera J, Ayala-Burgos A, Sandoval-Castro C, Solís-Pérez G (2012) Productive performance and urinary excretion of mimosine metabolites by hair sheep grazing in a silvopastoral system with high densities of Leucaena leucocephala. Trop Anim Health Prod 44(8):1873–1878

Barros-Rodríguez M, Solorio-Sánchez J, Sandoval-Castro C, Klieve AV, Briceño-Poot E, Rojas-Herrera R (2013) Effects of two intake levels of Leucaena leucocephala on rumen function of sheep. Tropical Grasslands-Forrajes Tropicales 1(1):55–57

Blokhuis HJ, Jones RB, Geers R, Miele M, Veissier I (2003) Measuring and monitoring animal welfare: transparency in the food product quality chain. Anim Welfare 12:445–455

Broom DM, Fraser AF (2007) Domestic animal behaviour and welfare, 4th edn. CABI, Wallingford

Chaudhri R, Timmer CP (2003) The impact of changing affluence on diet and demand patterns for agricultural commodities. World Bank working paper # 785

Cruz P (1997) Effect of shade on the growth and mineral nutrition of a C_4 perennial grass under field conditions. Plant Soil 188(2):227–237

Delgado C (2003) Rising consumption of meat and milk in developing countries has created a new food revolution. J Nutr. Supplement 2907S–3910S

FAO (2002) World agriculture towards 2015/2030. FAO, Rome

FAO (2006) Livestock's long shadow—environmental issues and options. Food and Agriculture Organization, Rome

FAO (2012) World agriculture towards 2030/2050. ESA Working Paper No. 12-03. Food and Agriculture Organization, Rome

FAO (2014) http://www.fao.org/livestock-environment/en/. Accessed 19 Sept 2014

FAOSTAT (2015) Food and agriculture organization of the United Nation, Statistics Division. http://faostat3.fao.org/download/FB/CL/E
Ferguson BG, Diemont SAW, Alfaro-Arguello R, Martin JF, Nahed-Toral J, Alvarez-Solís D, Pinto-Ruiz R (2013) Sustainability of holistic and conventional cattle ranching in the seasonally tropics of Chiapas, Mexico. Agri Syst 120:38–48
Finch E (2014) Suppliers struggle to meet demand for popular organic milk. CBS New York Business, http://newyork.cbslocal.com/2014/05/27/suppliers-struggle-to-meet-demand-for-popular-organic-milk/ Accessed 18 Sept 2014
Fukase A, Martin W (2014) Who will feed China in the 21st Century? Income growth and food demand and supply in China. The World Bank Development Research group Agriculture and rural development team. Policy Research Working Paper 6926
Galindo F, Olea R, Suzán G (2013) Animal welfare and sustainability. International Workshop on Farm Animal Welfare, São Paulo. http://www.workshopdebemestaranimal.com.br/indexen.html
Garnett T (2009) Livestock-related GHG emissions: impacts and options for policy makers. Environ Sci Policy 12:491–503
Gerbens-Leenes PW, Nonhebel S (2002) Consumption patterns and their effects on land required for food. Ecol Econ 42(1):185–199
Gerbens-Leenes PW, Nonhebel S (2005) Food and land use. The influence of consumption patterns on the use of agricultural resources. Appetite 45:24–31
Hogh Jensen H (2011) To meet future food demands we need to change from annual grain legumes to multipurpose semi perennial legumes. In: Food production approaches, Challenges and task, Tech-open access Publisher
Ibrahim M, Villanueva C, Mora J, Mosquera-Losada MR, Rigueiro-Rodríguez A, McAdam J (2005) Traditional and improved silvopastoral systems and their importance in sustainability of livestock farms. In: Silvopastoralism and sustainable land management. Proceedings of an international congress on silvopastoralism and sustainable management held in Lugo, Spain, April 2004, CABI Publishing, pp 13–18
Jank L, Valle CD, Resende RMS (2005) Grass and forage plant improvement in the tropics and sub-tropics. Grassland: a global resource. Wageningen Academic Publishers, Wageningen, pp 69–81
Kahn D (2014) As California record drought strengthens, organic milk faces financial crisis. Climate ware, http://www.scientificamerican.com/article/as-californias-record-drought-strengthens-organic-milk-faces-financial-crisis/ Accessed 18 Sept 2014
Lustig RH, Schmidt LA, Brindis CD (2012) The toxic truth about sugar. Nature 482:27–29
Mahecha L, Murgueitio M, Angulo J, Olivera M, Zapata A, Cuartas C, Naranjo J, y Murgueitio E (2012) Ceba de bovinos doble propósito pastoreando en sistemas silvopastoriles intensivos. In: Solorio-Sánchez FJ, Sánchez-Brito y C, Ku-Vera J (eds), Memorias del IV congreso internacional sobre sistemas silvopastoriles intensivos. Morelia, México, Fundación Produce Michoacán, Universidad Autónoma de Yucatán
Mayo Eusebio R (2014) Ganancia de peso de bovinos en crecimiento en un sistema silvopastoril intensivo y en un sistema en confinamiento en Michoacán, México, MSc. Thesis in Agricultural Sciences. Faculty of Veterinary Medicine and Animal Science, University of Yucatan, Mexico
McMichael A, Powles JW, Butler CD, Uauy R (2007) Energy and health 5. "Food, livestock production, energy, climate change and health". Lancet 370:1253–1263
Murgueitio E, Giraldo C (2009) Sistemas silvopastoriles y control de parásitos. Revista Carta Fedegán 115:60–63
Murgueitio E, Ibrahim M (2008) Ganadería y medio ambiente en América Latina. In: Murgueitio E, Cuartas CA, Naranjo JF (eds) Ganadería del Futuro. Fundación CIPAV, Cali, pp 19–39
Nahed-Toral J, Valdivieso-Pérez A, Aguilar-Jiménez R, Cámara-Cordova J, Grande-Cano D (2013) Silvopastoral systems with traditional management in southeastern Mexico: a prototype of livestock agroforestry for cleaner production. J Clean Prod 57:266–279

Ocampo A, Cardozo A, Tarazona A, Ceballos M, Murgueitio E (2011) La investigación participativa en bienestar y comportamiento animal en el trópico de América: oportunidades para nuevo conocimiento aplicada. Revista Colombiana Ciencias Pecuarias 24:332–346

Rosegrant MW, Valmonte-Santos RA, Cline SA, Ringler C, Li W (2005) Water resources, agriculture and pasture: implications of growing demand and increasing scarcity. In: Grassland: a global resource. Proceedings 2005 International Grassland Congress, vol 26. Dublin, pp 227–249)

Ruiz-González A (2013) Balance de nitrógeno y composición de la leche de vacas alimentadas con *Leucaena leucocephala*. MSc Thesis in Agricultural Sciences. Faculty of Veterinary Medicine and Animal Science, University of Yucatan, Mexico, p 63

Salman M, New Jr JC, Bailey M, Brown C, Detwiler L, Galligan D, Hall C, Kennedy M, Lonergan G, Mann L, Renter D, Saeed M, White B. Zika S (2008) Global food systems and public health: production methods and animal husbandry. A National Commission on Industrial Farm Animal Production Report, Pew Commission on Industrial Farm Animal Production. Comparative Medicine Publications and Other Work, http://trace.tennessee.edu/utkcompmedpubs/32 Retrieve 1 Oct 2011

Sarabia-Salgado L (2013) Efecto de la frecuencia de poda en *Leucaena leucocephala* y *Panicum maximum* sobre la fijación y transferencia de nitrógeno atmosférico en sistemas silvopastoriles intensivos. MSc. Thesis in Agricultural Sciences. Faculty of Veterinary Medicine and Animal Science, University of Yucatan, Mexico

Shelton M, Dalzell S (2007) Production, economic and environmental benefits of leucaena pastures. Trop Grasslands 41:174–190

Suttie JM, Reynolds SG & Batello C (2005) Grasslands of the World. Plant production and protection, 34. FAO. Rome, p 514

Thornton PK (2010) Livestock production: recent trends, future prospects. Phil Trans R Soc 365:2853–2867

United Nations Environmental Program (2005) Life cycle approaches. The road from analysis to practice, UNEP/SETAC, Life cycle initiative. 89 pp

Utrilla-García MP (2013) Conducta ingestiva e indicadores fisiológicos de toros en crecimiento en un sistema silvopastoril con Leucaena leucocephala asociado con Panicum máximum. MSc. Thesis in Agricultural Sciences. Faculty of Veterinary Medicine and Animal Science, University of Yucatan, Mexico

White T (2000) Diet and the distribution of environmental impact. Ecol Econ 34(234):145–153

World Health Organization (2014) Global status report on alcohol and health-2014.

Chapter 9
Role of Legumes for and as Horticultural Crops in Sustainable Agriculture

Matthew W. Blair, Xingbo Wu, Devendra Bhandari, Xiaoyan Zhang and Junjie Hao

Abstract Legumes are the second largest plant family on earth and arguably the second group of importance to current and past agricultural systems and human nutrition. Despite differences among legumes, their variability as early, medium, to late maturity annual crops that fix nitrogen and survive shading by larger adjacent plants, makes them very versatile in agronomics and horticultural cropping systems. In this chapter, we describe the importance of four vegetable legumes (garden peas, purple-hulled peas, snap beans, and yard-long beans) and a range of more minor legume crops as vegetables in today's world. Each crop is highlighted for its value in the local diets of peoples of different regions and the cropping systems to which they belong. We follow this by providing a large number of examples where vegetable and non-vegetable legumes can be used as intercrops between cereal crops such as corn or sorghum, between vegetables from the tomato/pepper and eggplant or cabbage/broccoli and cauliflower family or fruit tree seedlings and saplings that are being established. All of this shows that legumes are an amazingly diverse group of vegetable species which are advantageous to intensive horticultural systems.

Keywords Vegetable legumes · Garden peas · Snap beans · Intercropping

M.W. Blair (✉) · X. Wu · D. Bhandari
Department of Agriculture and Environmental Science, College of Agriculture, Human and Natural Sciences, Tennessee State University, 3500 John a Merritt Blvd., Nashville, Tennessee 37209 USA
e-mail: mblair@tnstate.edu; mwbeans@gmail.com

X. Zhang · J. Hao
Qingdao Academy of Agricultural Sciences, Qingdao 266100, Shandong, China

D. Nandwani (ed.), *Organic Farming for Sustainable Agriculture*, Sustainable Development and Biodiversity 9,
DOI 10.1007/978-3-319-26803-3_9

9.1 Introduction

Among the oldest group of crops to be domesticated, the vegetable and pulse legumes are multi-functional plant species useful for many horticultural settings and agronomic rotations (Lewis et al. 2005). The legume family is the second largest family of plants (Cannon and May 2009; Gepts et al. 2005; Weeden 2007; Young and Mudge 2003), with many species that have been successful in natural adaptation across many regions of the world.

Each Vavilovian center (Hummer and Hancock 2015) of origin for agriculture has a major legume or a suite of major and minor legumes which complement the production of cereals, such as corn, rice, and wheat (Teshome and Brown 2001). The Papillinoid subfamily has given rise to the majority of pulses and fresh peas and beans, while the two other subfamilies of legumes are only known for a large number of lumber or forestry species along with a few fruit crops (Sandhu and Singh 2007; Dwivedi et al. 2006). This review will concentrate on the horticultural legumes or a few grain legume species that are currently in cultivation with other vegetable crops.

Legumes can be consumed at physiological maturity, dried down like cereal grains but also in a fresh form (Kay 1979; Nielsen and Ohler 1997). Therefore, unlike cereals, legumes are often consumed green at immature physiological stages. Growing legume plants are often harvested for their fresh pods, fresh green seeds, and even leaves. This means that legumes are often horticultural crops early in their growth stages, but also serve as fully formed pulses (food legumes) when harvested dry at maturity. Because of this characteristic for many legume species, there is a horticultural and corresponding agronomic crop.

Take as examples, the garden (immature) and the field (mature) pea (*Pisum sativum* L.), the green (immature) and dry (mature) bean (*Phaseolus vulgaris* L.), the green versus white lima bean (*P. lunatus* L.), the fresh versus roasted faba bean (*Vicia fava* L.), or the yard-long (*Vigna unguiculata* L (Walp) var. *sesquipedilis*), and dry grain (*V. u.* var. *unguiculata*) cowpeas (Rubatzky and Yamaguchi 2012).

Unlike cereals, legumes are for the most part not consumed as flour or malt; although legume can be ground to a fine protein-rich flour which can be mixed with cereal flour and used in flat breads, chapatis or to enrich leavened breads (Shehata et al. 1988). Small grain cereals cannot be eaten fresh because of hard hulls around their kernels although they can be fermented.

Legumes due to their high protein are not normally fermented but can be extracted for a high protein, cheese like tofu (Messina 1999). Cereals meanwhile have low protein levels and cannot be turned into tofus but can produce starchy drinks. Among the cereals, only sweet corn is a major vegetable since the corn cobs and their sweet kernels on sweet corn ears can be consumed as a fresh vegetable when the leafy husks are removed (Tracy and Hallauer 1994).

In another difference with cereals, the legumes are nitrogen-fixing crops that can add to soil fertility thus increasing the productivity of adjacent or subsequent crops (Heichel and Helsel 1987). Legumes increase soil-nutrient availability and soil resource pools through host microbe interactions with bacterial Rhizobia species and Mycorhizal fungi which assist in the fixation, uptake, and utilization of nitrogen and phosphorus, respectively (Young and Johnston 1989). Given their access to nitrogen fixation and phosphorus uptake through plant symbiosis with bacterial and fungal species, legumes are a critical tool for the creation and maintenance of sustainable agricultural systems (Bohlool et al. 1992; Peoples and Ladha 1995).

The dedication of significant carbon resources that legumes make to their rhizospheres in the form of organic acids, root, and nodule biomass or nutrient uptake mechanisms benefits the roots of intercropped plants often even more than the legume itself. The smaller harvest index of legumes compared to cereals means that legumes can provide significant carbon and organic matter to the soil even compared to the higher biomass in cereals (Izaurralde and McGill 1992; Mucheru-Muna et al. 2010; Ngwira and Aune 2012).

In addition, legumes provide a valuable balance of nitrogen and other nutrients compared to carbon levels allowing microbial degradation to function to its greatest extent (Harinikumar et al. 1990). The result is high levels of nitrogen and phosphorus elements in the organic matter remaining in soils that have been planted to legumes. Recent studies show that phytates, for example, accumulate in the legume root zone and are available to bacterial phytase enzymes for release as free phosphorus over time (Hinsinger et al. 2015; Trouillefou et al. 2015).

Nodules that slough off of legume roots are a rich source of nitrogen, iron, and phosphorus for an intercrop or for a rotation crop in a subsequent season. The multi-branched root systems of legume species, many of which are deep rooting and high in organic matter investments, break up soil layers, and soil conglomerates increasing the porosity and water holding capacity of soils (Cong et al. 2015).

More than many other plant families, the legumes create a flourishing community underground microbial (Sugiyama and Yazaki 2012). Even in flooded situations, the advantages of a previous legume crop can assist the following crop to grow. It is no wonder that legumes have been a pillar of sustainable agriculture for centuries and are planted in almost any agro-ecosystem around the world that relies on a permanent farm site. The legume-cereal relay system, intercrop, or rotation is the historical basis for many civilizations starting with the oldest in Eurasia and near the Fertile Crescent (Buddenhagen 1990).

China, with its farmers of forty centuries, has relied on wheat or rice followed by mung bean, rice bean, and soybean to sustain a large farming population across a huge area of rugged mountains, vast valleys, steppes, and coastal plains and develop a preeminent unified state. Europe also has grown wheat in rotation with peas, faba beans, or forage legumes for many years.

The Indian Subcontinent developed a range of legumes such as the black, green, and red grams (*Vigna* spp.) or pigeonpea (*Cajanus cajan* L.) to follow rice and wheat in large areas of irrigation from the Indus and Ganges rivers.

Middle Eastern societies very early on developed wheat and barley crops that rotated with chickpeas (*Cicer arietum*), lentils (*Lens culinaris* L.), and peas (*Pisum sativum*). This cropping system spread around the Mediterranean through North

Africa and across the Bosporus during the time of Phoenician and Roman civilizations (Kassam et al. 2009).

In sub-Saharan Africa, cowpea was intercropped with millets and sorghum in the Sahel and along the rift valley and great lakes regions all the way to Southern Africa (Mortimore et al. 1997). Yams (*Dioscorea* spp.) may also have been intercropped with cowpeas, as this is a widely adaptable crop (Amusa et al. 1996).

Societies and civilizations of what we now know as Latin America, including the Aztec, Inca, and Maya developed and perfected the corn and common bean intercrop to a large extent (Abawi and Corrales 1990). In the high Andes Mountains, potatoes and other indigenous root and tuber crops were rotated with lupine species, while along the South American coast lima beans were grown together with sweet potatoes (Nieto-Cabrera et al. 1997). In the Bolivian Chaco, peanuts were domesticated and throughout the Brazilian hinterlands peanuts, and common beans were common together or following cassava production.

The North American "Three Sisters" system was based on common bean for nitrogen fixation, corn for cereal grain and stalks on which the common beans twined together with squash to suppress weeds (Ngouajio et al. 1997).

After the great Columbian agricultural exchange, the crops of the New World became established in the Old World and vice versa, expanding the repertoire of legumes available to societies around the world in a very sudden and abrupt manner. The result was the spread of corn–common bean intercrops into Europe (Santalla et al. 2001), Eastern and Central Africa (Blair et al. 2010b) and parts of middle and western China (Zhang and Blair 2008). Peanuts made the exchange with the crop becoming an extremely important food in Africa.

Reciprocally, cowpeas arrived from Africa and became established in the New World, especially southeastern USA and spreading to the Caribbean (Hummer and Hancock 2015). Peas came to North America with the European colonists. During World War II, soybean became an established crop of the USA and since then has spread to South America especially Argentina and Brazil in the 1970s (Fushan 1994), while lentils and peas have become important in the great plains provinces of Canada since about the same time (Sandhu and Singh 2007).

9.2 Legumes as an Intercrop in Organic Agriculture

Almost all civilizations over time have depended on legume crops in combination with cereals and sometimes with roots and tubers, to provide for a sustainable rotation or intercrop. From three sisters corn–bean–pumpkin intercrops to rice–mung bean relays, traditional agriculture has relied on a range of over 10 or more legume crops. Unfortunately, these cropping systems have been under threat since the introduction of inorganic fertilizers which substitute the nitrogen fixation of legumes. This chapter will review actual and potential agro-ecosystems that use or obtain the benefits of legumes.

The advantages of growing legumes amidst horticultural crops include nitrogen fixation, insect pest reduction, disease prevention in both the legume and the alternative crop as well as increases in economic yield per area (Majumdar 2011). Therefore, legumes should be considered as an intercrop or relay crop in most farming systems from organic farms to traditional monocropping farms that rotate across seasons in non-organic farming systems.

This chapter will have a global reach but will emphasize agricultural rotations and intercrops that work well in the climate of the southeastern USA and middle China, respectively, some of the newest and oldest agricultural lands in the temperate world. Applications to African, Australian, European, Latin American, and other Asian agricultures will be discussed, but the plethora of possible crops precludes a detailed discussion of these systems. While we will concentrate on major agricultural systems and horticultural legumes, even new agro-ecosystems such as beans grown with tomatoes and their relatives have the potential to be the principal method of growing solanaceous vegetables.

In our writing, we have centered the discussion in this chapter on the production of garden peas, snap beans, and yard-long beans as examples of horticultural legumes along with potential systems for intercropping sweet corn, tomatoes, peppers, and subtropical or exotic vegetables with any of these legumes. Other fresh vegetable legumes include the bean tree (*Erythrina edulis*) in the neotropics, the pigeonpea in the Indian subtropics, or faba beans in temperate zones; however, these are either not very common compared to snap peas, cowpeas and beans or alternatively consumed mostly for their dry seeds.

A few legume species, such as the Tamarind and Inga species, are eaten for their sweet pod linings instead of the pod wall or the seed itself, but these are more fruits than vegetables and will not be discussed further. Where appropriate we discuss the application of legume relays, rotations, and intercropping to organic production of both horticultural legumes and other vegetable crops. We also mention non-organic systems of production that are intensively based on legume, especially those based on legumes as horticultural crops as described below.

9.3 Types of Horticultural Legumes

9.3.1 Cowpeas or Purple-Hulled Peas

Cowpeas are mostly consumed as dry grain, but fresh pods and seed from them are known as purple-hulled peas which are a popular dish in the southern USA and parts of West Africa such as Senegal. In addition, fresh or dried leaves (in many parts of Asia and Africa) (Ahenkora et al. 1998), and fresh green pods (humid regions of Asia and in the Caribbean) are of importance in some localities or agro-economic/cultural situation (Ehlers and Hall 1997). In the USA, purple-hulled peas are also known as southern peas and were brought to the New World through exchanges between West Africa, the Caribbean, and the America.

The fresh consumption of green-shelled cowpeas became common place in the southern states and many varieties were named and new dishes created. In this case, the pods are picked and then manually or mechanically opened to shell out the immature or almost-mature seed. Since the pods are often a dark purple color, these fresh peas are called purple-hulled peas. In the south, purple-hulled peas have certain quality characteristics for the southern market that are considered better than most western grown California Black-Eyed peas, especially for canning. Splitting of the dry grain before canning is a concern for desert-grown seed but not for grain grown in the south that is processed quickly. A stew of southern peas is known as "Hopping John" in many southern states and is traditional on New Year's Day.

Varieties of cowpeas grown in the USA are often classified into black-eyed pea types (white seed coats with black hilum ring) or crowder peas, which are squarer rather than oblong because of their "crowding" in the pods. Crowder types are usually light yellow, tan, or cream in color. Black-eyed peas are usually light cream in color but vary by size. Some red or black cowpeas are also found but usually are not used as green-shelled peas.

Although low in consumption for North American consumers, cowpeas are still a very important food legume for traditional diets and cropping systems in the semiarid tropics (Benchasri and Bairaman 2010). They are grown in a wide area covering Sahelian West Africa, Eastern and Southern Africa along the rift valley and beyond into Botswana, Mozambique, and Namibia. Cowpeas are also grown in parts of Southern Asia and certain countries of Central and South America, such as Colombia, Haiti, and Venezuela (Mortimore et al. 1997; Singh and Chambliss 1997; Van Ek et al. 1997).

World cowpea production was 8,336,226 tons in 2012, and the top 5 producing countries were Nigeria (5,146,000 tons), Niger (1,329,514 tons), Burkino Faso (598,524 tons), Myanmar (180,000 tons), and Tanzania (179,570 tons). Based on FAO data (http://www.fao.org) and feedback from national programs in Sub-Saharan Africa, the estimated area worldwide under cowpea production is about 14 million ha. Average yields vary per year but generally have been low due to a lack of new varieties and the marginal conditions in which the crop is grown. For example, world averages were 732 kg/ha in 2012 but lower at 594 kg/ha in 2010, 522 kg/ha in 2013 and 515 kg/ha in 2009. Yield in the USA is reaching closer to potential yields of over 2,500 kg/ha.

Intercropping of black-eyed peas with corn and other crops in the Southern USA was traditional up to the 1940s. Cowpea roots are excellent for soil fertility management because of high rates of nitrogen fixation and good symbiotic properties with fungal mycorrhiza (Kwapata and Hall 1985). Furthermore, they tolerate a high range of pH in the soil as compared to other legumes and can be grown on marginal lands of low fertility or very sandy soils (Fery 1990).

9.3.2 Garden Peas

The garden pea is consumed as a vegetable and the field pea as a pulse crop. Together this species is the fourth most important legume in the world by volume of consumption (Zheng et al. 1997). The cultivated pea is an annual plant, with a life cycle of one season which can vary from early peas that mature in 75 days to late peas that take twice as long (Zheng et al. 1997). Peas are a cool season crop grown in many parts of the world but mainly in the temperate zones or at high altitudes in the tropics (e.g., in the Andes of South America) and subtropics (e.g., in the Himalayas).

The planting of peas in tropical regions can be at any time of the year when rainfall is adequate. In temperate regions, pea planting can take place from winter to early summer depending on the location and severity of winter conditions especially the minimum night temperatures and prevalence of wind chills (Lambert and Linck 1958; Smith and Herath 1973).

Many varieties of peas are grown primarily for their quality as a fresh vegetable (Edwards and Lee 1986). Immature pea seeds can be consumed boiled as a green vegetable as they develop chlorophyll early on during seed development. This allows them to be harvested early when the seed cotyledons are still tender and easy to cook. These young seeds are generally known as garden peas and can be either sweet or non-sweet. The immature peas are also known in some countries as snow peas because they can be harvested late in the fall.

The consumption of fresh immature green peas is popular in North America and Western Europe especially, notably in France where "petit pois" became a dish for all social classes. Today, the garden pea is found throughout all of Europe and North America and is popular in some parts of South America and Africa. Garden peas are a very common vegetable in many Asian countries becoming a staple for vegetarian dishes in India and Nepal as well as a major ingredient in China for stir fries and soups. In all these setting, the garden pea can provide abundant protein, dietary fiber, and vitamins to people.

In addition to the fresh green seed serving as a vegetable, the tender pod is also used as a vegetable. The pods can be harvested before the seed has swelled the pod wall or at slightly later maturity stages. These pods can be steamed, blanched in butter, or stir-fried in hot oil. Snap peas are low in fiber and easily eaten without a need for chewing (de Almeida Costa et al. 2006). This quality makes them like some snap beans, lima beans or yard-long beans, and other stringless legumes which can be eaten raw, steamed, or lightly boiled to consume as a typical vegetable.

The consumable products of garden peas whether fresh seed or snap pea pods can be eaten fresh in salads, boiled as a cooked vegetable or processed into cans. In addition, they can be cooked and frozen, making garden peas a major industrial vegetable useful in commercializing as imports or exports. When canned, garden peas can be found all over the world including off season and environments that otherwise would not have fresh peas available.

The pea plant's origin and spread trace the establishment of ancient civilizations around the Mediterranean Sea (Zheng et al. 1997). The earliest archaeological finds of peas date from the late Neolithic and were in the area of current-day Greece, Syria, Turkey and Jordan. Cultivated peas were domesticated from wild peas which are restricted to the Mediterranean basin and the Near East. In Egypt, early archeological findings date pea cultivation to ca. 4800–4400 BC in the Nile delta area, and from ca. 3800–3600 BC in Upper Egypt. The pea was also present in Georgia in the 5th millennium BC. In the second half of the 2nd millennium BC, this pulse crop appears in the Gangetic basin and southern India (Zheng and Wang 1997).

The spread of peas throughout Europe probably coincided with increasing production of cereals among early farmers in that region. Today, according to FAO, there are 97 countries in the world that produce field peas and 81 which produce garden peas. In total, 15 million acres produce a total of 10 million tons of field peas and 5.5 million acres produce 16.9 million tons of garden peas. The highest pea production statistics in the decade from 2001 to 2011 are from Canada, Russia, India, China, and Australia. The garden pea is one of the major vegetable crops spread throughout China and adapted to many agricultural niches across a large range of environments from north to south.

Crop rotation model of wheat-pea, maize-pea, and oat-pea are very popular in many parts of China and parallel early systems in Europe. Like other legume species, pea has the nitrogen fixation trait which enhances its role in sustainable agriculture. Thus, pea has an important role in food supply and sustainable agriculture for these two areas.

As a well-adapted legume, the pea can be planted from temperate zones in northern China to subtropical high land area in southern China. Landraces are divided into those for summer production in the north and winter production in the south since peas need a typical cold season crop. The southwest Chinese province of Sichuan, with high humidity and cool temperatures, is among the best planting environment for peas in Asia.

9.3.3 Snap Beans

Snap beans are one of the most important and commonly consumed horticultural products in the world (Myers and Baggett 1999). Unlike dry beans, snap beans are consumed for their edible, whole pods rather than dry seeds. They, therefore, have been selected as a vegetable to have succulent pod walls and low pod wall fiber. In the case of "stringless" beans, very little fiber is found at the pod suture as well (Silbernagel 1986). Synonyms of snap beans are "French bean," "Garden bean," or "Haricot bean," while "String bean" refers to many older varieties that have fiber at the pod suture but that are still consumed as a vegetable where the strings are removed manually before cooking. Snap beans are important sources of essential vitamins (A, B12 and C) and dietary fiber but are low in calories, which make them a healthy food.

Total world production of snap beans is around 9 million tons, with China, Turkey, India, Spain, France, and the USA being among the biggest producers and consumers. Marked preference and intense commercialization of snap beans occur in developed countries of North America and Europe with many seed and food processing companies intensively involved in the product chain (Silbernagel 1986).

Snap beans are of growing importance to developing countries, both as an export crop and as a local product (Henry and Jansen 1992). In terms of export, trade between Central America and the USA or East Africa and Europe produce important income streams for countries like Guatemala and Kenya. Meanwhile, as wages have gone up in countries such as Colombia or India, the markets for snap beans have also increased (Pachico 1987). The demand for a constant supply of fresh beans means that production in the northern hemisphere migrates with the seasons: winter or spring production is found in the very southern edges of Europe and North America (e.g., Florida, Mexico, or Spain) followed by summer production for fresh market and processing further north in both hemispheres (Henry and Jansen 1992). The production system in the USA is particularly mobile from south to north (Myers and Baggett 1999).

Snap beans require a short season of only sixty days to produce a good harvest of fresh pods, but do not resist drought as well as dry beans. Therefore, most snap beans in the USA are grown in rainier sections east of the Mississippi rather than in the west. Irrigated snap beans are grown in Mexico and California. Snap bean seed is usually purchased new by farmers every year because saved seed from vegetable production areas has poor germination and may carry disease. The harvest of the pods at an early reproductive growth stage precludes the seed saving.

Like other sorts of common beans, the primary center of diversity for snap beans is believed to be in the Americas; however, snap beans have a wide distribution and are very diverse in various regions around the world especially in Europe (Métais et al. 2002) and Asia (Zhang et al. 2008). Snap beans are thought to have been mainly selected for in Europe (Myers and Baggett 1999), but perhaps this theory is based on the lack of an archeological record for green bean pods in the Americas. Preservation of fresh pods in the dig sites was impossible even while dry seeds of common bean are found in early agricultural sites in a long arc from the USA through Mexico and Central America to the Andes (Kaplan and Kaplan 1988).

Therefore, it is thought that pre-Colombian societies of the Americas consumed only a few fresh green pods. Further to this issue of snap bean selections being from outside or inside the Americas, indigenous terms for snap beans exist in some Amerindian languages, which may show a long-term knowledge of snap beans that was not introduced or re-introduced as modified dry bean germplasm from outside the region. For example, in Quechua, the term Chaucha (interpreted as tender or light green/yellow) is used to refer to snap beans and show a probable original use of green pods in the Andean region.

A common misconception has been that most snap beans were derived from the Andean gene pool of common beans, one of two major gene pools of common bean, given that early analysis showed that bush-type snap beans generally had Andean-type seed protein (phaseolin) patterns (Brown et al. 1982; Gepts et al. 1986). However, some studies have suggested that certain groups of snap beans, such as the blue lake series, are actually from the Mesoamerican gene pool, the other major gene pool of common bean (Skroch and Nienhuis 1995; Cunha and Hintz 2004; Davis and Myers 2002; Blair et al. 2010a).

Various classes of snap beans exist such as large-sieve green beans, wax types, flat-podded Romano types, and fine-sieve types. Snap beans are especially diverse in Europe where pod types go from extra-fine to round with variability in pod color as well as other characteristics (Métais et al. 2002). In addition to their diversity for pod types, snap beans are very diverse in growth habit and plant ideotype, with type I erect bush beans to type IV (indeterminate) pole or climbing beans (Myers and Baggett 1999).

In the tropics, snap bean characteristics are somewhat less developed than in temperate production zones, and most genotypes are indeterminate climbing beans, except when export quality is required in which case imported bush bean varieties are produced. In temperate regions, pole bean-type snap beans are grown in home gardens but are not used for extensive mechanized production and the majority of snap beans are of bush type. Bush-type snap beans predominate in Europe and North America but are less well developed in China where snap beans are often climbing types.

9.3.4 Yard-Long Beans

Yard-long bean is characterized by its very long succulent pods of 30–90 cm in length (Verdcourt 1970). Unlike cowpea to which it is related as a subspecies, yard-long bean is grown exclusively as a vegetable and only produces seeds that are very narrow being about 2–3 mm wide and usually 8–12 mm long, with corresponding low 100 seed weight of 8–12 g. In comparison, cowpea grown for grain and fodder has large seed that has up to 20 g per 100 seed weight and is usually kidney shaped, round or square.

The origin of the yard-long bean as a subselection of cowpeas is uncertain as to geographical source. One possibility is that yard-long beans were selected from dry cowpeas with longer and longer pods over many generations when cowpeas were taken from Africa to Asia during trans-oceanic trade across the Indian Ocean. Worldwide production of yard-long bean is now centered in Eastern, Southern and Southeast Asia with minor production in Africa and North America.

Like cowpeas, the yard-long beans are for the most part a self-pollinated crop due to their cleistogamous flowers. Also similar to cowpeas, the yard-long bean is known by a variety of names. For example, yard long beans are also called asparagus bean, Chinese long bean, pea bean and snake bean (in English), Judea esparrago (Spanish), haricot asperge (French), Taao-hla-chao (Hmong); jurokusasagemae (Japanese), dow gauk (Chinese) and sitaw (Filipino).

The growth habit of the yard-long bean is of a trailing, climbing legume that often reaches heights of 9–12 feet, although some varieties have more moderate growth of 6 or 7 feet. Bush-type yard-long beans are not known since the short stature would not give the pods enough room to grow without touching the ground. yard-long beans therefore need a trellis to grow.

In terms of agronomic management, yard-long bean can be grown year round in the tropics but only in the summertime in temperate climates. The seed of yard-long bean is planted into warm soil (above 18 C) and the plants are staked or trellised at 25 days after planting. Plant flowering in yard-long bean occurs at about 40 days after planting. Since yard-long bean is primarily grown for its crisp, younger, and tender pods, the pods can be harvested two to four times starting at about 50 days after planting and at biweekly intervals (Kongjaimun et al. 2012a).

After harvest, the 60-to-90-cm-long beans are graded by length and tenderness and bundled together, often tied by one of the actual pods used as a long string. Although most yard-long beans are consumed fresh and tender, the mature pods can also be dried, stored, and the dry seed later cooked as a pulse or used as bean sprouts by soaking in water and allowing them to sprout. The green and dry vines are also used for feeding livestock in South Asia. Yard-long bean and cowpea are consumed in local and global marketplaces but mostly in developing countries (Net 2006; Benchasri 2009).

As mentioned above, the selection and domestication of yard-long bean is likely to be a derivative of cowpeas, but the exact site of this vegetable's development is uncertain. West Africa is known to be the major center of diversity of cultivated forms of cowpea for dry grain (Ng and Padulosi 1988), and the crop was postulated to have been domesticated either in this region (Ba and Pasquet 2004) or in northeastern Africa (Coulibaly et al. 2002), although some introgression of southern African wild cowpeas is also thought to have occurred (Pasquet 1999).

The yard-long bean is one of five cultivated cowpea subspecies groups determined mainly by pod and seed characteristics (Pasquet 1999; Steward 1969): In this case sbsp. *Sesquipedalis* or yard-long bean is differentiated from other cowpeas, because of their long pods (Kongjaimun et al. 2012a, b). Production of cowpeas is known to be over 8 M metric tons of dry grain, but the amount of vegetable cowpea harvested is unknown. Despite the lack of statistics, yard-long bean production is known to be very substantial in Bangladesh, Bhutan, Cambodia, China, Laos, India, Indonesia, Malaysia, Myanmar, Nepal, Philippines, Taiwan, Thailand, and Vietnam (Rachie 1985).

Yard-long bean pods have many advantages for the consumer. They are a flexible food that can be consumed both fresh and cooked (Kongjaimun et al. 2012a, b). When fresh, the pods are easily snapped into smaller portions due to their low fiber quantity. Tender green pods of vegetable cowpea can be boiled, steamed, or stir-fried. Pod tenderness, low fiber, and sweetness are key factors for the commercial acceptability of the yard-long bean.

The crop is tolerant to low soil fertility because of their high nitrogen fixation capacity (Elowad and Hall 1987). They can grow on soil with more high sand content, low organic matter, and low levels of phosphorus (Kolawole and Tian

2000; Sanginga and Lyasse 2000). While vegetable cowpea production in some Southeast Asian countries is constrained by diseases and insect pests (Fery 2002; Sarutayophat et al. 2007), however for the most part yard-long beans have few significant abiotic constraints and are heat and drought tolerant.

Unfortunately, vegetable cowpeas are not well known in the USA, although they have the potential to be an important vegetable adapted to the long hot summers of the Southeast region. One problem is that vegetable cowpeas are almost exclusively pole types and staking or trellises are essential. This requirement limits the production to backyard and community gardens or perhaps some small farms, since the hand labor for wide scale production would be too expensive. The seed market for home gardening is large and yard-long bean is certainly adapted to the hot summer climate faced by many American gardeners.

In addition to its heat tolerance for the USA, no fungicides are required other than for powdery mildew making it a promising organic vegetable. Currently, no insect problems are noted for vegetable cowpea which makes it more favorable than snap bean or purple-hulled peas as alternative legumes. Multiple pickings are possible with yard-long bean pods throughout the season and the crop has good marketing opportunities in ethnic restaurants and grocery stores.

9.3.5 Minor Vegetable Legumes

Faba beans are usually a dry pulse but in some areas are consumed as a freshly shelled green bean (Zong et al. 2009). The crop is a minor horticultural legume grown in the Andes Mountains of South America, southern China, the Himalayas of South Asia, and around the Mediterranean Sea. Faba beans grown in northern China as a summer crop, or the Indo-gangetic plain, or parts of Africa as a cool season crop are usually harvested dry and are not consumed as a vegetable.

The temperature regime in which faba beans are grown is usually cool to temperate. Therefore, faba beans can be found in the winter season in Southern China, North Africa, and Southern Europe but not north of these regions due to the risk of heavy freezes. Meanwhile, faba bean production in more tropical and subtropical regions is grown at high-altitude locations in countries such as Argentina, Bolivia, Colombia, Ecuador, and Peru in South America or in Bhutan, Nepal, Northern India, and Yunnan province of China in Asia. Abiotic stresses and especially fungal or viral diseases are major limitation to production in faba bean.

Lima bean (*P. lunatus* L.) is another minor vegetable species grown in a few specialized temperate and tropical growing environments. The USA, France, and parts of the Caribbean and South America are the only significant producers of Lima bean. Even within the USA, production is limited to a few states such as Delaware where commercial canners are located and several southeastern states where the crop is grown for home-consumption. Southern Colombia is an area where lima beans are cooked as fresh shelled beans. Shelled lima beans are also popular as vegetables in Haiti and the Dominican Republic. Drought tolerance is

Table 9.1 Diseases in legumes caused by fungus, bacteria, and virus

	Fungal Pathogens						
	Anthracnose	*Fusarium oxysporium*	*Fusarium solani*	Leaf spot	Powdery Mildew	*Rhizoctonia solani*	Rust
Garden pea	√	√	√		√	√	
Snap bean	√	√	√	√		√	√
Yard-long bean	√	√	√	√	√	√	√
Faba bean	√	√	√	√	√	√	√
Cowpea	√	√	√	√	√	√	√
	Bacterial and viral Pathogens						
	Bacterial blight	Bacterial leaf spots	Halo blight	Bacterial wilt	Potyviruses	Comoviruses	Geminiviruses
Garden pea					√		
Snap bean	√	√		√	√	√	√
Yard-long bean	√	√	√	√	√	√	√
Faba bean	√	√			√		
Cowpea	√	√	√	√	√	√	√

very high in lima beans, and their wide diversity of germplasm allows breeding improvements. This, combined with high salt and heat tolerance and a variety of architectural growth habits, makes lima bean a promising crop for agricultural systems facing climate change.

The worldwide average yield of faba bean is less than 1 MT/ha dries grain and 2 MT of green faba beans which is below the yields of most tropical legumes which grow in long summer days (common beans, purple-hulled peas from black-eyed peas, mung beans and lima beans). Variability in the geographical statistics is perhaps due to variability in yield and production potential of each vegetable.

The typical biotic and abiotic stresses of each crop described above, from black-eyed peas to garden peas and faba beans to snap beans are summarized in Table 9.1 based on the compendiums of diseases for each crop (Hagedorn 1984; Lin and Rios 1985; Hall 1991; Emechebe and Florini 1997; Sillero et al. 2010). The principal biotic stresses include bacterial, fungal, and viral pathogens affecting the legumes. Meanwhile the abiotic stresses include cold, drought, high heat, low fertility, and salinity stresses. The pathogen species for each of the specific legume crops are variable but the general names, such as leaf blights, root rots, and wilts apply even with different bacterial or fungal disease species. Nematodes are also a universal problem for legumes and can be of the root knot or cyst types.

9.4 Vegetable Legumes as Intercrop or Rotation with Cereal Crops

Improving crop yield has been a core issue in modern agriculture. One way to increase crop yields on a per acre basis is intercropping of cereals with legumes. Intercropping is an ancient but mostly abandoned practice of simultaneous cultivation of crops. Legume intercropping with maize (e.g., with bean) is considered to be good alternatives to mono-cropping for an efficient use of space and for improvements in soil nitrogen. The increased or equivalent yield of maize besides the bonus yield of legume results in greater productivity per unit time and space and higher net returns of intercropping compared to monoculture.

Among the vegetable legumes, the most common grown as an intercrop with cereals are the purple-hulled pea or black-eyed pea grown with corn or millets in West Africa or the bush-type and climbing-type common beans grown with corn, sorghums, or bananas in the East Africa or Southern Africa highlands and some parts of Central and South America. Faba beans are also often grown in mixed systems at high elevations in equatorial regions of the Andes. Climbing beans can be grown either simultaneously with corn or in a relay with corn so that the legume has support above the ground.

Climbing beans usually yield most of their pods at least half a meter up the corn stalks. Some climbing beans have been bred to avoid lodging corn in Ecuador, though most intercropped corn are tall varieties. Climbing beans can also sometimes be staked with bamboo poles and these are called pole beans. Pole snap beans are grown in the Andes of South America, China, Southern Europe, and parts of the United State. US varieties include Blue Lake Pole, Kentucky Wonder, Provider, and Mountain Runner (Fig. 9.1).

Fig. 9.1 Intercropping of vegetable legumes and horticultural crops in China, **a** double row of summer okra grown between maturing winter peas in Hebei province; **b** single row bush common beans intercropped with eggplant rows in Shandong province; **c** spaced winter cabbage grown after trellised summer–fall snap bean crop in a greenhouse near Beijing

9.5 Vegetable and Non-vegetable Legumes as Intercrop or Rotation with Horticultural Crops

Horticultural crops are a high-value commodity that can be useful for generating small farm income. In this scenario, investments are high and land is used for a crop that is mostly valuable in the marketplace rather than for food security. Under these conditions, the inclusion of a legume, whether it should be also for vegetable and harvest early or a non-vegetable legume use for dry grain can reduce the risks associated with intensive vegetable farming.

Combined yield advantage in intercropping can occur because component crops differ in their use of soil and carbon resources in such a way that when they are grown together they are complementary and so make better overall use of soil and fertility resources than when grown separately in monocrops. An alternative is for the legume to be grown in rotation (and in a previous season) to or with the vegetable leaving the advantages of fixed nitrogen and crop residues in or on the soil making the next crop grow in a healthier manner.

Examples of legume rotation with vegetable horticultural crops include bush beans proceeding and or following winter vegetables. Garden Peas following summer tomatoes, garlic, and onion crops as a winter crop. Snap beans can be grown between tomatoes and empirical evidence shows that this reduces insect attack on susceptible tomatoes: although more study is needed in this area. The tomato/snap bean intercrop is common in trellises where post-digging, wiring, and netting or weaves for vertical growth require a heavy investment and yields must be maximized with minimum risks. Intercrop systems have also been developed for high tunnel agriculture especially between tomato and green beans using trellising to the hoop house structure. In this case, polypropylene string and a Florida or Georgia weave system can maintain the plants in row and growing all the way to or almost to the hoop house ceiling. Another advantage of this system is nematode control since most vegetable crops are of greater susceptibility than the legume crops that grow around or with, before or after the vegetables (Powers and McSorley 1993).

Winter-grown legumes like garden peas and faba beans can easily be worked into an early season crop before vegetable production and this is the most common system in China (Fig. 9.2). However, summer legumes including black-eyed peas, fresh shelled garden beans, purple-hull peas, snap beans, or yard-long beans are also useful, either in greenhouses, hoop houses, or in the field with summer production. In the southeastern USA, purple-hulled peas as a summer heat tolerant crop followed by canola which can be planted later than winter vegetables are also a possibility. Another *Vigna* species, the mung bean, is ideal for cropping with various summer vegetables. Many examples exist in China of mung beans beneficial nutritional, protective, and restorative role in the intercropping of vegetables which can easily be decimated by diseases and insects when grown in monocultures, as is so often done in the USA and Europe.

Fig. 9.2 Intercropping of vegetable legumes and fruit tree crops in China, **a** rows of common bean grown between pear trees during orchard establishment in Shandong province, **b** a wide row of mung bean bordered by peppers and ornamental plum trees growing as transplant trees in the field near Beijing

In small-holder agriculture in Latin American and African countries, many vegetables are grown among legume species for home-consumption. These legumes can be either vegetables themselves or used for dry grain but are ultimately important as an intercrop with the commercial crop of vegetables. Intercropping system has been found to have potential for small farm and community based agriculture in various regions of the world where it has been abandoned.

A recent study in Tennessee showed that legumes can be advantageous to a vegetable cropping system. Specifically, the first authors of this chapter carried out an experiment at the Tennessee State University farm using plots of fertilized and unfertilized sweet corn with or without a legume intercrop. Overall, the goal was to evaluate the effect of planting mung bean intercropping on yield and yield components of sweet corn compared under the different conditions The mung bean intercrop of sweet corn was found to increase the yield of corn cobs and their weight by 10 % despite a lack of fertilization and was comparable to urea fertilization treatment of the economically important sweet corn crop. In this case, legumes contribute significant nitrogen to the sweet corn through symbiotic nitrogen fixation (SNF) in nodules which resulted in a net gain of N into the cropping system. Sweet corn yields could be increased even more by using mung bean double rows in intercropping between sweet corn rows planted with conventional tillage.

Intercropping would have added advantage of even greater weed control by the mung bean which can shade out many sweet corn competitors. The use of legume intercropping may reduce the reliance on synthetic fertilizers and result in less greenhouse gas emissions, thus helping to mitigate climate change. Organic farming or home gardening pairing legumes with vegetables has spurred the interest in crop diversification and use of cover crops to provide nitrogen and weed control; however, this remains an under-researched area.

9.6 Vegetable and Non-vegetable Legumes Intercrops with Fruit Trees

An extremely valuable intercropping system of legumes intercropped between young fruit trees has been developed extensively in China but is not used in most other countries. This difference in agro-ecosystems is very notable in the differences between Chinese and Western fruit tree production. The use of legumes as an intercrop is a method of sustainable intensification of land use providing a second crop in the alleys between fruit trees.

The legume provides advantages of nitrogen fixation and diversity in the fruit orchard. Fruit tree leaves drop and provide ground cover for the legumes reducing weed pressure and providing a source of phosphorus from the tree root zone to the legumes. Meanwhile, tillage for legumes and the microbial community of legumes provides a control for diseases transmitted through leaf litter. Some fruit orchards are established for short seasons especially in the tropics.

Fruit–legume intercrops are very common in Asia where many fruit species from apples to oranges and persimmons were domesticated and cultivated. Chinese farmers have the most expertise in using legumes in fruit orchards and have made this sustainable intensification technique one of the cornerstones of China's intensive agriculture. Winter production of peas between dormant fruit trees is a popular system of intercropping and makes valuable use of some of the best alluvial soils of China.

For example, in Shandong province plum trees are spaced in rows about 4 m apart which provides room for six rows of peas. In areas near the bigger cities of China, this fruit intercrop system has evolved into an ornamental horticultural system, whereby flowering trees and shade trees are grown from liners into bur-lapped transplantable trees within a field double cropped to legumes. Peas in these tree intercrops can be followed by common beans which are for dry grain or snap bean production (Fig. 9.3).

In addition to growing legumes that are mostly for vegetables among fruit trees, some grain legumes are also widely used in Chinese fruit tree producing areas. One of these legumes that are most flexible for production purposes is the mung bean. As mentioned above, mung beans can be grown in double rows and are ideal in plant height to grow between vegetables or fruit trees (Fig. 9.4). Common beans, faba beans and garden or field peas can all be grown between fruit trees in the same way with different seasonality typical of each of these crops compared to the more heat tolerant mung bean. Faba beans as winter crop, garden or field peas as a spring and fall crop, and common bean as an early summer crop are all feasible productive systems for growing between fruit trees during the first few years of orchard establishment. After this period, legume fodders such as clovers and medicagos can continue to be grown between the trees along with shade-tolerant grasses.

Legumes between fruit trees are highly productive in leaf and stem biomass making them an ideal cover crop for the alleys between almost all fruit species. Since most legumes are flexible in plant spacing plans and grown to between 0.5 m

Fig. 9.3 Corn intercropped with common bean or mung bean with different spatial arrangements and at different development stages in China, **a** single row of common bean and single row of corn; **b** double row of common bean and single row of corn; **c** double row of common bean and double row of corn; **d** four row mung bean and single row of corn

Fig. 9.4 Close-ups of corn intercropping with legumes, **a** bush snap bean over shadowed by corn; **b** maturing bush beans surrounded by newly planted corn in tight spacing; **c** within row intercropping of common beans and corn; **d** single row of mung bean relay planted four weeks after a single row of corn

Fig. 9.5 Use of faba bean as legume for intercropping with various cereals in China, **a** double row or faba bean grown between wheat as spring crops in East–West orientation; **b** maturing faba beans flanked by planted corn for summer production

and 1.5 m in height, they form a dense cover that shades out any weeds and therefore makes for a clean orchard floor almost to the base of the trees. Unlike soybeans which do not branch much, the mung bean can cover into the spaces between trees within a row by slanting their growth architecture toward any empty spaces (Fig. 9.5). Cowpeas will also cover the ground between fruit trees efficiently. All the legumes from the *Vigna* family are high biomass producers which add organic matter and nitrogen to the soil allowing for a reduction in fertilization.

Examples found in China commonly are garden peas in the winter between rows of deciduous stone fruit trees or snap beans and mung beans between young apple, nectarine, peach, pear, and plum trees (Fig. 9.6). Chinese agriculture emphasizes intensive land use so much that even ornamental trees are grown between legumes. These ornamental trees can be grown closely together from liners in rows since they will be transplanted when approximately 3 m high, allowing a continuous rotation of legume and vegetable crops in these fruit and flowering tree nurseries. Other regions of the world have similar legume/fruit tree intercrops but mostly in traditional systems found in Europe or around the Mediterranean, with few current examples in North America and South America.

Solanaceous fruit crops such as cape gooseberry or lulo intercropped are commonly intercropped with common bean and garden peas in Colombia. These crops can be grown for a year and a half or less but still allows three or four legume rotations. Yields of legumes allow the generation of income early in the production cycle and permit expenses to be covered for the establishment and fertilization of the solanaceous crops. This is extremely valuable since both cape gooseberry and lulo have to be staked and strung from trellises. The legumes also require staking and provide an early economic yield to justify the planting and staking efforts for the fruit species (Figs. 9.7 and 9.8). Corn and coffee are alternative alley crops that can be grown with the legumes.

There is a large potential for growth in the use of vegetable or grain legumes in intensive agriculture especially around urban areas where land is at a premium and

Fig. 9.6 Garden peas grown for rotation/intercropping/companion crop in China, **a** garden peas as a winter cover crop with corn seedlings grown as wind breaks for summer vegetables; **b** alley cropping of multiple rows of garden peas between rows of mature pear trees; **c** garden peas and faba beans intercropped for winter production and in maturity during spring months; **d** garden peas between cement trellis carrying fruit grapes for early production under plastic

Fig. 9.7 Legumes as rotation/intercrop/companion crops of tropical solanaceous species in Colombia, **a** garden peas trellised with cape gooseberry (uchuva), *Physalis peruvianum* in Granada, Cundinamarca department; **b** garden peas grown as a relay with common beans in Fusagasuga, Cundinamarca department; in preparation for new fruit plantings and **c** alley cropping of common beans between coffee and old tamarillo (tomate de arbol) bushes, *Solanum betaceum* in Darien, Valle department

Fig. 9.8 Common bean for a rotation/intercrop/companion crop in Colombia: **a** alley cropping of multiple rows of common beans growing between newly planted naranjilla (or lulo in Spanish) bushes (*Physalis peruvianum_* in Restrepo, Valle department, **b** climbing and bush bean varieties grown in a relay on dry corn stalks in Calima, Valle department

less likely to be used for extensive cereal or industrial crops. The examples given above of production systems where legumes are grown together, in rotation or between, in intercrop with vegetables or fruit trees can be iterated into many permutations of intensive agricultural systems with the adaptability of legume to fit into any rotation.

9.7 Advantages of Intercropping with Vegetable Legumes in Organic Agriculture

Organic agriculture as discussed in other chapters of this book requires the careful use of labor, land, and water resources within a given field and the recycling of nutrients for the fields' soils. Legumes are very valuable for this purpose because they are capable of symbiotic nitrogen fixation and therefore usually contribute a net gain of this nutrient to the soil. In some cases, where biomass of the legume is low, the legume crop can use up soil N, but it does so sparingly compared to cereals that are heavy feeders and utilizers of N.

Therefore, legumes are a much better cover crop for most situations in organic agriculture than a cereal crop or grass family member, which are the other main type of cover crop used in organic agriculture. Examples of grass-based covers are fescues, oats, rye, and wheat, while legumes include all the crops discussed above from the genera *Phaseolus*, *Vicia*, and *Vigna* as well as the *Medicago* and *Trifolium* genera. Most of these rotations have not been well studied, although the use of soybean in organic rotations with winter wheat has been evaluated at the Rodale Institute New Farm near Kutztown, Pennsylvania in the USA. In this system, the winter wheat is mechanically harvested with both grain and hay crops removed. The soybean is then drill planted into the wheat stubble and is allowed to grow until maturity for a second crop late in the season, providing two crops in one year. This system is a very practical part of low-input sustainable agriculture (LISA) for US farmers in the mid-Atlantic and some parts of the mid-south and has become fairly widespread.

Modifications of the winter crop/summer legume system for other regions would be valuable but have been of low priority for USDA funding. One possibility being studied in Tennessee State University is to follow winter canola with cowpea, mung bean, or soybean planted in May or June for summer production for the southeast, but this requires the use of early maturing canola varieties. The over-wintering and build-up of insect pests in the canola crop is one concern for this system which might require the use of non-organic chemical pesticide inputs. However, even in this system many of the canola pests can be controlled by pyrethroids which are allowed in organic farming. Economic analysis is required to determine the profitability of organic systems which do not allow genetically modified organism (GMO) crops. Most southern farmers would be likely to use GMO canola and soybean although some non-transgenic canola and soybean varieties exist and are significantly less expensive in terms of seed costs.

Likewise, legume intercropping or rotation systems of farmers considering organic production with field corn must analyze the advantages or disadvantages of growing GMO or non-GMO corn varieties in terms of weed (round up ready corn) and insect (Bt corn) control.

Apart from their advantages in nitrogen fixation and as cover crops, legumes are a valuable mechanism for increasing microbial biodiversity in the soil which have other advantages to plant and soil health in a more natural farming system such as organic agriculture. Like many plants, legumes can augment the growth of plant grown promoting bacteria around their root systems. These bacteria can be of benefit to the legume itself or to following crops. The center for agricultural research in Rothamsted, England, has had long-term plots with legumes where soil health has been monitored over decades,

While most work in organic agriculture of legumes has been with grain legumes, the vegetable legumes obviously have an important role in organic systems. Above we have discussed four major vegetable legumes and two minor vegetable legumes along with several non-vegetable legumes giving their potential use in various horticultural cropping systems. Intercropping systems lend themselves to organic agriculture since multi-cropping usually reduces the pest problems on any of the individual crops in the system through physical interference or more active pest defense systems.

Intercropping is very important for vegetable legume production especially in Asia where the majority of snap bean and garden pea production occurs. These two crops are also widely grown in Europe at similar latitudes to their production regions in East Asia, but with less intensity due to the differences in landholdings and the use of greenhouses. North America is a producer of snap beans and some garden peas but usually in mono-cropping systems unless grown by backyard gardeners. Faba beans are also grown widely in Asia and the Middle East as well as Europe, but the production figures are not well reported. Most of these systems are currently not organic and do use chemical pesticide, herbicide, and fertilizer inputs but could be rapidly converted to organic systems if placed on certified land for organic agriculture. Non-organic inputs could be reduced in any of the systems to make them more self-reliant.

Among the potentials of legumes for organic agriculture, perhaps most promising is the use of legumes in organic fruit production. Many eco-physiological advantages are present for the legume grown in the fruit orchard. For examples, most legumes can withstand a lot of shade and therefore are ideal for the shady conditions of the fruit orchard. The cooler conditions under the cover of orchard trees are often beneficial to legumes which as C3 plants require shadier conditions than grasses which are the alternative pasture intercrop systems for fruit production. Although little studied, the reduction in light intensity typical of an orchard can actually help legumes establish themselves and grown better. Finally, the reduction in disease transmission and pest buildup from a partitioned intercropped landscape can reduce or eliminate the use of pesticides and help in the conversion of orchards from conventional to organic production. In summary, the legume and fruit tree intercrop represent a win-win situation for both crop species and can lead to organic agriculture.

Apart from the economic advantage of a conversion to organic production under the current societal paradigms for organic production, there are multiple advantages to intercropping legumes among fruit trees. For example, the economic advantages of legume–fruit tree intercropping are spatial and temporal as well as economic, environmental, and physiological. Legumes that cover the ground can prevent the emergence of overwintering pathogens that cannot get up into the canopy if a crop is established beneath it. Aeration of the root zone by a legume crop can counteract the compaction from heavy machinery that is pulled through the orchard.

In economic terms, the legume can provide a quick return on investment that adds rather than detracting from the profits generated by a fruit orchard. While fruit trees are being established, a farmer can use legumes to cover the ground and prevent weeds from growing but also gaining an income from the sale of legumes as vegetables or dry grain. The use of legumes in vineyard establishment is an area that should be pursued for organic wine production.

In organic orchard production, where growth is not accelerated by fertilization, legumes play an important role both agroecologically and economically. First the legume can be used as a nitrogen-fixing cover or soil enhancement crop. In this case, the legume is often the main source of nitrogen for the fruit trees in the orchard and can be used to increase soil fertility and avoid weed growth in the alleys between trees. Economically, the legumes are important if harvested. During the first two or three years of growth of fruit trees, the legumes can be the sole source of profit for the orchard. After the fruit trees start to yield fruit, the legumes can be grown on a more limited scale. For example, legumes can be grown as six-row swaths through the typical spacing of most stone fruits (almonds, apricots, peaches, and plums) or as eight-row strips with larger trees such as apples and pears that are spaced further apart.

Acknowledgement The Evans Allen program from the USDA and State of Tennessee is recognized for graduate student assistantship support and funding for the corresponding author. We also thank the Chinese Agriculture Research System for collecting and providing the pictures in this book.

References

Abawi GS, MAP Corrales (1990) Root rots of beans in Latin America and Africa: diagnosis, research methodologies, and management strategies. CIAT

Ahenkora K, Dapaah HKA, Agyemang A (1998) Selected nutritional components and sensory attributes of cowpea (*Vigna unguiculata* [L.] Walp) leaves. Plant Foods Hum Nutr 52:221–229

Amusa AN, Ikotun T, Bankole JO (1996) Short communication: survey of leaf spot-causing microorganisms on yams. Afr Crop Sci J 4:111–113

Ba FS, Pasquet RS, Gepts P (2004) Genetic diversity in cowpea [*Vigna unguiculata* (L.) Walp.] as revealed by RAPD markers. Genet Resour Crop Evol 51:539–550

Benchasri S (2009) Inheritance of cowpea aphid (*Aphis craccivora* Koch) resistance and preliminary study of DNA Markers associated with cowpea aphid resistant genes in yard long bean and cowpea. Diss. PhD thesis. Department of Plant Science. PSU. Songkhla, Thailand

Benchasri S, Bairaman C (2010) Evaluation of yield, yield components and consumers' satisfaction towards yard long bean and cowpea in agricultural organic system. Bulg J Agric Sci 16:705–712

Blair MW, Chaves A, Tofiño A, Calderón JF, Palacio JD (2010a) Extensive diversity and inter-genepool introgression in a world-wide collection of indeterminate snap bean accessions. Theor Appl Genet 120:1381–1391

Blair MW, González LF, Kimani PM, Butare L (2010b) Genetic diversity, inter-gene pool introgression and nutritional quality of common beans (*Phaseolus vulgaris* L.) from Central Africa. Theor Appl Genet 121:237–248

Bohlool BB, JK Ladha, DP Garrity and T George (1992) Biological nitrogen fixation for sustainable agriculture: a perspective. In: Biological nitrogen fixation for sustainable agriculture, Springer, Berlin, pp 1–11

Brown JW, TC Osborn, FA Bliss and TC Hall (1982) Bean lectins: Part 2: Relationship between qualitative lectin variation in *Phaseolus vulgaris* L. and previous observations on purified bean lectins. Theoretical applied genetics. Theoretische und angewandte Genetik 62(4):361–367

Buddenhagen IW (1990) Legumes in farming systems in Mediterranean climates. In: The role of legumes in the farming systems of the Mediterranean areas. Springer, Berlin, pp 3–29

Cannon SB, May GD, Jackson SA (2009) Three sequenced legume genomes and many crop species: rich opportunities for translational genomics. Plant Physiol 151:970–977

Cong WF, Hoffland E, Li L, Six J, Sun JH, Bao XG, Zhang FS, Van Der Werf W (2015) Intercropping enhances soil carbon and nitrogen. Glob Change Biol 2:1715–1726

Coulibaly S, Pasquet RS, Papa R, Gepts P (2002) AFLP analysis of the phenetic organization and genetic diversity of *Vigna unguiculata* L. Walp. reveals extensive gene flow between wild and domesticated types. Theor Appl Genet 104:358–366

Cunha C, Hintz T, Griffiths P (2004) Genetic diversity of snap bean cultivars determined using randomly amplified polymorphic DNA (RAPD) markers. HortScience 39:81–484

Davis J, Myers JR (2002) Phylogenetic analysis of snap beans using RAPD markers. Ann Rep Bean Impr Coop 45:16–17

de Almeida Costa GE, da Silva Queiroz-Monici K, Reis SMPM, de Oliveira AC (2006) Chemical composition, dietary fiber and resistant starch contents of raw and cooked pea, common bean, chickpea and lentil legumes. Food Chem 94:327–330

Dwivedi SL, Blair MW, Upadhyaya HD, Serraj R, Balaji J, Buhariwalla HK, Ortiz R, Crouch JH (2006) Using genomics to exploit grain legume biodiversity in crop improvement. Plant Breed Rev 26:171

Edwards CG, Lee CY (1986) Measurement of provitamin A carotenoids in fresh and canned carrots and green peas. J Food Sci 51:534–535

Ehlers JD, Hall AE (1997) Cowpea (*Vigna unguiculata* L. Walp.). Field Crops Res 53:187–204

Elowad HOA, Hall AE (1987) Influences of early and late nitrogen fertilization on yield and nitrogen fixation of cowpea under well-watered and dry field conditions. Field Crop Res 15:229–244

Emechebe AM, Florini DA (1997) Shoot and pod diseases of cowpea induced by fungi and bacteria. Advances in cowpea research. International Institute of Tropical Agriculture (IITA), Ibadan, Nigeria, pp 176–192

Fery RL (1990) The cowpea: production, utilization, and research in the United States. Hortic Rev 12:197–222

Fery RL (2002) New opportunities in *Vigna*. Trends in new crops and new uses, pp 424–428

Fushan L (1994) Study on origin and evolution of soybean. Soybean Science 01

Gepts P, Osborn TC, Rashka K, Bliss FA (1986) Phaseolin-protein variability in wild forms and landraces of the common bean (*Phaseolus vulgaris*): evidence for multiple centers of domestication. Econ Bot 40:451–468

Gepts P, Beavis WD, Brummer EC, Shoemaker RC, Stalker HT, Weeden NF, Young ND (2005) Legumes as a model plant family. Genomics for food and feed report of the cross-legume advances through genomics conference. Plant Physiol 137:1228–1235

Hagedorn DJ (1984) Compendium of pea diseases. American Phytopathological Society

Hall R (1991) Compendium of bean diseases. American Phytopathological Society

Harinikumar KM, Bagyaraj DJ, Mallesha BC (1990) Effect of intercropping and organic soil amendments on native VA mycorrhizal fungi in an oxisol. Arid Land Res Manag 4:193–197

Heichel GH, Helsel ZR (1987) Legume nitrogen: symbiotic fixation and recovery by subsequent crops. Energy in plant nutrition and pest control. Energy World Agric 2:63–80

Henry G, Jansen W (1992) Snap Beans in the developing world. Centro internacional de agricultura tropical, Calí 366p

Hinsinger P, Herrmann L, Lesueur D, Robin A, Trap J, Waithaisong K, Plassard C (2015) Impact of roots, microorganisms and microfauna on the fate of soil phosphorus in the rhizosphere. Ann Plant Rev Phosphorus Metab Plants 48:377

Hummer KE, Hancock JF (2015) Vavilovian centers of plant diversity: implications and impacts. HortScience 50:780–783

Izaurralde RC, McGill WB, Juma NG (1992) Nitrogen fixation efficiency, interspecies N transfer, and root growth in barley-field pea intercrop on a Black Chernozemic soil. Biol Fertil Soils 13:11–160

Kaplan L, Kaplan LN (1988) Phaseolus in archaeology. In: Genetic resources of *Phaseolus* beans, Springer, Berlin, pp 125–142

Kassam A, Friedrich T, Shaxson F, Pretty J (2009) The spread of conservation agriculture: justification, sustainability and uptake. Int J Agric Sustain 7:292–320

Kay DE (1979) Food legumes. TPI Crops and Product Digest, Tropical Products Institute

Kolawole GO, Tian G, Singh BB (2000) Differential response of cowpea lines to aluminum and phosphorus application. J Plant Nutr 23:731–740

Kongjaimun A, Kaga A, Tomooka N, Somta P, Shimizu T, Shu Y, Isemura T, Vaughan DA, Srinives P, Scoles G (2012a) An SSR-based linkage map of yard long bean (*Vigna unguiculata* (L.) Walp. subsp. *unguiculata Sesquipedalis* Group) and QTL analysis of pod length. Genome 55:81–92

Kongjaimun A, Kaga A, Tomooka N, Somta P, Vaughan DA, Srinives P (2012b) The genetics of domestication of yardlong bean, *Vigna unguiculata* (L.) Walp. ssp. *unguiculata* cv.-gr. *sesquipedalis*. Ann Bot 109:1185–1200

Kwapata MB, Hall AE (1985) Effects of moisture regime and phosphorus on mycorrhizal infection, nutrient uptake, and growth of cowpeas (*Vigna unguiculata* (L.) Walp.). Field Crops Res 12:241–250

Lambert RG, Linck AJ (1958) Effects of high temperature on yield of peas. Plant Physiol 33:347

Lewis GP, Schrire B, Mackinder B, Loc M (2005) Legumes of the World. Royal Botanic Gardens Kew

Lin MT, Rios GP (1985) Cowpea diseases and their prevalence in Latin America. Cowpea research, production and utilization. IITA, Ibadan, pp 189–204

Majumdar DK (2011) Pulse crop production: principles and technologies. PHI Learning Pvt, Ltd

Messina MJ (1999) Legumes and soybeans: overview of their nutritional profiles and health effects. Am J Clinical Nutr 7:439s–450s

Métais I, Hamon B, Jalouzot R, Peltier D (2002) Structure and level of genetic diversity in various bean types evidenced with microsatellite markers isolated from a genomic enriched library. Theor Appl Genet 104:1346–1352

Mortimore MJ, B Singh, F Harris and SF Blade (1997) Cowpea in traditional cropping systems. Advances in Cowpea Research, pp 99

Mucheru-Muna M, Pypers P, Mugendi D, Kung'u J, Mugwe J, Merckx R, Vanlauwe B (2010) A staggered maize–legume intercrop arrangement robustly increases crop yields and economic returns in the highlands of Central Kenya. Field Crops Res 115:132–139

Myers JR and JR Baggett (1999) Improvement of snap bean. In: Common bean improvement in the twenty-first century. Springer, Berlin, pp 289–329

Net G (2006) Earth Net Foundation, 2006. Organic Agriculture Situation in Thailand

Ng NQ, Padulosi S (1988) Cowpea genepool distribution and crop improvement. IITA

Ngouajio M, Foko J, Fouejio D (1997) The critical period of weed control in common bean (*Phaseolus vulgaris* L.) in Cameroon. Crop Protection 16:127–133

Ngwira AR, Aune JB, Mkwinda S (2012) On-farm evaluation of yield and economic benefit of short term maize legume intercropping systems under conservation agriculture in Malawi. Field Crops Res 132:149–157

Nielsen SS, Ohler TA, Mitchell CA (1997) Cowpea leaves for human consumption: production, utilization, and nutrient composition. Adv Cowpea Res 326

Nieto-Cabrera C, Francis C, Caicedo C, Gutiérrez PF, Rivera M (1997) Response of four Andean crops to rotation and fertilization. Mountain Res Dev 273–282

Pachico D (1987) The potential of snap beans as a crop for small farmers in the tropics. Davis J, Janssen W (comps). El mejoramiento genético de la habichuela en América Latina: Memorias de un taller, CIAT, Cali, Colombia, 11 al 15 de mayo de 1987

Pasquet RS (1999) Genetic relationships among subspecies of *Vigna unguiculata* (L.) Walp. based on allozyme variation. Theor Appl Genet 98:1104–1119

Peoples MB, DF Herridge and JK Ladha (1995) Biological nitrogen fixation: an efficient source of nitrogen for sustainable agricultural production? Plant Soil 174:3–28

Powers LE, McSorley R, Dunn RA (1993) Effects of mixed cropping on a soil nematode community in Honduras. J Nematol 25:666

Rachie KO (1985) Introduction. In: Singh SR, Rachie KO (eds) Cowpea research, production and utilization. Wiley, New York, pp i–iv

Rubatzky VE, Yamaguchi M (2012) World vegetables: principles, production, and nutritive values. Springer, Science & Business Media

Sandhu JS, Singh S (2007) History and origin. In: Lentil. Springer, Berlin, pp 1–9

Sanginga N, Lyasse O, Singh BB (2000) Phosphorus use efficiency and nitrogen balance of cowpea breeding lines in a low P soil of the derived savanna zone in West Africa. Plant Soil 220:119–128

Santalla M, Rodino AP, Casquero PA, De Ron AM (2001) Interactions of bush bean intercropped with field and sweet maize. Eur J Agron 15:185–196

Sarutayophat T, C Nualsri, Q Santipracha and V. Saereeprasert (2007) Characterization and genetic relatedness among 37 yard long bean and cowpea accessions based on morphological characters and RAPD analysis. Songklanakarin J Sci Technol (Thailand)

Shehata NA, Darwish N, El-Nahry F, Abdel Razek FA (1988) Supplementation of wheat flour with some local legumes. Food/Nahrung 32:1–8

Silbernagel MJ (1986) Snap bean breeding. Breeding vegetable crops. Avi Publishing Company Inc, Westport, pp 234–282

Sillero JC, Villegas-Fernández AM, Thomas J, Rojas-Molina MM, Emeran AA, Fernández-Aparicio M, Rubiales D (2010) Faba bean breeding for disease resistance. Field Crops Res 115:297–307

Singh BB, Tarawali SA (1997) Cowpea and its improvement: key to sustainable mixed crop/livestock farming systems in West Africa. In: Renard C (eds) Crop residues in sustainble mixed crop/livestock farming systems. Commonwealth Agricultural Bureau International (CABI), Wallingford, pp 79–100

Skroch PW, Nienhuis J (1995) Qualitative and quantitative characterization of RAPD variation among snap bean (*Phaseolus vulgaris*) genotypes. Theor Appl Genet 9:1078–1085

Smith BN, Herath HMW, Chase JB (1973) Effect of growth temperature on carbon isotopic ratios in barley, pea and rape. Plant Cell Physiol 14:177–182

Steward FC (1969) Tropical crops: dicotyledons. In: Purseglove JW (ed). Wiley, New York, 1968, 2 vols, 719 pp. American Association for the Advancement of Science

Sugiyama A, Yazaki K (2012) Root exudates of legume plants and their involvement in interactions with soil microbes. In Secretions and exudates in biological systems. Springer, Berlin, pp 27-48

Teshome A, Brown AHD, Hodgkin T (2001) Diversity in landraces of cereal and legume crops. Plant Breed Rev 21:221–262

Tracy WF, Hallauer AR (1994) Sweet corn. Specialty corns, pp 147–187

Trouillefou CM, Le Cadre E, Cacciaguerra T, Cunin F, Plassard C, Belamie E (2015) Protected activity of a phytase immobilized in mesoporous silica with benefits to plant phosphorus nutrition. J Sol-Gel Sci Technol 74:55–65

Van Ek GA, Henriet J, Blade SF, Singh BB (1997) Quantitative assessment of traditional cropping systems in the Sudan savanna of northern Nigeria. II. Management and productivity of major cropping systems. Samaru J Agric Res 14:47–60

Verdcourt B (1970) Studies in the Leguminosae-*Papilionoïdeae* for the ‘Flora of Tropical East Africa’: IV. Kew Bulletin, pp 507–569

Weeden NF (2007) Genetic changes accompanying the domestication of *Pisum sativum*: is there a common genetic basis to the ‘domestication syndrome’ for legumes? Ann Bot 100:1017–1025

Young JPW, AWB Johnston (1989) The evolution of specificity in the legume-Rhizobium symbiosis. Trends Ecol Evol 4:341–349

Young ND, Mudge J, Ellis THN (2003) Legume genomes: more than peas in a pod. Curr Opin Plant Biol 6:199–204

Zhang X, Blair MW, Wang S (2008) Genetic diversity of Chinese common bean (*Phaseolus vulgaris* L.) landraces assessed with simple sequence repeat markers. Theor Appl Genet 117:629–640

Zheng Z, Wang S, Zong X (1997) Food legume crops in China. China Agriculture Press, Beijing, pp P53–P92

Zong X, Liu X, Guan J, Wang S, Liu Q, Paull JG, Redden R (2009) Molecular variation among Chinese and global winter faba bean germplasm. Theor Appl Genet 118:971–978

Chapter 10
Principles of Vermitechnology in Sustainable Organic Farming with Special Reference to Bangladesh

A.A. Ansari, S. Jaikishun, M.A. Islam, S.K. Kuri, K. Fiedler and D. Nandwani

Abstract Organic farming is an important concept toward sustainable development that involves the use of organic input for the production of crops. Biofertilizers such as vermicompost and vermiwash increase the organic matter content necessary for the maintenance of soil properties, which is beneficial for long-term sustainability and crop productivity. Biofertilizers are enriched with micronutrients and beneficial microbes that enhance the soil quality and aid in slow release of nutrients required for the healthy growth of plants. Various small- and large-scale experiments on field crops such as wheat, sugarcane, paddy along with vegetables such as tomato, okra, and eggplant have been successful in terms of productivity and quality of produce. These technologies have also been adopted successfully by food growers across the globe resulting in substantial markets for organic produce. Organic agriculture is still in its infancy in Bangladesh. Very little cultivated land is being used for organic vegetable production, primarily because of lack of consumer awareness and demand. Though demand in-country in developing country is low, exporting certified safe vegetables can add to the country's economy. Establishing effective and reliable organic vegetable producers in Bangladesh

A.A. Ansari (✉) · S. Jaikishun
Department of Biology, University of Guyana, Georgetown, Guyana
e-mail: abdullah.ansari@uog.edu.gy

M.A. Islam
Department of Horticulture, Faculty of Agriculture, Bangladesh Agricultural University (BAU), Mymensingh 2202, Bangladesh
e-mail: ashrafulmi@bau.edu.bd

S.K. Kuri
Department of Agricultural Extension Education, Faculty of Agriculture, Bangladesh Agricultural University (BAU), Mymensingh 2202, Bangladesh

K. Fiedler
Fulbright Foundation and International Institute of Education, New York, NY, USA

D. Nandwani
Department of Agricultural and Environmental Sciences, College of Agriculture, Human and Natural Sciences, Tennessee State University, Nashville, TN 37209, USA
e-mail: dnandwan@tnstate.edu

D. Nandwani (ed.), *Organic Farming for Sustainable Agriculture*,
Sustainable Development and Biodiversity 9,
DOI 10.1007/978-3-319-26803-3_10

would enable the country to prepare for the potential growing market demand by establishing technology and infrastructure. Such technologies are way forward to sustainable organic farming that contributes to the development in terms of green economy.

Keywords Biofertilizer · Vermicompost · Vermiwash · Soil improvement · Crop productivity

10.1 Introduction

Chemical fertilizers and pesticides make it possible to produce multiple crops per year on the same land, which stresses the soil and leads to a depleted nutrients. This cycle of chemical inputs is continuously required to continue high levels of production that is required in many developing countries such as Bangladesh. Soil fertility is declining, and structure is no longer ideal for sustainable farming. Bangladesh has great potential for organic agriculture that could improve the suitability of soils and improve the economy of farmers with organic produce exports to other countries.

There is an increasing desire for organic produce locally in urban centers such as Dhaka, which is primarily hub of commercial activities. As a result, grocery stores are starting to carry organic foods in the capital city, Dhaka, such as Probortana, PROSHIKA, Meena Bazar, Nandan, Agora, and others that are produced by several NGOs. Organic tea products, as well as vegetables and herbs, are also available. The organic products are produced by farmers in rural areas of Bangladesh supported by NGOs. They are encouraged to produce organic products, but there are gaps between production and marketing.

The frequent use of synthetic fertilizers reduces soil quality, primarily soil organic matter. The conversion period is a transitional time from conventional agriculture management to an organic agricultural management system. Considering this, conversion periods are extended until the improvement of soil fertility, and it is a necessary component of organic certification. In Europe, the transition period can last two to three years of production before organic certification is accomplished. In Bangladesh, lands managed by different organizations such as PROSHIKA, BARI, BAU and Kazi have taken 7–10 years to convert the land. Kazi tea plantations required to 5–7 years in the Dinajpur district of Bangladesh. This location required unique strategies to prevent chemical contamination from neighboring land by installing a boundary canal and legume plantation.

To maintain sustainable organic practices in Bangladesh, nutrient sources need to be applied strategically. Nutrient management strategies for organic crop production result in higher crop productivity and regeneration of soil with improved fertility. Productive soil should be enriched with organic matter. In Bangladesh, some soils have less than 1 % organic matter, though low organic matter content is variable based on elevation. A large component of nutrient management in organic agriculture practices is the use of compost and manure, though availability of commercially licensed products is limited or nonexistent.

Earth has diverse plants that have been providing food, promoting health and some form of shelter toward civilization throughout the years. The soil is considered to be a major component for plant growth that helps to provide homes for many organisms (Ismail 2005). Soil microbiology influences above-ground ecosystem by contributing to plant nutrition, health, soil structure, and fertility. They also play a pivotal role in various biogeochemical cycles and cycling of organic compounds (Kirk et al. 2004). Plant growth is improved when beneficial microbes increase nutrient availability and stimulate plant growth (Ismail 2005).

Biofertilizers referred to the use of soil microorganisms to increase the availability and uptake of mineral nutrients for plant (Ansari 2008). Also, they are substance added to the soil to enhance the microorganisms in order to increase the nutrient status. Vermicompost is one of the biofertilizers that helps to promote humification, increased microbial activity and enzyme production, which subsequently helps to increase the aggregate stability of soil particles resulting in better aeration when applied to the soil. The material has excellent structure, porosity, aeration drainage, and moisture holding capacity and helps to improve the physical, chemical, and biological properties of the soil (Ansari 2008).

The biocomposting method is made up of two phases such as breakdown and buildup phase. In the breakdown phase, biodegradable wastes are decomposed into smaller particles. Proteins are broken down into amino acids and finally to ammonia, nitrates, and free nitrogen. Similarly, urea, uric acids, and other non-protein nitrogen-containing compounds are reduced to form different plant nutrients. In the buildup phase, there is the resynthesis of simple compounds into complex humic substances. The organisms responsible for transformation to humus are aerobic and facultative aerobic, sporing and nonsporing, and nitrogen-fixing bacteria of the *Azotobacter* and *Nitrosomonas* group. Similarly, *Actinomycetes* also play an important role. There are two major reasons why vermicomposting is better. Waste is converted faster. Conventional composting takes weeks to months to convert organic matter to compost and is very labor intensive. By using earthworms, waste is rapidly turned into vermicompost. The vermicompost is far superior to conventional compost. The worm castings in the vermicompost have nutrients that are highly utilizable by plants, and the castings have a mucous coating which allows the nutrients to "time release." Vermicompost forms fine stable granular organic matter that assists in the aeration, released mucus that is hygroscopic absorbs water and prevents water logging as well as improves water holding capacity. Vermicompost added to the soil releases nutrient slowly and consistently and enables the plant to absorb these nutrients more readily. Soils enriched with vermicompost provide additional substances that are not found in the chemicals (Ansari and Ismail 2001; Kale 1998). Biofertilizers contribute both macro- and micronutrients in amounts that are required by the plant and upon application have emphatic effect on plant growth parameters and production.

Organic waste possesses a serious environmental problem globally. This can be solved by Vermitechnology including Vermiwash and vermicompost, and also biodynamic preparation (500), which is essential component of biodynamic farming. Many researches over the years have been conducted, whereby solid waste were used and recycled to produce organic fertilizers using different technologies. In many developing countries, there is a serious organic solid waste problem; preparing these organic fertilizers will be cost-effective and beneficial for farming (Ansari 2008). The use of organic processes and materials in agriculture also helps to prevent environmental hazards, soil damage, and nutrients loss due to the excess use of toxic chemical fertilizers and pesticides (Nath et al. 2009).

10.2 Earthworms in the Soil

Earthworms play a vital role in maintaining soil fertility and in bringing of efficient nutrient cycling. Earthworms assist in recycling of organic nutrients for the efficient growth of plants. Earthworms not only inhabit the soil, but contribute to the physical and chemical alterations in the soil, leading to soil fertility and plant growth. Soils inhabited by earthworms have casts which is enriched by microorganisms. The role of earthworms in such a process as an indicator and biomanager is critically important. Soils could be sustained through the use of organic amendments such as vermicompost and inoculation of earthworms which facilitate humus formation and prevent leaching of nutrients from the soil by their slow release compared with conventional farming using chemical fertilizers (Rao 1994; Thampan 1995; Kale 1996).

Most earthworms are terrestrial organisms which live in the soil and are generally classified as saprophages. Based on their feeding habit, they are classified in detritivores and geophages. Detritivores feed at or near the soil surface. They feed mainly on plant litter or dead roots and other plant debris in the organic matter-rich surface soil horizon or on mammalian dung. These worms are called humus formers and comprise the epigeic and anecic forms. Geophagous worms, feeding beneath the surface, ingest largest quantities of organically rich soil. These are generally called humus feeders and comprise the endogeic earthworms (Ismail 2005).

Epigeics (*Eisenia fetida, Eudrilus eugeniae*) are surface dwellers serving as efficient agents of comminuting and fragmentation of leaf litter. They are phytophagous and generally have no effect on the soil structure as they cannot dig into the soil. Anecics (*Lampito mauritii*) feed on the leaf litter mixed with the soil of the upper layers and are said to be geophytophagous. They may also produce surface casts generally depending on the bulk density of the soil. Endogeic earthworms (*Octochaetona thurstoni*) are geophagous and live within the soil deriving nutrition from the organically rich soil they ingest (Ismail 2005).

10.3 Vermitechnology

Vermitechnology is a method of converting all the biodegradable wastes into useful product i.e., vermicompost, through the action of earthworms. Vermicompost is a sustainable biofertilizer regenerated from organic wastes using earthworm which contains 1.2–6.1 % more nitrogen, 1.8–2.0 % more phosphate, and 0.5–0.75 % more potassium compared to farmyard manure. It also contains hormones such as auxins and cytokinins, enzymes, vitamins, and useful microorganisms such as bacteria, actinomycetes, protozoans, and fungi (Ansari and Ismail 2001). This process of decomposition results in the production of vermicompost. Vermicompost, or castings, is worm manure. It is considered by many in farming arena to be the very good soil improver. The nutrient content of castings is dependent on the material fed to the worms, and worms are commonly fed materials with high nutrient content (Ismail 1997). It is the worm castings that provide these nutrients in a form that is readily available to plants. The biology of the worm's gut facilitates the growth of fungus and bacteria that are beneficial to plant growth.

10.4 Vermicomposting

Vermicomposting is a simple biotechnological process of composting, in which epigeic species of earthworms are used to enhance the process of waste conversion and produce a better end product. Vermicompost is a nutrient-rich organic soil conditioner which can be applied to improve soil conditions for a wide range of soil types. The use of earthworms is very essential in this process, and the worms act for the composing of organic matter into a stable nontoxic material with good structure which has a potential of the high economic value. Also, it acts as soil conditioner for plant growth. Vermicomposting has many environmental benefits that are proven to be an easy way of getting rid of garbage waste. This technique is also beneficial to the soil and results in a lower use of synthetic fertilizers.

10.5 Setting up a Vermicomposting Unit

Vermicomposting units can be set up on many ways. This system can be set up in a large box, a bucket, a bin or a cement bin, a basket, and even in a pit in the soil. It is very important to keep in mind that a vermicomposting unit should be more than 1 m in depth, but may be as long as preferred in width. This structure should be set up in the shade. Organic matter that is added to the unit should be dried to prevent an increase of temperature in the unit. The unit should be kept moist; therefore, watering is very essential. The amount of materials which are layered during the building of the unit depends on the size of the unit which is set up.

The basic layering of a vermicompost in bin or in cemented bin is as follows (Figs. 10.1 and 10.2):

- The basal layer of the vermibed comprises of broken bricks followed by a layer of coarse sand (10 cm thick) in order to ensure proper drainage.
- A layer (10 cm) of loamy soil should be placed at the top. 100 locally collected earthworms should be introduced into the soil.
- Fresh cattle dung is scattered over the soil, and then, it was covered with a 10 cm layer of dried grasses.
- Water is sprinkled on the unit in order to keep it moist.
- The dried grasses along with cattle dung are turned once a week.

Fig. 10.1 Layering in the vermicomposting unit (*Source* Ansari 2012)

Fig. 10.2 Vermicompost preparation procedure in cement bin: **a** site selection, **b** cemented bin placement, **c** collection of earthworms, **d** placement of processed cattle dung on cemented bin and creation of pore in middle to release gas, **e** sieve use for sieving, and **f** final product—vermicompost (*Photo* Md. Ashraful Islam)

- After 60 days, vermicompost units are regularized for the harvesting of vermicompost every 45 days.
- When the layering is completed, the unit should be covered with dried leaves and left for 60 days.
- During the period of these 60 days, organic material and cow dung should be added on a weekly basis, while watering every other day, depending on the moisture content of the material in the bin.

10.6 Harvesting of Vermicompost

Vermicompost should be ready for harvesting in maximum 40–45 days. When the organic material in the unit is changed completely in structure and smells soil-like, it is ready for harvest. The compost should be pressed in the hand to check on moisture content (Fig. 10.3). Before harvesting, no water should be added to the unit for 3–4 days and a heap of the compost should be formed after harvesting. These actions will derive the earthworms in the deeper layers of the unit where the moisture content is slightly higher. The fourth day, the compost can be harvested and is ready to be used for agricultural purpose. This compost can be used directly in the soil and can be stored for 3 months if disposed well in a plastic bag.

10.7 Benefits of Vermicompost

Vermicompost has many benefits on the soil, but has also many economic benefits which are as follows:

1. Source of plant nutrients to the soil
   (a) Improves its physical structure.
   (b) Enriches soil with microorganisms (adding enzymes such as phosphatase and cellulase).

Fig. 10.3 Harvested vermicompost (*Source* Ansari 2012)

(c) Microbial activity in worm castings is 10–20 times higher than in the soil and organic matter that the worm ingests.
(d) Attracts deep-burrowing earthworms already present in the soil.
(e) Improves water holding capacity.

2. Improving crop growth and yield (plant growth)

(a) Vermicompost plays a major role in improving growth and yield of different field crops, vegetables, flower, and fruit crops.
(b) Enhances germination, plant growth, and crop yield.
(c) Improves root growth and structure (rhizosphere).
(d) Enriches soil with microorganisms (adding plant hormones such as auxins and gibberellic acid).

3. Economic benefits

(a) Biowastes conversion reduces waste flow to landfills.
(b) Elimination of biowastes from the waste stream reduces contamination of other recyclables collected in a single bin (a common problem in communities practicing single-stream recycling).
(c) Boost to rural economy.
(d) Less waste land formation.
(e) Low capital investment and relatively simple technologies make vermicomposting practical for less-developed agricultural regions.
(f) It creates the employment opportunity.

4. Eco-friendly environmental factors

(a) Good quality organic soil additives enhance the water holding capacity and nutrient-supplying capacity of soil and also the development of resistance in plants to pests and diseases, thereby providing a sustainable environment in the soil.
(b) Wastes create no pollution, as they become valuable raw materials for enhancing soil health.
(c) Helps to close the "metabolic gap" through recycling waste on-site.
(d) Reduction in greenhouse gas emissions such as methane and nitric oxide (produced in landfills or incinerators when not composted or through methane harvest).

10.8 Nutrient Quality of Vermicompost

Nutrient status of vermicompost (Table 10.1) produced from the organic waste (Shinde et al. 1992) is an excellent biofertilizer, which has been investigated to have favorable influence on the growth and yield parameters of several crops such as paddy, sugarcane, tomato, brinjal, and okra (Ismail 1997, Ansari and Sukhraj

Table 10.1 Physicochemical properties of vermicompost (*Source* Ansari 2012)

Parameters	Vermicompost
pH	6.12
Total salts (ppm)	3148.67
Total nitrogen (%)	1.11
Organic carbon (%)	9.77
C/N ratio	8.80
Available phosphate (ppm)	597.67
Calcium (ppm)	322.33
Magnesium (ppm	137.33
Potassium (ppm)	2428.33
Manganese (ppm)	0.69
Iron (ppm)	0.11
Copper (ppm)	0.01
Zinc (ppm)	2.13

2010). Vermicompost contributes to the supply of essential micronutrients (Kale 1998) and moreover contains growth-promoting substances such as auxins and cytokinins (Krishnamoorthy and Vajranabhiah 1986).

10.9 Vermiwash

Vermiwash is one of the materials produced by vermicomposting which is an "eco-biotechnological process that transforms energy-rich and complex organic substances into a stabilized vermicomposts" primarily through the action of earthworms but with the support of other microorganisms. Vermiwash contains the soluble nutrients that were released in the vermicomposting process (Nath et al. 2009). Organic fertilizers such as vermiwash provide a relatively cost-effective and safe alternative to chemical fertilizers. According to Ansari and Sukhraj (2010), the use of chemical fertilizers is very common in many developing countries. On the other hand, the use of chemical fertilizers can lead to soil damage and reduced soil health and production levels while increasing the incidence of pests and disease and environmental pollution.

10.10 Vermiwash Production

Vermiwash is a liquid that is obtained when water is left to flow slowly through a vermicomposting-like unit. Vermiwash has fertilizing abilities and has also been proven to have a pesticidal action when applied as a foliar spray. The layering of a vermiwash bin is the same as a vermicomposting unit (Fig. 10.4). The exception is

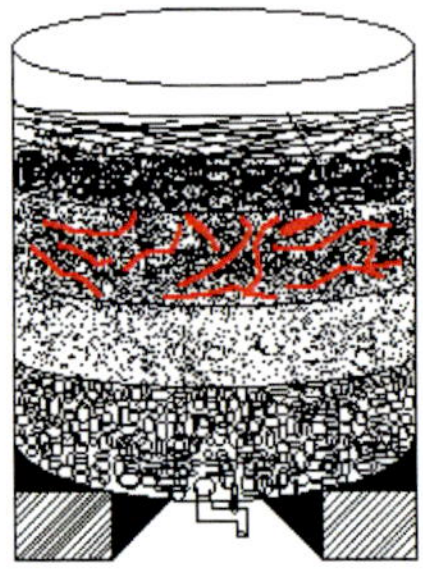

Fig. 10.4 A detailed design of a vermiwash unit (*Source* Ansari 2012)

that this unit consists of a bucket to which a tap is attached at the lowest point to collect the vermiwash when ready. The organic matter that is added to this unit varies from ordinary grass clippings to plant material with pesticidal properties. The organic matter should be dried for 3–4 days to accelerate the composting action and regulate the temperature in the bin.

- The vermiwash unit is set up using buckets.
- A tap is fixed on the lower side of each bucket.
- The bucket is placed on a stand to facilitate collection of vermiwash.
- 5 cm of broken pebbles are placed at the bottom of the buckets followed by 5 cm layer of coarse sand.
- Water is then allowed to flow through these layers to enable the settling of the basic filter unit.
- A 15 cm layer of loamy soil is placed on top of the filter bed.
- Approximately 300 earthworms are introduced into the soil.
- Dried grass and cattle dung are placed on top of the soil.
- The vermiwash unit is left to regularize after 60 days for collection of vermiwash every day.
- Approximately 0.5 l can be collected on a daily basis.

After layering the different materials to the bin, the unit is left for 60 days to regulate with the tap open. Organic matter and cattle dung should be added on a weekly basis as needed. The unit should be watered every other day depending on the moisture content in the bin. Access water should be left to flow through the open tap. Vermiwash will be ready to collect when the liquid flowing through the tap gets pale yellow in color. After that, the tap should be closed and water should be allowed to drip through the unit overnight. The following day, the tap should be opened and the vermiwash should be collected in a plastic container. The color intensity of the vermiwash will differ according to the organic material that is added to the bin. After the first collection, vermiwash can be collected on a daily basis by repeating the same process of adding water to the unit. The vermiwash which is collected can be kept in store for 3 months in plastic containers. Vermiwash can be used by a dilution of 10 % of the vermiwash with water and spray to the desired plant/crop.

10.11 Nutrient Status of Vermiwash

Vermiwash, a liquid fertilizer (Table 10.2) produced by the action of earthworms, contains soluble plant nutrients, some organic acids, mucus, and microbes, which has proved to be effective, both as a biological fertilizer (as a foliar spray) as well as a pesticide (Pramoth 1995; Ismail 1997; Kale 1998).

10.12 Organic Amendments and Impact on Soil

Organic amendments such as vermicompost promote humification, increased microbial activity and enzyme production. Ultimately, it increases the aggregate stability of soil particles, resulting in better aeration (Tisdale and Oades 1982; Dong et al. 1983; Haynes and Swift 1990; Perucci 1990). Organic matter has a property of binding mineral particles such as calcium, magnesium, and potassium in the form of colloids of humus and clay, facilitating stable aggregates of soil particles for desired porosity to sustain plant growth (Haynes 1986). Soil microbial biomass and enzyme activity are important indicators of soil improvement as a result of addition of organic matter (Perucci 1990). Apart from these, earthworm castings are reported to contain plant growth promoters, such as auxins and cytokinins (Krishnamoorthy and Vajranabhaiah 1986).

The high content of organic matter in compost and the resultant effects of the organic matter on the humic fractions and nutrients in soil effectively increase the microbial population, activity, and enzyme production, which in turn increases the aggregation of stability (Tisdale and Oades 1982; Dong et al. 1983; Haynes and Swift 1990; Perucci 1990). Humic acid and fulvic acid are important as persistent

Table 10.2 Physicochemical properties of vermiwash (*Source* Ansari 2012)

Parameters	Vermiwash
pH	7.11
Total salts (ppm)	9841.67
Total nitrogen (%)	0.02
Organic carbon (%)	0.18
Available phosphate (ppm)	48.86
Calcium (ppm)	192.4
Magnesium (ppm	142.53
Potassium (ppm)	245.67
Manganese (ppm)	0.04
Iron (ppm)	2.21 ± 0.04
Copper (ppm)	0.35 ± 0.01
Zinc (ppm)	0.03 ± 0.01

binding agents in mineral organic complexes, and 52–92 % of soil organic matter may be involved in these complexes (Edwards and Bremner 1967; Hamblin 1977). Increased plant litter incorporation improves aggregation, better aeration, and water relationships. Also, the development of mull characteristics can be observed through soils amended with organic inputs. These improvements in soil structure are confirmed by soil morphological studies which are described by Rogaar and Boswinkel (1978). On the contrary, there was a reduction in organic carbon in plots treated with chemical fertilizers, and it may be due to the negligible organic matter as input. Moreover, chemical inputs cause degradation of the soil structure resulting in unfavorable conditions for crop growth and development of that degraded soil (Pagliai et al. 1983a, b; Shipitalo and Protz 1988).

Vermicompost, one of the important types of compost, contains earthworm casts that are reported to be higher in available nitrogen (de Vleeschauwer and Lal 1981) which enhances the activity and number of microorganisms (Satchell and Martin 1984; Satchell et al. 1992). Soil nitrogen becomes higher through the application of vermicompost which is likely to be due to stimulation of microbial activity specifically through increase in the colonization of nitrogen fixers and actinomycetes (Kale 1998; Borken et al. 2002). Much of the effect of application of compost on crop yield and productivity is derived from the plant nutrients, particularly nitrogen in composts (Woodbury 1992; Maynard 1993; Ozores-Hampton et al. 1994).

It is indicated in the several reports that the adequate quantities of phosphorus and potassium are met up to the crop by compost application to the soil (Smith 1992; Maynard 1993; Ozores-Hampton et al. 1994; de Vleeschauwer and Lal 1981). Ultimately, it enhances the activity and number of microorganisms producing acid phosphatases in the soil (Satchell and Martin 1984; Satchell et al. 1984). Synergistically, these specific effects appear to raise phosphorus availability in soils amended with vermicompost (Buchanan and Gliessman 1990).

Vermicompost application in the wheat–paddy cropping system has been reported to increase the crop yield (Sharma and Mittra 1991; Ismail 1997). This is because of nutrients present in vermicompost are readily available to the plants (Ismail 2005; Rajkhowa et al. 2000). The effect of application of organic amendments such as vermicompost on crop yield and production is derived from the plant nutrients, particularly nitrogen (Woodbury 1992; Maynard 1993; Ozores-Hampton et al. 1994). Organic phosphorus solubilized by microbial activity in composts such as the vermicompost is more effective for plant absorption (Mishra and Banger 1986; Singh et al. 1987).

The reduced cost of cultivation, less cost–benefit ratio, and higher net income have been recorded in wheat and paddy cultivation through vermitech compared with the use of chemical fertilizers along with the other economically important crops such as peanut (*Archis hypogaea*) and brinjal (*Solanum melongena*) by organic methods (Ismail 1997). Organic farming has been proved to be environment friendly, sustainable, and cost-effective (Reganold et al. 2001).

Significant effect of yields on different crops such as tomato (*Lycopersicum esculentum*), brinjal (*Solanum melongena*), and okra (*Abelmoschus esculentus*) has been found in the experiments on the effect of earthworms and vermicompost application (Ismail 1997, 2005). Application of composts such as vermicompost could contribute to increase the availability of food (Ouédraogo et al. 2001). This is attributed to better growth of plants and higher yield by slow release of nutrients for absorption with additional nutrients such as gibberellin, cytokinin, and auxins, by the application of organic inputs such as vermicompost in combination with the vermiwash (Raviv et al. 1998; Subler et al. 1998; Lalitha et al. 2000). The yield of potato and the average weight of potato tubers were significantly higher in plots treated with vermicompost. This may be attributed to the increased bioavailability of phosphorus by the application of organic amendment in the form of vermicompost (Erich et al. 2002).

Addition of organic manures such as vermicompost and vermiwash to soil augments the crop growth and yield (Lalitha et al. 2000). The yields of spinach and onion in response to diluted vermiwash along with vermicompost were highly significant which may be due to increased availability of more exchangeable nutrients in the soil by the application of vermiwash along with vermicompost (Ponomareva 1950; Finck 1952; Nijhawan and Kanwar 1952; Nye 1955; Atlavinyte and Vanagas 1973, 1982; Czerwinski et al. 1974; Watanabe 1975; Cook et al. 1980; Tiwari et al. 1989).

Concern about the environment and the economic and social impacts of chemical or conventional agriculture has led to many thinking groups seeking alternative practices that will make agriculture more suitable. Organic farming practices and systems have shown promise in mitigating some of the detrimental effects of chemical-dependent conventional agriculture on the environment (Reganold et al. 1993).

10.13 Fact Sheet on Bangladesh

Compost is one kind of organic manure which is prepared by mixing plant matter and animal waste available. There are few large-scale compost producers in Bangladesh, including Annapurna Agro Service and Waste Concern. NGOs such as Grameen Shakti and Grameen Krishok Shohayak Sangstha; autonomous organizations such as Rural Development Academy; and research/academic institutes such as Bangladesh Agricultural University (BAU) and Bangladesh Agricultural Research Institute (BARI) are also involved in research and production of different types of composts. Composts are also produced by farmers through traditional methods, but mainly use cattle dung and ashes (Rashid 2011).

Two types of composting methods practiced in Bangladesh are heap and pit methods. The preferable size of heap is 1.5–2 m wide and <1.5 m height. During compost preparation, water, air, and quick starter, such as urine and cattle dung, are essential materials. Moisture is maintained at 15–20 % water content depending on the weather conditions (FAO 2011). Naturally occurring microorganisms such as fungi, bacteria, worms, and insects actively turn the waste materials into compost. The surface should be covered to prevent drying of material and to maintain consistent moisture levels. After one week, a bamboo stick or pole is inserted to determine the compost progress, using temperature, and smells as indicators. The excessive heat produced from composting destroys the majority of weed seeds, pathogens, and pests. The whole procedure requires the duration of 2–3 months. The fungus *Trichoderma harzianum* is commonly mixed with materials for rapid composting which called trichocompost.

Green manure is also used in Bangladesh, which involves green plants or leaves (at vegetative stage before flowering) incorporated into soil for decomposition. It adds organic matter and improves physical structure of the soil. Either leguminous (*Sesbania sp.,* mung bean, cowpea, grasspea, lentil, sun hemp, Ipilipil, blackgram) or nonleguminous (mustard, wheat, radish, carrot, jowar, sunflower) crops are used. Legume crops are popular as a green manure crop because they have the ability to fix nitrogen from the air with root nodule bacteria. After incorporation, the next crop is planted after 3–4 weeks of decomposition.

Farmers in Bangladesh have produced vegetables traditionally for many centuries, which by default was organic production. Traditional practices have been almost completely lost through the introduction of the green revolution and the need to feed the ever-growing population. Excessive use of synthetic fertilizer, pesticide, and other chemicals is destroying the lands. Subsequently, health hazards and unsafe environment now require growers to seek alternate agricultural options. In this era, people seek out new innovative technology which will help increase production as well assure human health and the environment. The following strategies can be followed for establishment of organic food production.

- Increase the awareness organic vegetable among producers and consumers.
- Develop an organic agricultural value chain and improve the marketing strategies for organic production.
- Develop national organic policy and government organic certification and regulation system. It will help to improve the marketing system domestically and abroad.
- Fortify the research on high value crops and minor crops for the promotion of domestic and export marketing of organic products.
- Coordinate the collaboration of national and international scientists for organic production and build up the infrastructure in different Bangladesh organizations and institutes.
- Create policy to register and patent biopesticides/biofungicides/organic fertilizers and standardize their application rate for higher production and available to develop the value chain market.

10.14 Conclusion

Soils are fundamental to the well-being and productivity of both agriculture and natural ecosystems. Soil is an integral system which can be maintained for sustainable agriculture. The continuous worldwide soil degradation by erosion, chemicals, acidification, and physical abuse requires management in terms of soil quality. The use of organic amendments augmented with vermitechnology could be adopted as a means for crop production and soil stability.

The use of combinations of organic amendments such as vermiwash and vermicompost can effectively bring about an improvement in soil quality. Also, it can increase in microbial population and enhance crop productivity which would be beneficial in the long term for the stability of crop production. Finally, vermitechnology could be applied for the successful of sustainable soil productivity considering all aspects such as studies on soil, soil health, yield of crops, and cost effectiveness of vermitechnology.

References

Ansari AA (2008) Effect of vermicompost on the productivity of potato (*Solanum tuberosum*) Spinach (*Spinach oleracea*) and Turnip (*Brassica campestris*). World J Agric Sci 4(3):333–336

Ansari AA (2012) Permutations and combinations of organic waste-Vermitechnology. Lambert Academic publishing, Germany, pp 76

Ansari AA, Ismail SA (2001) A case study on organic farming in Uttar Pradesh. J Soil Biol Ecol 27:25–27

Ansari AA, Sukhraj K (2010) Effect of vermiwash and vermicompost on soil parameters and productivity of okra (*Abelmoschus esculentus*) in Guyana. Pak J Agric Res 23(3–4):137–142

Atlavinyte O, Vanagas J (1973) Mobility of nutritive substances in relation to earthworm numbers in the soil. Pedobiologia 13:344–352

Atlavinyte O, Vanagas J (1982) The effect of earthworms on the quality of barley and rye and grain. Pedobiologia 23:256–262

Borken W, Muhs A, Beese F (2002) Changes in microbial and soil properties following compost treatment of degraded temperate forest soils. Soil Biol Biochem 34:403–412

Buchanan RA, Gliessman SR (1990) The influence of conventional and compost fertilization on phosphorus use efficiency by broccoli in a phosphorus deficient soil. Am J Altern Agric 5:38

Cook AG, Critchley BR, Critchley U (1980) Effects of cultivation and DDT on earthworm activity in a forest soil in the sub-humid tropics. J Appl Ecol 17:21–29

Czerwinski Z, Jakubczyk H, Nowak E (1974) Analysis of sheep pasture ecosystem in the Pieniny Mountains (The Carpathians) XII. The effect of earthworms on pasture soil. Pol J Ecol 22:635–650

De Vleeschauwer D, Lal R (1981) Properties of worm casts under secondary tropical forest regrowth. Soil Sci 132:171–181

Dong A, Chester G, Simsiman GV (1983) Soil dispersibility. J Soil Sci 136:208

Edwards AP, Bremmer JM (1967) Microaggregates in soils. J Soil Sci 18:64

Erich MS, Fitzgerald CB, Porter GA (2002) The effect of organic amendments on phosphorus chemistry in a potato cropping system. Agric Ecosyst Environ 88:79–88

FAO (2011) How to make and use compost in Climate change and food systems resilience in Sub-Saharan Africa. Edn Lim Li Ching Sue Edwards and Nadia El-Hage Scialabba Itally 385–436
Finck A (1952) Ökologische und Bodenkundliche Studien über die Leistungen der Regenwürmer für die Bodenfruchtbarkeit. Z PflErnähr Düng 58:120–145
Hamblin AP (1977) Structural features of aggregates in some East Anglian silt soils. J Soil Sci 28:23
Haynes RJ (1986) The decomposition process mineralization, immobilisation, humus formation and degradation. In: Haynes RJ (eds) Mineral nitrogen in the plant-soil system. Academic Press, NewYork
Haynes RJ, Swift RS (1990) Stability of soil aggregates in relation to organic constituents and soil water content. J Soil Sci 41:73
Ismail SA (1997) Vermicology: the biology of earthworms. Orient Longman Press, Hyderabad, 92 pp
Ismail SA (2005) The Earthworm book. Other India Press, Mapusa Goa, 101 pp
Kale RD (1996) Earthworms-the significant contribution to organic farming and sustainable agriculture. In: Proceeding of national seminar of organic farming and sustainable agriculture, pp 52–57
Kale RD (1998) Earthworm cinderella of organic farming. Prism Book Pvt Ltd, Bangalore, p 88
Kirk JL, Beandette LA, Hart M, Moutoglis P, Klironomos JN, Lee H, Trevors JT (2004) Methods of studying soil microbial diversity. J Microbiol Methods 58:169–188
Krishnamoorthy RV, Vajranabhaiah SN (1986) Biological activity of earthworm casts: an assessment of plant growth promoter levels in the casts. Proc Indian Acad Sci (Anim Sci) 95:341–351
Lalitha R, Fathima K, Ismail SA (2000) Impact of biopesticides and microbial fertilizers on productivity and growth of *Abelmoschus esculentus*. Vasundhara Earth 1 & 2:4–9
Maynard A (1993) Evaluating the suitability of MSW compost as a soil amendment in field growth tomatoes. Part A: yield of tomatoes. Compost Sci Utilization 1:34
Mishra MM, Banger KC (1986) Rock phosphate comprising: transformation of phosphorus forms and mechanisms of solubilization. Biol Agric Hortic 3:331
Nath G, Singh K, Singh D (2009) Chemical analysis of vermicomposts/vermiwash of different combinations of animal agro and kitchen wastes. Aust J Basic Appl Sci 3(4):3671–3676
Nijhawan SD, Kanwar JS (1952) Physico-chemical properties of earthworm castings and their effect on the productivity of soil. Indian J Agric Sci 22:357–373
Nye PH (1955) Some soil-forming processes in the humid tropics. IV the action of soil fauna. J Soil Sci 6:78
Ouédraogo E, Mando A, Zombré NP (2001) Use of compost to improve soil properties and crop productivity under low input agricultural system in West Africa. Agric Ecosyst Environ 84:259–266
Ozores-Hampton M, Schaffer B, Bryan HH, Hanlon EA (1994) Nutrient concentrations, growth and yield of tomato and squash in municipal solid-waste-amended soil. Hortic Sci 29:785
Pagliai M, Bisdom EBA, Ledin S (1983a) Changes in surface structure (crusting) after application of sewage sludges and pig slurry to cultivated agricultural soils in northern Italy. Geoderma 30:35–53
Pagliai M, LaMarca M, Lucamante G (1983b) Micromorphometric and micromorphological investigations of a clay loam soil in viticulture under zero and conventional tillage. J Soil Sci 34:391–403
Perucci P (1990) Effect of the addition of municipal solid-waste compost on microbial biomass and enzyme activities in soil. Biol Fertil Soils 10:221
Ponomareva SI (1950) The role of earthworms in the creation of a stable structure in ley rotations. Pochvovedenie 476–486
Pramoth A (1995) Vermiwash-A potent bio-organic liquid "Ferticide". M.Sc dissertation University of Madras, 29 pp

Rajkhowa DJ, Gogoi AK, Kandal R, Rajkhowa KM (2000) Effect of vermicompost on Greengram nutrition. J Indian Soc Soil Sci 48:207–208

Rao NS (1994) Model scheme on financing vermicompost. National Bank Newsl 5:1–5

Rashid SZ (2011) Composting and use of compost for organic agriculture in Bangladesh. In: Proceedings of the 4th international conference for the development of integrated pest management in Asia and Africa, pp 20–22

Raviv M, Zaidman BZ, Kapulnik Y (1998) Compost Sci Utilization 6:46–52

Reganold JP, Palmer AS, Lockhart JC, Macgrogor AN (1993) Soil quality and financial performance of biodynamic and conventional farms in New Zealand. Science 260:344–349

Reganold JP, Glover JD, Andrews PK, Hinman HR (2001) Sustainability of three apple production systems. Nature 410:926–930

Rogaar H, Boswinkel JA (1978) Some soil morphological effects of earthworm activity, field data and X-ray radiography. Neth J Agric Sci 26:145–160

Satchell JE, Martin K (1984) Phosphatase activity in earthworm species. Soil Biol Biochem 16:191

Satchell JE, Martin K, Krishnamoorthy RV (1992) Stimulation of microbial phosphatase production by earthworm activity. Soil Biol Biochem 16:195

Sharma AR, Mittra BN (1991) Effect of different rates of application of organic and nitrogen fertilisers in a rice- based cropping system. J Agric Sci 117:313

Shinde PH, Naik RK, Nazirkar RB, Kadam SK, Khaire VM (1992) Evaluation of vermicompost. In: Proceeding of national seminar of organic farming. MPKV, Pune

Shipitalo MJ, Protz R (1988) Factors influencing the dispersibility of clay in worm casts. Soil Sci Soc Am J 52:764–769

Singh CP, Singh YP, Singh M (1987) Effect of different carbonaceous compounds on the transformation of soil nutrients. II. Immobilization and mineralization of phosphorus. Biol Agric Hortic 4:301

Smith SR (1992) Sewage sludge and refuse composts as peat alternatives for conditioning impoverished soils: effects on the growth response and mineral status of Petunia grandiflora. J Hortic Sci 67:703

Subler S, Edwards CA, Metzer J (1998) Biocycle 39:63–66

Thampan PK (1995) Perspectives on organic agriculture. In: Thampan PK (ed) organic agriculture. Peekay Tree Crops Development Foundation, Cochin, pp 1–38

Tisdale JL, Oades JM (1982) Organic matter and water-stable aggregates in soil. J Soil Sci 33:141

Tiwari SC, Tiwari BK, Mishra RR (1989) Microbial populations, enzyme activities and nitrogen-phosphorus-potassium enrichments in earthworm casts and in the surrounding soil of a pineapple plantation. Biol Fertil Soils 8:178–182

Watanabe H (1975) On the amount of cast production by the megascolecid earthworm Pheretima hupeiensis. Pedobioligia 15:20–28

Woodbury PB (1992) Trace elements in municipal solid waste composts: a review of potential detrimental effects on plants, soil biota, and water quality. Biomass Bioenergy 3:239

Chapter 11
Composting, Crop Rotation, and Cover Crop Practices in Organic Vegetable Production

Ajay Nair and Kathleen Delate

Abstract For nearly a decade, there has been an increased awareness toward food quality, health standards, and global environmental issues in our communities. In that context, adoption of organic production practices has been increasing rapidly in vegetable production. Organic farming is grounded in a holistic view of agriculture that aims to reflect a profound interrelationship between on-farm living biota, farm production, and the overall environment. Organic agriculture has emerged as a powerful tool in re-establishing production practices that are self-sufficient, promote biodiversity, and support practices that conserve soil, water, and the environment. Organic production systems utilize practices such as composting, crop rotation, and use of cover crops, all of which have a positive impact on soil physical, chemical, and biological properties. Although these practices are widely used, there is still uncertainty among growers when it comes to the actual process of composting, compost nutrient concentration and availability, use of compost in transplant mixes, and application rates. Similarly, other areas that need attention are crop rotation, sequence of crops within a rotation, and integration of cover crops in these rotations. Cover crops have an important role in reducing soil erosion, suppressing weeds, improving soil structure and water holding capacity, and increasing soil organic matter. This chapter will highlight the role of composting, use of compost, crop rotation, and cover crops in organic vegetable production systems. This chapter will discuss in detail the composting process, raw materials used, composting methods, quality assessment of compost, and potential avenues where compost can be used in organic vegetable production. The crop rotation portion of this chapter will highlight various crop rotation plans and strategies that growers could utilize to improve soil quality, break pest and disease cycles, and increase yields. The chapter will also provide information on cover crop types, their planting, management, benefits, and challenges in organic vegetable cropping systems. Organic production systems are complex and dynamic. Understanding techniques and practices that directly influence soil is critical in building a production system that is self-sustaining, strong, and

A. Nair (✉) · K. Delate
Department of Horticulture, Iowa State University, 106 Horticulture Hall, Ames, IA 50011, USA
e-mail: nairajay@iastate.edu

D. Nandwani (ed.), *Organic Farming for Sustainable Agriculture*, Sustainable Development and Biodiversity 9,
DOI 10.1007/978-3-319-26803-3_11

resilient. A better understanding of such practices is of paramount importance to build, strengthen, and support organic vegetable production.

Keywords C:N ratio · Composting · Conservation tillage · Cover crops · Crop rotation · Green manure · Soil quality · Vegetable production

11.1 Introduction

As agriculture evolved from agrarian to industrial methods, heavy reliance on synthetic fertilizers and pesticides followed. Uncontrolled and indiscriminate use of those resources led to increases in the number of issues related to environmental pollution, habitat destruction, and risks to human health. Since the last decade, there has been an increase in awareness, among growers and consumers, toward food quality, health standards, and global environmental issues. Coupled with environmental concerns, rising energy costs and shrinking profit margins have motivated growers to transition and adopt environmentally sound production practices. Organic agriculture has thus emerged as a powerful tool in re-establishing production practices that are self-sufficient and biodiverse and support practices that conserve soil and water. Specific benefits of organic agriculture include increased biodiversity; improved soil physical, chemical, and biological properties; lower levels of soil erosion; reduced nitrate pollution; reduced amount of pesticides in the environment; and enhanced water quality. The US organic acreage has increased from 1.5 M acres in 1995 to 5.1 M in 2011 (USDA-NASS, 2011), with annual organic sales at 11 % in 2014 (OTA,2015). Demand-driven organic food sales increased from $3.6 billion in 1997 to $39.1 billion in 2011 (Organic Trade Association,2015).

Soil fertility and health are the foundation of organic production. One of the most aggressively sought after means of increasing soil fertility and improving soil health is through increasing soil organic matter. The nutrient sources most often used in organic production to increase soil organic matter are manure, compost, and cover crops. Manure and compost not only add organic matter, but they also supply nutrients for crop production, including micronutrients. Proper use of manure and compost in cropping systems is essential from both a crop production and environmental standpoint. Compost that is immature (not properly done) can detrimentally affect plant growth by tying up nitrogen and producing harmful compounds that can stunt the growth or even kill sensitive plant species. Applying too low rates can lead to nutrient deficiency and reduced yields. On the other hand, excessive applications can lead to nitrate leaching, phosphorus runoff, and excessive vegetative growth of some crops (Rosen and Bierman 2005). Thus, understanding the process of composting, compost nutrient composition and availability, and proper handling techniques is important when utilizing compost as a major source of nutrients. Sections below will discuss in detail the composting process, raw materials, composting methods, storage and handling, quality assessment of compost, and potential avenues where compost can be used in organic vegetable production. Another important tenet/principle of this chapter is the focus on crop rotation, which is critical to build healthy soils, break pest and diseases cycles, reduce weed pressure, and recycle soil nutrients. Integration of cover crops in vegetable cropping systems, with respect to crop rotation, will also be discussed.

11.2 Composting

Composting can be defined as decomposition of organic matter under aerobic conditions into humus-like substances and minerals by the action of microorganisms combined with chemical and physical reactions (Piegne and Girardin 2004). Composting stabilizes nutrient content of manures and other organic materials and releases nutrients slowly, minimizing nutrient loss and potential environmental contamination (Evanylo et al. 2008). Depending on the raw material used, the time required to produce a mature batch of compost could range anywhere from 6 to 8 months. In addition, the process utilized for composting also dictates the length of the composting process. To be successful, composting process must be carefully managed from the mixing of the initial ingredients through the high-temperature phase to the maturation phase when the compost is deemed ready for use. Preparation of high-quality compost requires appropriate raw materials, proper temperature and moisture management, and a good understanding of the science behind the composting process.

11.2.1 Composting Process

Composting is predominantly an aerobic or oxygen-requiring process in which microorganisms consume oxygen while feeding on the organic matter. In doing so, they produce carbon dioxide, water, heat, miscellaneous gaseous by-products, and compost (Fig. 11.1) (Stofella and Kahn 2001). As soon as the appropriate raw materials and water are mixed and brought together in a pile, composting process starts. In the presence of oxygen, microorganisms consume and start decomposing the organic matter (Fig. 11.1). The major group of microorganisms that participate in composting are bacteria, fungi, and actinomycetes. In terms of proportion, bacteria are about 100 times more prevalent than fungi (Poincelet 1977). Major genera of bacteria include *Bacillus, Pseudomonas,* and *Arthrobacter.*

The microbial population of bacteria, fungi, and actinomycetes change during the composting process. Bacteria tend to flourish during the early stages of composting. Within bacteria, initially there is higher population of mesophilic bacteria that grow actively in the 35–45 °C range. Due to rapid growth of bacteria, there is generation of large quantities of metabolic heat energy which raises the temperature of the compost pile. As the temperature crosses 45 °C, it favors the growth of

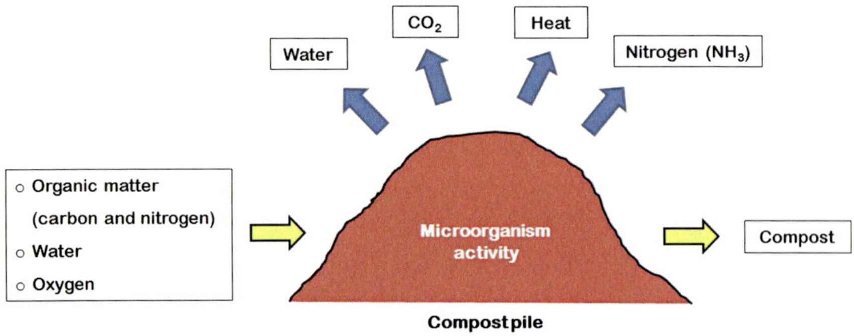

Fig. 11.1 Schematic of composting process

thermophilic bacteria (heat loving bacteria). Increased microbial activity of thermophilic bacteria raises the temperature of the pile to 65–70 °C. As the oxygen levels deplete, microbial activity reduces and the temperature of the pile falls. At this point, the compost pile should be agitated or mixed to infuse oxygen and for the bacteria cycle to restart. After successive agitations and depletion of easily degradable compounds, bacterial activity and population decreases. The pile enters final or maturation phase and is inhabited by mesophilic bacteria. Also, population of fungi and actinomycetes increase as they feed on resistant organic materials that remain in the pile. The most commonly observed fungi include: *Aspergillus*, *Penicillum*, *Fusarium*, and *Trichoderma* (Bhardwaj 1995). The final phase of composting is the curing phase where the pile no longer reheats after agitation. Curing phase furthers the colonization of the compost pile with common microorganisms, protozoa, worms, insects, and other large organisms that feed on microorganisms and organic matter. The concentration of nitrate-nitrogen also increases as the pile cools down and enters the maturation and curing phase. The curing time for a pile varies based on the length of the active composting phase which in turn depends on raw material, composting conditions, and management of the pile. Usually, the recommended time for curing is around 30 days; however, a longer period is necessary if active composting was not completed (Rynk 1992).

11.2.2 Composting Methods

Aeration is crucial element of the composting process as it facilitates aerobic decomposition of the compost pile. Agitation of the compost pile to maintain the presence of oxygen for rapid decomposition of organic material is desirable. Aerobic decomposition is not the only driving mechanism in the composting process. Anaerobic decomposition also takes place in the compost pile and contributes to the overall decomposition process (Stofella and Kahn 2001). Depending upon how the compost pile is handled, composting can be broadly classified into three methods: passive composting, windrow composting, and aerated static pile. There are other methods used commercially, but passive composting, windrow composting, and aerated static pile methods are common on organic vegetable production farms.

11.2.2.1 Passive Composting

In passive composting, there is minimal disturbance made to the compost pile. In this method, organic materials are placed in a pile and left for extended period of time to decompose. Usually, growers do not maintain the correct pile temperature and they usually make no attempt to adjust the moisture content of the pile. Aeration, which is a critical factor for composting process, happens passively by diffusion, natural air movement, and thermal convection. Under outdoor conditions, wind can facilitate oxygen transfer to the pile, although it is not a major driving force for aerating the pile (Haug 1993). Thermal convection plays a significant role in passive composting (Lynch and Cherry 1996). As the temperature of the pile rises, gases in the pile also heat up and rise. This creates a vacuum in the pile resulting in the movement of cool fresh air from the surrounding (Rynk et al. 1992). This process could take place for a few days after the pile is set up, but due to lack of further mixing or agitation of the pile, oxygen movement to the pile is restricted and the pile starts decomposing anaerobically. In addition, the moisture content of such a pile exceeds levels that are required to maintain a porous structure of the pile for air movement. All these factors lead to low temperature, slow decomposition, and release of malodorous gases including hydrogen sulfide from the pile. Passive composting method is not approved for certified organic production (USDA 2010). To meet National Organic Program (NOP) requirements, piles that are passively composted must be aerated to sustain microbial activity and adequate temperatures. Growers often install perforated pipes at the base of the pile and, in some cases, install blowers to force air through the pile. This is called the aerated static pile method which is explained later in the chapter.

11.2.2.2 Windrow Composting

This is the most commonly practiced composting method in organic vegetable production. In this method, the mixture of raw material is placed in long narrow piles or windrows that are agitated or turned on a regular basis (Fig. 11.2). Windrow dimensions vary with the materials being composted and the turning equipment available at the farm. Typically, windrow height range from 4 to 5 m for fluffy material such as leaves and from 1 to 2 m for dense material such as manure (Rynk et al. 1992). It is important to maintain the right height for the windrow as large windrows can develop anaerobic zones near its center, which is undesirable for earlier mentioned reasons. Windrows that are too small also pose problems as they lose heat quickly and are not able to maintain high temperatures that are needed to

Fig. 11.2 Windrow composting consists of long and narrow piles that are agitated or turned on a regular basis

evaporate moisture and kill harmful pathogens. Windrow width varies from 3 to 6 m. The windrow is turned or mixed on a periodic basis to provide oxygen throughout the pile. Turning of the windrow also helps to rebuild the pore space in the pile that is lost due to decomposition and settling of the organic material, and releases trapped heat, water vapor, and gases. Turning also distributes water, nutrients, and microorganisms throughout the windrow.

Aeration of the windrow happens at the time of turning and also by diffusion, wind, and convection between turnings. The number of times a windrow pile is turned is determined by many factors, including the pile temperature, moisture content, and porosity of the pile. Weather conditions at the farm also influence the turning schedule. For certified organic production, NOP stipulates the number of times windrow compost piles must be turned when composting plant and animal materials. According to the NOP, compost piles in the windrow composting system should maintain temperatures between 55 and 77 °C for 15 days, during which the materials must be turned a minimum of five times (NOP 2010). A common strategy adopted by growers is to turn the windrow based on temperature patterns. Generally, a pile should be turned when its interior temperature falls below 50 °C. This results in the turning of the windrow every 2–3 days for the first 2–3 weeks, followed by weekly turnings for another 6–8 weeks. Growers often use thermometers with long stem, 1–1.5 m, to measure temperatures at 50–75 ft intervals along the windrow. Turning of the windrow in small to moderate scale farms usually takes place using a front end loader or a bucket loader. Specialized equipment such as tractor-assisted windrow turners can also be used as they are highly efficient and save time. Organic vegetable production farms mostly use the windrow composting method. It is a proven, simple, and successful method of composting that easily accommodates wide range of feedstock, equipment, farm size, and management strategies.

11.2.2.3 Aerated Static Pile

This method is an improvement of the passive composting system. As the name indicates, in this method, a static pile is aerated using a combination of pipes and fans/blowers to enhance the pace of decomposition (Stoffella and Kahn 2001). Fans or blowers either blow air into the pile or suck air out of the pile. Also, the base of the compost pile is built with wood chips, chopped straw, etc., to provide porosity to the pile. Underneath the base is a perforated aeration pipe that provides oxygen and removes water vapor, carbon dioxide, and other products of decomposition. No turning or agitation of the pile occurs after the pile has been set up. Growers also cover the static pile with a layer of finished compost, straw, or wood chips to insulate the pile and retain heat. The feedstock material for the compost pile needs to be well mixed before being placed on the pile as there is no future turning or agitation of the pile. A common practice among growers is to add wood chips as an amendment to the feedstock.

Size of the aerated static pile varies with feedstock, equipment available to make the pile, and the capacity of the fan/blower. Aerated static piles range in height from 2 to 4 m. The length of the pile depends on the efficiency and uniformity of air distribution of the pipe and ducts. Typically, the length of the pile ranges from 30 to 70 m. One of the advantages of this method over the windrow method is that it requires less area and the efficiency of aeration is higher, although uniform distribution of forced air largely depends on the porosity of the pile and how well feedstock was mixed. This method is also a proven method of composting and can produce high-quality compost in 6–8 weeks.

11.2.3 Raw Materials

Raw materials for composting are organic by-products or waste materials. These materials come in different shape, sizes, and chemical composition. Some commonly used raw materials for composting include animal manure, crop residues, bedding, and processing wastes. The elemental composition of the final compost largely depends on the chemical composition of the feedstock. Of all the elements, two most important elements are carbon and nitrogen. During the composting process, nitrogen concentration of the pile directly influences microbial population growth and carbon serves as the energy source. The most important aspect of the feedstock is the carbon to nitrogen ratio. Higher C:N ratio raw materials (more than 40:1) can immobilize nitrogen and slows the composting process (Coyne and Thompson 2006) and lower ratios lead to the loss of N as ammonia, although higher and lower ratios are debatable. The most accepted and agreed upon C:N ratio of the feedstock is between 25 and 35 (Hamoda et al. 1998). Table 11.1 outlines C:N ratios of commonly used feedstocks for composting.

Table 11.1 C:N ratios of common composting materials

Raw material	C:N ratio	Raw material	C:N ratio
Crop residue and fruit/vegetable		*Manures*	
Coffee grounds	20:1	Broiler litter	14:1
Corn stalks	60–73:1	Dairy manure	19:1
Cull potatoes	18:1	Horse manure	30:1
Fruit wastes	18–49:1	Poultry	6:1
Hay	15–32:1	Sheep	16:1
Straw	48–150:1	Swine	14:1
Vegetable wastes	11–13:1		
Wood and yard waste			
Bark–hardwood	223:1		
Bark–softwood	496:1		
Grass clippings	17:1		
Leaves	54:1		
Wood chips	600:1		

Source Rynk et al. (1992). On-Farm Composting Handbook, NRAES 54

In addition to considering the C:N ratio of the feedstock, other factors that need to be considered include the degradability and odor potential of the feedstock. Not all organic materials degrade equally. Biodegradability and bioavailability of organic materials vary considerably based on the source material (Tuomela 2000). For example, the biodegradability of wood chips could vary drastically depending on their source (Allison 1965). Biodegradability largely depends on the form in which carbon exists in the pile. He et al. (1995) investigated the C content of compost and categorized C into three classes: total extractable organic carbon, carbonate carbon, and residual carbon and estimated their distribution to be 20, 8, and 72 %, respectively. Feedstocks and raw materials also impact the odor potential of the pile. Considering ways to mitigate or reduce odor during the composting process is critical. It is advisable to distance composting sites from schools, hospitals, or residential neighborhoods. The source of odor stems from three things: odorous raw materials, ammonia generated during composting, and anaerobic conditions within the compost pile. Feedstocks that have higher odor potential include fish wastes, swine manure, and other forms of liquid manure. Materials such as crop residue, leaves, and sawdust present little or no odor issues. A good mix of feedstock, appropriate moisture content, and frequent turning or agitation reduce odor problems. In aerated static piles, lining the surface with peat moss or finished compost helps to trap odor-forming gases (Rynk et al. 1992).

11.2.4 Maturity and Quality Assessment

Composting is considered complete after the temperature of the pile does not rise, even after turning, and subsides to near-ambient levels. Fully mature compost is well decomposed, stable, and has an earthy smell. Assessing compost maturity and quality is necessary before it is incorporated in the soil and potting mix or used as landscape mulch. Maturity is a general term that describes fitness of the compost, depending on the end use of the product. There are other physical, chemical, and biological parameters that can be tested to ascertain compost maturity. Because of the diversity in origin and type of feedstock in the compost, it is difficult to use a single method to evaluate the maturity of given compost. Below are certain key aspects that help with the assessment of maturity and quality of compost.

11.2.4.1 Sensory Assessment

Although not a rigorous assessment, growers often test maturity of the compost based on its color and odor. A pile that is not fully composted usually smells foul and is considered immature. Leege and Thompson (1997) developed a standardized matrix including color and odor for evaluating the maturity of the compost (Table 11.2; Method 9.03A). For on-farm use of compost, the sensory method could be considered as viable indicator but compost that is used in potting mixes or sold to the general public needs to be scrutinized harder for maturity and quality.

11.2.4.2 Carbon and Nitrogen Ratio

Although carbon and nitrogen concentrations cannot be used as an indicator of compost maturity, their ratio is often taken into account for practical reasons. A compost with a high C:N ratio (usually higher than 30:1) could lead to soil immobilization of N and detrimentally affect plant growth. Ideal compost feedstock mixture contains a C:N ratio of 30:1 which gradually comes down to 20:1 or lower by the end of the composting process. After composting is over, the rest period or the curing phase, which usually lasts for a month, plays a critical role in affecting

Table 11.2 Compost maturity assessment based on sensory and chemical indicators

Method	TMECC[a] method number	Value for matured compost
Color	9.03A	Black to very dark
Odor	9.03A	Earthy, soil like, no odor
C:N ratio	9.02A	15–20:1
Inorganic N	9.02C	More nitrate-N than ammonium-N

[a]*TMECC* Tests for the Examination of Composting and Compost (Leege and Thompson 1997)
Source Stoffella and Kahn (2001)

the form of nitrogen in the pile. During the curing phase, C:N ratio decreases, the pH of the pile shifts toward neutral, and conversion of NH_4–N to NO_3–N occurs. Plants absorb both forms of nitrogen, but NH_4–N, at higher concentrations, can cause temporary stunting and burning of the foliage of susceptible young plants. Vegetable crops absorb most of their N in NO_3–N form although in their younger stages they will absorb NH_4–N form as well. Fully mature compost has gone through the curing phase and contains more of the nitrate (NO_3–N) than ammonium form (NH_4–N) of nitrogen.

11.2.4.3 Electrical Conductivity and pH

Measuring the electrical conductivity (EC) (soluble salts concentration) and pH is another tool growers often use to assess compost maturity. The pH range for most finished compost is between 6.0 and 8.0 (Sullivan and Miller 2001). The final pH of the finished compost largely depends upon feedstock used and the management technique during composting. Electrical conductivity provides information on total salt concentration. Similar to pH, EC largely depends on feedstock used. Compost with high salt concentrations can affect seed germination and stunt root growth. Usually, compost with EC higher than 3 dS/m is considered phytotoxic for seed germination; however, compost EC can be in the 8–10 dS/m range if the intended use of compost is soil incorporation in the field. In addition, certain vegetable crops are more susceptible to higher salt concentration than others. For example, crops such as onions and beans are highly sensitive to high salt concentrations (Bischoff and Werner 1999). Table 11.3 lists salt tolerance of few vegetable crops.

Table 11.3 Sensitivity rating of vegetable crops to soluble salts

Crop	Sensitivity rating[a]	Crop	Sensitivity rating
Asparagus	T	Muskmelon	MS
Bean	S	Onion	S
Broccoli	MS	Pepper	MS
Cabbage	MS	Potato	MS
Carrot	S	Pumpkin	MS
Corn, sweet	MS	Spinach	MS
Cucumber	MS	Squash, zucchini	MT
Eggplant	MS	Sweet potato	MS
Lettuce	MS	Tomato	MS

[a]*MS* Moderately Sensitive, *MT* Moderately Tolerant, *S* Sensitive
Source Bischoff and Werner (1999)

11.2.4.4 Respiration and Phytotoxicity Tests

Tests that measure oxygen consumed and carbon dioxide released from a finished compost pile can help determine the maturity of the pile. The rate of compost respiration determined over three days by carbon dioxide respirometry at 37 °C is a standard method of measuring compost stability by the US Composting Council (Thompson et al. 2002; TMECC 05.08-B). These tests are robust and provide valuable information but require preconditioning of the sample and need to be conducted under laboratory conditions. At small- and medium-scale organic vegetable farms, conducting these tests can be challenging. A rapid semiquantitative test called the Solvita® test is now being used by many of the states in USA (Woods End Research Laboratory 1999). This test involves the use of colorimetric pads sensitive to carbon dioxide and ammonia in a jar that contains a fixed volume compost sample. The pads are left in the jar for 4 h; they absorb carbon dioxide and ammonia and change color. The color change on the pad surface is visually compared to a precalibrated coded color chart. Solvita test accuracy of carbon dioxide and ammonia measurements using Solvita test were not accurate when compared to respirometry tests, but the test is a simple, inexpensive relative method of estimating compost stability and extremely useful for on-farm applications.

Cooperband (2002) suggested a more broad and general approach when it comes to testing compost quality for on-farm use (Table 11.4). Growers often use phytotoxicity tests to evaluate compost maturity. The premise is that growth of most plant species is inhibited with immature and unfinished composts (Garcia et al. 1992). A common example is the germination test conducted by seeding cucumber in pots containing finished compost. These tests seem to be a practical hands-on approach to test compost maturity but have been subjected to considerable controversy ever since it was proposed by Zucconi et al. in 1981 (Warman 1999). The choice of plant species is the most confounding factor when it comes to reliability of phytotoxicity tests. Moreover, even stable composts can exhibit high salt concentration and inhibit seed germination (Sullivan and Miller 2001). Thus, it is not advisable to solely depend on phytotoxicity tests to verify maturity of compost.

Table 11.4 Suggested optimum qualities of compost for on-farm use

Compost attribute	Optimum
Organic matter	Should range between 40 and 60 %
C:N ratio	10–15:1
pH	6–8
Electrical conductivity	Below 10 dS/m
Phytotoxicity	Seed germination >85 %
Weeds	No or few seeds

Source Cooperband (2002)

11.2.5 Use of Compost in Organic Vegetable Production

Organic systems heavily rely on organic matter-based amendments such as manure and compost to meet crop nutrient demand (Lammerts van Bueren et al. 2010; Russo and Webber 2007). Composts and manures are applied to agricultural lands as a source of essential microbes, as plant nutrients, and as a source of organic matter (He et al. 2001; Schroder 2005). Composts have also been successfully used in organic vegetable transplant production (Nair et al. 2011). Organic fertility amendments such as compost and manures have been shown to improve soil physical, chemical, and biological properties and produce yields equivalent to conventional cropping systems (Bulluck et al. 2002; Drinkwater et al. 1995; Ozores-Hampton et al. 1998). Compost amendments in organic vegetable production systems resulted in increased soil microbial biomass and enhanced microbial diversity (Nair and Ngouajio 2012). The USDA's NOP regulation 7 CFR 205.203(c), the soil fertility, and crop nutrient management practice standard set forth the requirements for management and application of plant and animal materials in organic production. Compost made in accordance with NOP rules may be applied in organic production systems without restriction on the time interval between application and crop harvest.

Numerous studies have shown benefits of using compost in organic vegetable production systems (Roe and Cornworth 1997; Martini et al. 2004). Studies conducted on organic pepper production have shown better growth and yields in compost-based organic fertilizer treatments than their conventional counter parts (Delate et al. 2008a, b). Similarly, study conducted in organic cucumber production showed enhanced crop growth and higher marketable yields in compost-amended soils (Nair and Ngouajio 2010). In another study, yields of pepper grown with dairy leaf compost produced similar yields as conventionally grown peppers (Hepperley et al. 2009). After 3 years of compost additions, yields of the three Spanish onion cultivars from the compost plots were significantly greater than from unamended plots (Maynard and Hill 2000). Compost application rates in organic vegetable production systems vary depending upon N content of the compost, N demands for the crop, and soil test results. It is not uncommon to see application rates ranging from 12 to 25 metric tonnes/ha. General guidelines suggest that 10–25 % of compost N will be plant available during the first year of application. Estimates for P and K availability in the first year are higher, 40 and 60 %, respectively.

11.3 Crop Rotations for Organic Vegetable Production

11.3.1 Introduction to Crop Rotations

Crop rotations involve a systematic farm plan, whereby the crop planted in one area or field on the farm changes every year or every season (Mohler and Johnson 2009).

Rotations occur more rapidly in warmer climates, with up to seven types of crops grown annually in one field, depending on the farm's climatic zone. Most growers rotate both crop type and crop variety, but for the purposes of this chapter, we refer to rotation based on crop type or species. The history of crop rotations in farming systems in Western civilization dates back to ancient Greece and Rome, where Pliny described the benefits of incorporating legume crops to enhance soil and crop quality in grain crop systems that included wheat, barley, and emmer. In the USA, George Washington in the 1700s established a seven-year crop rotation of grain and legume crops, along with carrots, cabbage, peas, potatoes, pumpkins, and turnips, to enhance soil quality on his Mount Vernon farm. According to Karlen et al. (1997), soil quality is a product of inherent parent material, climate and topography, and human-mitigated operations, including tillage and crop rotation. Crop rotations that are more diverse, or include more crops in the rotation, tend to have greater soil quality (Liebig and Doran 1999). Certified organic farmers, by law under the USDA-NOP, are required to practice crop rotation, according to CFR Title 7, Subtitle B, Chap. 1, Subchapter M, Part 205, §205.205—Crop rotation practice standard (USDA-NOP 2010). This standard dictates that the producer "must implement a crop rotation, including, but not limited to, sod, cover crops, green manure crops, and catch crops" that aid in soil quality and pest management. Among the many benefits of crop rotations, the USDA-NOP recognizes that crop rotations can lead to improving soil organic matter, supplying necessary plant nutrients, and providing erosion control. In the NOP standard on pest, weed, and disease management (§205.206), crop rotation is specifically stated as the first method used in managing pests, expressing the linkage between healthy soils and healthy plants. Because organic certification verifies that only organic practices were used for a minimum of 36 months prior to certification, farmers must complete an Organic System Plan (OSP), which provides information on the history of crop rotations planted in the last three years for every organic field. Thus, determining your crop rotations for the next few years will greatly assist the certification process and lead to better farm management overall.

11.3.2 Crop Rotations for Organic Vegetable Systems: Management Considerations

Vegetable growers around the world are cognizant of the need for separating or rotating fields of vegetable crops of the same family in order to avoid soilborne diseases prevalent in one particular family (e.g., a 3-year rotation to avoid Verticillium wilt in Solanaceae crops). In organic systems, this separation is even more critical, as synthetic fungicides are disallowed and prevention is the main factor utilized for disease management. Typically, organic farms maintain greater spatial and temporal diversity of crops than conventional counterparts, as green manure and perennial legume crops, in addition to vegetable crops, are often part of the OSP. Longer crop rotations have been shown to improve soil physical properties (Reganold 1988; Lal et al. 1994; Gerhart 1997), decrease erosion (Lockeretz et al. 1978; Reganold et al. 1987; Gantzer et al. 1991), reduce N leaching potential (Poudel et al. 2002; Kramer et al. 2006), improve soil organic matter (Lockeretz et al. 1981; Reganold et al. 1993; Clark et al. 1998; Drinkwater et al. 1998; Liebig and Doran 1999; Pulleman et al. 2000; Pimentel et al. 2005; Marriot and Wander 2006), and provide competitive crop yields (Delate and Cambardella 2004; Drinkwater et al. 1998; Teasdale et al. 2007). The next sections address specific benefits in relation to soil fertility, pest management, and economic considerations.

11.3.3 Crop Rotations for Enhancing Soil Fertility

Many organic farmers are striving for a closed, integrated organic farm, relying on on-farm or locally produced inputs/techniques, such as crop rotations, as much as possible to meet crop nutritional needs. Building or maintaining soil carbon (C) and nitrogen (N) pools for subsequent crop use is an important consideration in developing sustainable organic farming systems. Incorporation of crop residues from crop rotations and manure has been found to sequester C in soils, improve soil function, and mitigate erosion (Russelle and Franzluebbers 2007). Long-term organic farming practices, including crop rotations and manure application, were shown to enhance nutrient cycling and pest control by promoting soil quality and biodiversity (Birkhofer et al. 2008; Carpenter-Boggs et al. 2000; Pimentel et al. 2005). Liebig and Doran (1999) found that in four of five locations, soils on organic farms had higher soil quality, as represented by greater water holding capacity, higher microbial biomass C and N, enhanced soil respiration, and greater potentially mineralizable N relative to nitrate-N in the surface 30.5 cm, compared to conventional farms. This result was attributed to the use of diverse crop rotations that included cover crops and applications of organic-based amendments.

In organic production systems, high-N-demanding crops, such as sweet corn, are usually planted in a field following a soil-building crop. Soil-building crops in organic rotations in the Midwest include combinations of oats (*Avena sativa*), barley (*Hordeum vulgare*), rye (*Secale cereale*), wheat (*Triticum spp.*), hairy vetch (*Vicia villosa*), and red or white clover (*Trifolium incarnatum; T. repens*), due to their quick establishment, ability to overwinter, weed competitiveness, and ease of mechanical termination (Nelson et al. 1991; Creamer and Bennett 1997). Because of their ability to fix N, leguminous cover crops provide the greatest potential for improving yields, but cereal crops generally result in higher levels of soil organic matter helping suppress weeds, immobilize soil nitrogen, and reduce nitrate leaching during winter months (Snapp et al. 2005; Cherr et al. 2006). Planting small grains and N-fixing cover crops together may be an effective management strategy to increase soil C and improve soil N cycling processes, thereby reducing N leaching while maintaining robust yields. An example of a crop rotation plan from an organic farm in the northeast is shown in Fig. 11.3.

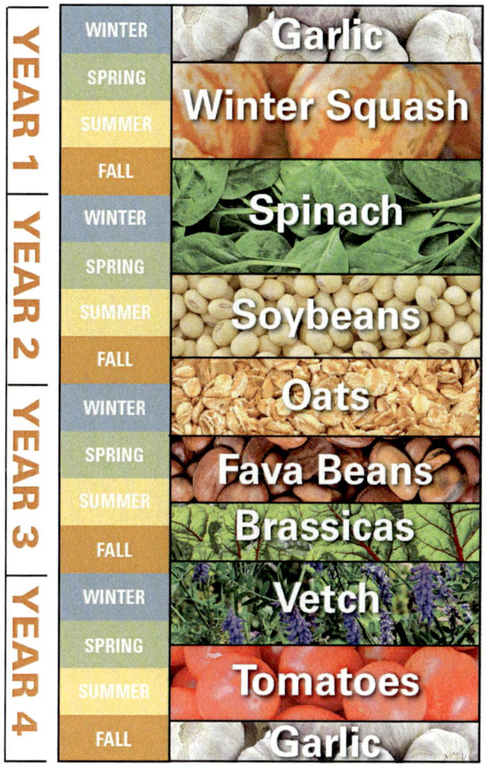

Fig. 11.3 Typical organic vegetable crop rotation on Northeast US farm (from Mohler and Johnson 2009: *Crop Rotation on Organic Farms: A Planning Manual*, NRAES 177)

While the legumes may provide a significant amount of N (20–120 lb/acre, depending on the species mixture), this contribution may not meet the complete needs of the cash crop. Soil testing in the fall following crop harvest can help determine the need for further amendments. Before planting in the spring, producers can supplement the soil with animal manure or a manure-based compost in an amount that will provide the full complement of N necessary for vigorous plant growth. Many NOP-compliant fertilizers, such as fish emulsion, humates, humic acids, surfactants, bioactivators, Biodynamic™ preparations, and others, can also be used. However, these amendments may be viewed as cost-prohibitive on a large scale and must be compatible with marketing requirements in order not to limit marketing options. Maintaining a soil pH of 6.0–7.0 is also critical for optimal crop production. Various agricultural liming materials can be used to neutralize the acidity of soils and to provide calcium and magnesium, but concern over soil magnesium buildup from dolomitic lime applications has led to the popularity of naturally mined calcium carbonate (limestone) in organic systems. Again, soil testing will help determine the need for lime and other rock mineral powders, such as rock phosphate. Hard rock phosphate varies considerably in soil reactivity, while soft rock or colloidal phosphate has greater applicability. On many organic farms, gypsum is used to supply calcium and sulfur, especially on high pH and sodic soils.

Fig. 11.4 Overview of the 44 fields of the Long-Term Agroecological Research (LTAR) experiment, Greenfield, Iowa, which examines biological and economic outcomes from five crops in four crop rotations over time

There are several organic-complaint commercial fertilizers and soil amendments that can be used for supplemental potassium, including sulfate of potash-magnesia (e.g., Sul-Po-Mag®) and naturally mined potassium sulfate, but all must be approved by a certification agency before application.

Research at the Long-Term Agroecological Research experiment (LTAR) in the Midwest (Fig. 11.4) has demonstrated excellent organic corn yields in the range of 120–209 bu/acre when rotations with soil-building legumes preceded corn crops (Delate and Cambardella 2004; Delate et al. 2008a, b). Soil quality has remained high in these systems even with multiple tillage operations (Delate et al. 2013). High yields have been achieved by preceding organic corn crops with legumes, such as alfalfa, and composted manure applications. Many organic farmers seek optimal yields, based on the limits of their farm's internal resources, as opposed to maximal yields, achieved through external inputs. Vegetable crops strictly relying on crop rotations, or cover crop residues, usually require additional compost applications to equal conventional yields. A systems' experiment from 1998 to 2003 comparing organic and conventional bell pepper production demonstrated similar growth and yield of conventional and organic crops, only when 100 lb/acre N was applied as compost (Delate et al. 2003a, b). Using a rotation of hairy vetch/rye preceding the pepper crop, without compost additions, resulted in reduced pepper yields compared to conventional yields 50 % of the time. Soil analysis revealed higher N in plots where cover crops were tilled compared to strip-tilled plots, leading to recommendations for side-dressing N in strip-tilled fields. Thus, most organic growers use a combination of crop rotations, cover crops, and compost to achieve the highest yields.

The issue of ground and surface water contamination from excess N applications has become increasingly critical for the future of farming. A significant proportion of the NO_3–N in the Mississippi River comes from agricultural land in the Midwest (Goolsby et al. 1999; Jaynes et al. 1999), and nitrate contamination of surface and groundwaters from ag lands flowing into a municipal water plant is the subject of a current lawsuit in Iowa (Des Moines Register, 2015). Accelerated NO_3–N loading in the Mississippi River has also been linked to the spread and increased severity of hypoxia within the Gulf of Mexico (Rabalais et al. 1996). Relying primarily on crop rotations (legumes) to supply N to the vegetable crop can assist with alleviating potential leaching problems associated with excess N applications, found even from manure sources. Using composted manure, which is in a more stable organic form, will help reduce leaching loss compared with fresh manure or synthetic nitrogen. A study evaluating nitrate-N leaching loss in an organic vegetable system amended with composted poultry litter demonstrated that nitrate-N concentrations in lysimeter leachates were generally below 10 mg L^{-1} during 52 months of monitoring (Evanylo et al. 2008). Thus, an ongoing challenge for organic growers is to be able to synchronize nutrient release from various crop residues and amendments with crop needs (Evanylo et al. 2008), which, in turn, will help reduce both N_2O emissions and NO_3–N leaching.

11.3.4 Crop Rotations for Pest Management

Recommendations for crop rotations to avoid plant pathogen carryover include 3 years between Solanaceae crops to prevent Verticillium wilt; 3–4 years for blackleg in brassicas; and up to 20 years for white rot in allium crops (Kuepper 2015). Variety selection should include cultivars designated as V, F, and N, which signifies resistance to Verticillium wilt, Fusarium wilt, and pathogenic nematodes. In a properly rotated organic field of the most resistant or tolerant vegetable cultivar available, planting at the proper time to permit quick germination and growth will generally keep disease and insect pests below economic injury levels. Because disease inocula can survive on infected crop residue, crop rotation can break the disease triangle (pathogen–host–environment) by changing to a non-host that does not support the growth of that particular pathogen. As with fertility regimes, a systems approach, including crop rotation and tillage, can limit continued spread of pathogenic organisms. Seedcorn maggots, *Hylemya cilicrura*, for example, the legless fly larvae that attack corn seeds particularly in cool, wet fields, can be avoided through the use of quality seed, crop rotation (especially away from previously infected fields) and planting when soils are warm (above 10 °C) to ensure quick germination.

Habitat diversification, which includes rotations of crop types across space and time, has been recommended as a strategy to enhance biological control and subsequent insect pest reduction, through either resource provisioning for natural enemies (Altieri 1994; Andow 1991), or spatial interference from a mixture of host crop and non-host crop species (Root 1973). As an example, corn rootworms (Northern and Western types of *Diabrotica* spp.) are not generally problematic on organic farms where three- to four-year crop rotations are practiced. There are also many natural enemies of prominent lepidopteran pests in vegetable systems, including predators that feed on eggs and larvae, such as lady beetles (various species), lacewings (various species), bigeyed bugs (*Geocorus* spp.), damsel bugs (*Nabis* spp.), minute pirate bugs (various spp.), and others. The most significant parasitic wasps against European corn borer are *Macrocentris grandii*, a braconid larval parasite, and *Trichogramma ostriniae,* an egg parasite. Pathogens of corn borer include *Nosema pyrausta* and *Beauveria bassiana.* A diverse habitat has been found to support natural enemies through provisioning of nectar, pollen, and insect pest (host) sources, as some host must be maintained for natural enemy survival (Chaplin-Kramer et al. 2011).

In areas where soil fertility is adequate, weeds are considered the greatest constraint in organic vegetable production. Weeds generally occupy the same ecological niche as the annual or perennial crop plants where they grow (Bullock 1992; Liebman and Dyck 1996) and thus can be reduced through crop rotations utilizing crops with different life cycles and management requirements, such as deep-rooted, perennial legumes with annual, shallow-rooted vegetable crops. Variety selection can also impact crop competitiveness over weed species, as quick-germinating, taller, and leafier plants tend to be more competitive in their resource utilization (Liebman et al. 2001). Longer crop rotations (3 years or longer) have been found to be instrumental in disrupting weed establishment and growth. In a study in Greenfield, Iowa, the shorter 2-year organic rotation of soybean–wheat had, on average, two to three times the weed population of the 3- and 4-year rotations rotating grain crops with oats/alfalfa (Delate et al. 2008a, b). Schreiber (1992) also found the greatest reduction of giant foxtail (*Setaria faberi*) in a soybean–wheat–corn rotation compared to a soybean–corn or corn–corn rotation.

With the focus in organic crop production on prevention of weed problems, establishment and growth of weed seeds can be greatly managed through both crop rotations and allelopathic cover crops. Rye (*Secale cereale*) is particularly important in crop rotations in helping prevent weed proliferation through its allelopathic properties (Bullock 1992; Delate and Hartzler 2003). Weed reductions, as high as 99 %, were observed for lambsquarter (*Chenopodium album*), when soybeans and sunflower were planted into killed rye compared with tilled plots with no mulch (Worsham 1984). In Iowa, weed populations in organic tomato plots with a rolled/crimped hairy vetch/rye mulch were lower or statistically equivalent to tilled plots with no mulch (Delate et al. 2012). Other fast-growing, high-biomass cover crops, such as sorghum–sudangrass and sunn hemp (warmer climes), when used in rotation with vegetable crops, can provide excellent weed control and are particularly useful when rotating out of sod crops such as bermuda grass or bahia grass (Kuepper 2015).

11.3.5 Economic Considerations of Crop Rotations

Both farm and field considerations are involved in determining the economic viability of specific crop rotations (Mohler and Johnson 2009). The balance between financial and biological considerations should be considered before long-range crop rotations plans are established. This includes both short-term (annual) and long-term (multi-year) farm management decisions. Several factors can often override rotation plans, including weather, market opportunities, and/or crop failure. Organic certification agencies have the authority to grant variances in crop rotation plans when unforeseeable events, such as extended cold and rainy weather causing failed germination in spring vegetable crops, could lead to a summer crop planting instead. Economic decisions are often based on growing the most profitable crop for the area, such as heirloom tomatoes, which can return $547.21 over all costs for a 4 × 100-ft bed (Chase 2006). Not rotating tomato crops, however, can be detrimental to the long-term viability of the farm if/when diseases, such as late blight, severely affect yields and profits. Chase et al. (2008) recommend growing 3–6 “signature vegetables” which provide the main income for the farm, but in rotation with other less-profitable vegetable crops and non-vegetable crops, which are useful for the ecosystem services they provide: N fertilization and other nutrients; beneficial insect habitat; weed management; and potential mitigation of greenhouse gases. Economic analysis shows higher returns in the longer crop rotations that include grain crops and legumes compared to a two-crop rotation (Delate et al. 2003a, b, 2013) and a general equivalency between vegetable crops grown with cover crops in rotation and those without cover crops (Delate et al. 2012). When crops in rotation are grown strictly for soil-building purposes, such as hairy vetch/rye, and nothing from this crop is marketed off the farm, the cost of cover crop seed must be offset by the additional gain in yield and/or, ideally, “green payments” for their carbon sequestration benefits (Singerman et al. 2011). The soil-building properties of these cover crops and other benefits they provide to the whole farm, however, can be considered a type of bank to support long-term farm viability.

11.3.6 Conclusion

Crop rotations are an essential component of sustainable, organic vegetable production. Because organic rules require the protection and/or enhancement of carbon and other nutrients in soil organic matter to maintain soil fertility and structure, organic farmers view crop rotations as the foundation of their OSP. Other aspects of the OSP include approved soil amendments and tillage to optimize production, but without a systematic crop rotation plan, farms risk losing ecological stability and may be denied certification. There are many types of organic vegetable crop rotations; these vary based on climatic zone, soil type, biological needs of the site (weed management, soil improvement, pest prevention), labor demands for specific crops, and desired markets. Most organic vegetable rotations follow a pattern of soil-building crops (cover crops), followed by high-nutrient-demanding vegetable crops (fruiting vegetables), then lesser demanding crops (bulb crops and leafy vegetables), before returning to soil building again. Separation of 3–4 years between crops of the same botanical family has been a standard practice to avoid disease issues. Because organic farmers utilize crop rotations, it has been found that organic soils sequester more carbon, and cycle and store nutrients better than conventional soils. The inclusion of legumes in the rotation, particularly perennial species, has been associated with greater soil quality improvements. High-tillering species, such as oats and rye, and crops providing extensive coverage, such as alfalfa and red clover, are also useful for reducing tillage and providing weed management, which, in turn, benefits soil health. In the overall analysis, the organic grower is striving for a crop rotation system that improves soil quality, and provides ecosystem services in the long term, while meeting market demand on an annual basis to sustain the economic viability of the farm.

11.4 Cover Crops in Organic Vegetable Production

Organic production systems heavily rely on organic inputs that improve soil fertility, quality, and health. In this respect, cover crops have profound impact as they add soil organic matter, enhance soil structure and fertility, improve water holding capacity, suppresse weed, and reduce soil erosion (Carrera et al. 2007; Clark 2007; Snapp et al. 2005). Cover crops help support diverse and active soil biotic communities that serve as a foundation for agricultural sustainability (Nair and Ngouajio 2012; Nair et al. 2013). Some of the key benefits of cover crops in organic vegetable cropping systems are mentioned below.

11.4.1 Nitrogen Fixation

Legume cover crops, in addition to adding organic matter, add nitrogen to soil by fixing atmospheric nitrogen through symbiotic relationship with soil bacterium (*Rhizobium* sp.) The bacteria, living in legume roots, absorbs nitrogen from the air and transforms it into forms that are used by the plant. The amount of nitrogen contributed by legumes varies by species. There are specific species of bacteria that form symbiotic relationship with individual legume cover crop species (Nair et al. 2015b). Growers should inoculate legume seeds accordingly with the proper nitrogen-fixing bacteria strain for efficient nitrogen fixation. The cost for the inoculum packet is $5–$10 and can usually treat 50 pounds of seeds. Research has shown significant increase in cover crop biomass and nitrogen-fixing potential in inoculated legume cover crop systems (Nair et al. 2015a).

11.4.2 Weed Suppression

Cover crops can be used to manage weeds in vegetable production systems. Cover crops can reduce weed germination and establishment by competing and/or producing allelochemicals, which suppress weed seed germination (Nair et al. 2014). Cover crops such as cereal grains and grasses establish quickly in the fall, cover the soil, and grow throughout the winter, thereby suppressing fall and winter weeds (Liebman and Dyck 1993). Small-seeded legumes that are seeded in the fall are sometimes not a good choice for weed suppression as they grow slowly during cold weather and can be outcompeted by weeds. Cover crops can influence weeds either in the form of living plants or as plant residue remaining after the cover crop is killed (Nair et al. 2014).

11.4.3 Soil Erosion and Water Quality

Most vegetable growers use cover crops as a strategy to reduce soil erosion in the fall and early spring. A cover crop provides vegetative cover during periods when a vegetable crop is not present and reduces the impact of falling raindrops, which otherwise would detach soil particles and increase erosion (Nair et al. 2015b). It also slows the rate of runoff, thus improving moisture infiltration into the soil. No tillage and other conservation tillage practices combined with cover crops have shown to significantly reduce runoff and soil erosion losses. Cover crops have also shown to improve water quality by suppressing nitrate leaching. Jokela and Nair (2014) found that nitrate leaching was reduced by 50 % using a rolled cover crop of cereal rye and hairy in organic broccoli and pepper production.

11.5 Conclusion

Cover crops can provide numerous ecological benefits in vegetable production systems. A systems approach to production is necessary to identify and understand the significance of the linkages between grower practices and their implications on crop growth, productivity, and the environment. They improve the sustainability of vegetable production systems although the diversity of the farming enterprise, size and scale of the farm, and climatic conditions provide unique opportunities and barriers to effectively integrate cover crops in vegetable cropping systems. Table 11.5 provides few examples and scenarios of how cover crops could be integrated into vegetable cropping systems. These examples are a starting point and can be modified to fit grower need, resources, and crop rotation plans.

Table 11.5 Examples for integrating cover crops in vegetable cropping systems

Year	Month*	Season	Example 1	Example 2	Example 3	Example 4	Example 5	Season	Month	Year
		Fall (previous year)	Oats + peas		Oilseed radish	Cowpea	Yellow mustard	Fall (previous year)		
	March		Winter-killed field peas	Cereal rye + hairy vetch	Winter-killed oilseed radish	Winter-killed cowpea	Winter-killed yellow mustard		March	
	April	Spring						Spring	April	
	May				Lettuce				May	
Year 1	June		Onion			Sweet corn	Muskmelon		June	Year 1
	July	Summer			Buckwheat			Summer	July	
	August			Pumpkin		Buckwheat			August	
	Sep				Cauliflower				Sep	
	Oct	Fall	Crimson clover					Fall	Oct	
	Nov						Cereal rye + hairy vetch		Nov	
	March		Winter-killed crimson clover	Cereal rye	Cereal rye				March	
	April	Spring				Garlic		Spring	April	
	May		Potato						May	
Year 2	June			Broccoli					June	Year 2
	July	Summer			Eggplant or pepper		Sweet potato	Summer	July	
	August								August	
	Sep		Sorghum sudangrass			Sorghum sudangrass			Sep	
	Oct	Fall		Buckwheat				Fall	Oct	
	Nov				Triticale		Triticale		Nov	
	March		Winter-killed sorghum sudangrass	Winter-killed buckwheat		Winter-killed sorghum sudangrass			March	
	April	Spring						Spring	April	
	May			Carrot			Cauliflower		May	
Year 3	June				Onion				June	Year 3
	July	Summer	Sweet potato				Buckwheat	Summer	July	
	August					Pumpkin or winter squash			August	
	Sep			Crimson clover			Lettuce or spinach		Sep	
	Oct	Fall			Oats + field peas			Fall	Oct	
	Nov					Cereal rye			Nov	
	March		Cereal rye	Winter-killed crimson clover	Winter-killed oats + Field peas		Cereal rye		March	
	April	Spring						Spring	April	
	May					Potato			May	
Year 4	June			Sweet corn	Cucumber		Pepper		June	Year 4
	July	Summer	Cucumber					Summer	July	
	August								August	
	Sep					Cowpea			Sep	
	Oct	Fall	Oats + field peas	Cereal rye + hairy vetch	Oilseed radish		Yellow mustard	Fall	Oct	
	Nov								Nov	
Year 5			Return to Year 1	Return to Year 1	Return to Year 1	Return to Year 1	Return to Year 1			Year 5

*Months indicate planting time for crops. Planting time within a month may vary based on weather conditions

Acknowledgements This work was supported by Iowa State University College of Agriculture and Life Sciences and the Leopold Center for Sustainable Agriculture.

References

Allison E (1965) Organic carbon. In: Black CA et al (eds) Methods of soil analysis, Part 2. American Society Agronomy, Madison, pp 1367–1378

Altieri MA (1995) Agroecology. Westview Press, Boulder, CO

Andow DA (1991) Vegetational diversity and arthropod response. Annu Rev Entomol 36:561–586

Bhardwaj KKR (1995) Improvements in microbial compost technology: a special reference to microbiology of composting. In: Khawna S, Mohan K (eds) Wealth from waste. Tata Energy Research Institute, New Delhi, pp 115–135

Birkhofer K, Fließbach A, Wise DH, Scheu S et al (2008) Generalist predators in organically and conventionally managed grassclover fields: implications for conservation biological control. Ann Appl Biol 153:271–280

Bischoff J, Werner H (1999) Salt/salinity tolerance of common horticultural crops in South Dakota. South Dakota Extension, Fact Sheet 904

Bullock DG (1992) Crop rotation. Crit Rev Plant Sci 11:309–326

Bulluck LR III, Brosius M, Evanylo GK, Ristaino JB et al (2002) Organic and synthetic fertility amendments influence soil microbial, physical and chemical properties on organic and conventional farms. Appl Soil Ecol 19:147–160

Carpenter-Boggs L, Kennedy AC, Reganold JP et al (2000) Organic and biodynamic management effects on soil biology. Soil Sci Soc Am J 64:1651–1659
Carrera LM, Buyer JS, Vinyard B et al (2007) Effects of cover crops, compost, and manure amendments on soil microbial community structure in tomato production systems. Appl Soil Ecol 37:247–255
Chaplin-Kramer R, O'Rourke ME, Blitzer EJ, Kremen C et al (2011) A meta-analysis of crop pest and natural enemy response to landscape complexity. Ecol Lett 14:922–932
Chase, C. 2006. Iowa Vegetable Production Budgets. PM 2017. Extension Communications, Iowa State University, Ames, IA
Chase C, Delate K, Liebman M, Leibold K et al (2008) Economic analysis of three Iowa rotations. PMR 1001. Iowa State University, Ames
Cherr CM, Scholberg JMS, McSorley R et al (2006) Green manure approaches to crop production: A synthesis. Agron J 98:302–319
Clark A (2007) Managing cover crops profitably, 3rd edn. Sustainable Agriculture Network, Beltsville
Clark MS, Horwath WR, Shennan C, Scow K et al (1998) Changes in soil chemical properties resulting from organic and low-input farming practices. Agron J 90:662–671
Cooperband L (2002) The art and science of composting. Center for Integrated Agricultural Systems, University of Wisconsin, Madison, Wisconsin. http://www.cias.wisc.edu/wp-content/uploads/2008/07/artofcompost.pdf. Accessed on 21 July 2015
Coyne MS, Thompson JA (2006) Math for soil scientists. Thomson Delmar Learning, Clifton Park, New York, pp 176–190 and 199–208
Creamer NG, Bennett MA (1997) Evaluation of cover crop mixtures for use in vegetable production systems. HortScience 32:866–870
Delate K, Cambardella C (2004) Agroecosystem performance during transition to certified organic grain production. Agron J 96:1288–1298
Delate K, Cambardella C, McKern A et al (2008) Effects of organic fertilization and cover crops on an organic pepper system. HortTechnology 18:215–226
Delate K, Cambardella C, Chase C, Johanns A, Turnbull R et al (2013) The long-term agroecological research (LTAR) experiment supports organic yields, soil quality, and economic performance in Iowa. Crop Manag. doi:10.1094/CM-2013-0429-02-RS
Delate K, Cambardella C, Chase C, Turnbull R et al (2008) Beneficial system outcomes in organic fields at the long-term agroecological research (LTAR) site, Greenfield, Iowa, USA. Organic E-prints: http://www.orgprints.org/12441/. Accessed 17 June 2015
Delate K, Cwach D, Chase C et al (2012) Organic no–till system effects on soybean, corn and irrigated tomato production and economic performance in Iowa, USA. Renew Agric Food Syst 27(1):49–59. doi:10.1017/S1742170511000524
Delate K, Duffy M, Chase C et al (2003a) An economic comparison of organic and conventional grain crops in a long-term agroecological research (LTAR) site in Iowa. Am J Altern Agric 18:1–11
Delate K, Friedrich H, Lawson V et al (2003b) Organic pepper production systems using compost and cover crops. Biol Agric Hortic 21:131–150
Delate K, Hartzler R (2003) Weed Management for Organic Farmers. PM 1883. Extension Communications, Iowa State University, Ames, IA
Des Moines Register. 2015. Des Moines Water Quality Suit Slated for Trial in 2016. http://www.desmoinesregister.com/story/money/agriculture/2015/07/15/des-moines-water-works-lawsuit-buena-vista-calhoun-sac-counties-water-quality/30191169. Accessed 11 August 2015
Drinkwater LE, Letourneau DK, Workneh F, VanBruggen AHC, Shennan C et al (1995) Fundamental differences between conventional and organic tomato agroecosystems in California. Ecol Appl 5:1098–1112
Drinkwater LE, Wagoner P, Sarrantonio M et al (1998) Legume-based cropping systems have reduced carbon and nitrogen losses. Nature 396: 262–265

Evanylo G, Sherony C, Spargo J, Starner D, Brosius M, Haering K (2008) Soil and water environmental effects of fertilizer, manure, and compost-based fertility practices in an organic vegetable cropping system. Agric Ecosyst Environ 127:50–58

Gantzer CJ, Anderson SH, Thompson AL, Brown JR et al (1991) Evaluation of soil loss after 100 years of soil and crop management. Agron J 83:74–77

Garcia C, Hernandez T, Costa F, Pascual J et al (1992) Phytotoxicity due to the agricultural use of urban wastes: germination experiments. J Sci Food Agric 59:313–319

Gerhardt RA (1997) A comparative analysis of the effects of organic and conventional farming systems on soil structure. Biol Agric Hort 14:139–157

Goolsby EA, Battaglin WA, Lawrence GB, Artz RS, Aulenbach BT, Hooper RP, Keeney DR, Stensland FJ et al (1999) Flux and sources of nutrients in the Mississippi-Atchafalaya river basin: Topic 3 Report for the Integrated Assessment of Hypoxia in the Gulf of Mexico

Hamoda MF, Abu Qdais HA, Newham J et al (1998) Evaluation of municipal solid waste composting kinetics. Resour Conserv Recycl 23:209–223

Haug RT (1993) The practical handbook of compost engineering. Lewis Publishers, Boca Raton, p 717

He XT, Logan TJ, Traina SJ et al (1995) Physical and chemical characteristics of selected U.S. municipal solid waste compost. J Environ Qual 24:543–552

He Z, Yang Z, Kahn BA, Stoffella PJ, Calvert DV et al (2001) Plant nutrition benefits of phosphorous, calcium, magnesium, and micronutrients from compost utilization. In: Stoffella PJ, Kahn BA (eds) Compost utilization in horticultural cropping systems. CRC Press LLC, Florida, pp 307–320

Hepperly P, Lotter D, Ziegler UC, Seidel R, Reider C et al (2009) Compost, manure and synthetic fertilizer influences crop yields, soil properties, nitrate leaching and crop nutrient content. Compost Sci Utilization 17:117–126

Jaynes DB, Hatfield JL, Meek DW et al (1999) Water quality in Walnut Creek watershed: Herbicides and nitrate in surface waters. J Environ Qual 28:45–59

Jokela D, Nair A (2014) Effects of reduced tillage and split fertilizer application in organic broccoli and pepper production systems. ISRF 14–36. http://farms.ag.iastate.edu/sites/default/files/EffectsReducedTillage.pdf

Karlen DL, Mausbach MJ, Doran JW, Kline RG, Harris RF, Schuman GE et al (1997) Soil quality, a concept, definition, and framework for evaluation. Soil Sci Soc Am J 61:4–10

Kramer SB, Reganold JP, Glover JD, Bohannan BJM, Mooney, HA et al (2006) Reduced nitrate leaching and enhanced denitrifier activity and efficiency in organically fertilized soils. PNAS 103:4522–4527

Kuepper G (2015) Market Farming with Rotations and Cover Crops: An Organic Bio-Extensive System. Kerr Center for Sustainable Agriculture, Poteau, OK. http://kerrcenter.com/wp-content/uploads/2015/02/Bioextensive-Model-report-2015-small.pdf. Accessed on 11 August 2015

Lal R, Mahboubi AA, Fausey NR et al (1994) Long-term tillage and rotation effects on properties of a central Ohio soil. Soil Sci Soc Am J 58:517–552

Lammerts van Bueren ET, Jones SS, Tamm L, Murphy KM, Myers JR, Leifert C, Messmer MM (2010) The need to breed crop varieties suitable for organic farming, using wheat, tomato and broccoli as examples: a review. J. Life Sci. NJAS-Wageningen. doi:10.1016/j.njas.2010.04.001

Leege PB, Thompson WH (1997) Test methods for the examinations of composting and compost. www.edaphos.com. Accessed on 24 July 2015

Liebig M, Doran JW (1999) Impact of organic production practices on soil quality indicators. J Environ Qual 28:1601–1609

Liebman M, Dyck E (1993) Crop rotation and intercropping strategies for weed management. Ecol Appl 3:92–122

Liebman M, Drummond FA, Corson S, Zhang J et al (1996) Tillage and rotation crop effects on weed dynamics in potato production systems. Agron J 88:18–26

Liebman M, Mohler CL, Staver CP et al (2001) Ecological management of agricultural weeds, Cambridge University Press, Cambridge

Lockeretz W, Shearer G, Klepper R, Sweeney S et al (1978) Field crop production on organic farms in the Midwest. J Soil Water Conserv 33:130–134

Lockeretz W, Shearer G, Kohl D et al (1981) Organic farming in the Corn Belt. Science 211:540–547

Lynch NJ, Cherry RS (1996) Winter composting using the passively aerated windrow system. Compost Sci Utilization 4:44–52

Marriot EM, Wander MM (2006) Total and labile soil organic matter in organic and conventional farming systems Soil Sci Soc Am J 70:950–959

Martini EA, Buyer JS, Bryant DC, Hartz TK, Denison RF et al (2004) Yield increases during the organic transition: improving soil quality or increasing experience? Field Crops Res 86:255–266

Maynard A, Hill D (2000) Cumulative effect of leaf compost on yield and size distribution in onions. Compost Sci Utilization 8:12–18

Mohler CL, Johnson SE (2009) Crop rotation on organic farms: a planning manual. NRAES 177, Ithaca

Nelson WA, Kahn BA, Roberts BW et al (1991) Screening cover crops for conservation tillage systems for vegetables following spring plowing. HortScience 6:860–862

Nair A, Carpenter B, Tillman J, Jokela D, Tester K (2013). ISRF 13–36. http://farms.ag.iastate.edu/sites/default/files/IntegratingCoverCrops.pdf. Accessed 26 July 2015

Nair A, Delate K, Artz G, Bregendahl C (2015a) Assessing nitrogen credits from clover cover crops and effects of seed inoculation. http://farms.ag.iastate.edu/sites/default/files/AssessingNitrogenCredits_0.pdf. Assessed 26 July 2015

Nair A, Jokela D, Tillman J (2014) Principles and practices of sustainable vegetable production systems. In: Nandwani D (ed) Sustainable horticultural systems. Springer, International Publishing, Switzerland, pp 51–78

Nair A, Kaspar T, Nonnecke G (2015b) Cover crops in vegetable production systems. HORT3026, Iowa State University. www.https://store.extension.iastate.edu/Product/Cover-Crops-in-Vegetable-Production-Systems. Accessed 26 July 2015

Nair A, Ngouajio M (2010) Integrating rowcovers and soil amendments for organic cucumber production: implications on crop growth, yield, and microclimate. HortScience 45:1–9

Nair A, Ngouajio M (2012) Soil microbial biomass, functional microbial diversity, and nematode community structure as affected by cover crops and compost in an organic vegetable production system. Appl Soil Ecol 58:45–55

Nair A, Ngouajio M, Biernbaum J et al (2011) Alfalfa-based organic amendment in peat-compost growing medium for organic tomato transplant production. HortScience 46:253–259

National Organic Program 5021 (2010) Draft Guidance: Compost and vermicompost in organic crop production. United States Department of Agriculture–Agricultural Marketing Service. http://www.ams.usda.gov/AMSv1.0/getfile?dDocName=STELPRDC5087122. Accessed 24 July 2015

Organic Trade Association (2015) Organic industry growth–2014. Organic Trade Association, Washington, D.C. http://ota.com/news/pressreleases/18061

Ozores-Hampton M, Obreza TA, Hochmuth G et al (1998) Using composted wastes on Florida vegetable crops. HortTechnology 8:130–137

Peigné J, Girardin P (2004) Environmental impacts of farm-scale composting practices. Water Air Soil Pollut 153:45–68

Pimentel D, Hepperly P, Hanson J, Douds D, Seidel R et al (2005) Environmental, energetic, and economic comparisons of organic and conventional farming systems. BioScience 55:573–582

Poincelet RP (1977) The biochemistry of composting. In: Composting of municipal sludges and wastes. Proceedings of the national conference. Rockville, Maryland, pp 33–39

Poudel DD, Horwath, WR, Lamini WT, Temple SR, van Bruggen AHC et al (2002) Comparison of soil N availability and leaching potential, crop yields and weeds in organic, low-input and conventional farming systems in northern California. Agric Ecosyst Environ 90:125–137

Pulleman MM, Bourma J, van Esssen EA, Meijles EW et al (2000) Soil organic matter content as a function of different land use history. Soil Sci Soc Am J 64:689–693

Rabalais NN, Wiseman WJ, Turner RE, SenGupta BK, Dortch Q et al (1996) Nutrient changes in the Mississippi River and system responses on the adjacent continental shelf. Estuaries 19:386–407

Reganold JP (1988) Comparison of soil properties as influenced by organic and conventional farming systems. Am J Alt Agric 3:144–155

Reganold JP, Elliott LF, Unger YL et al (1987) Long-term effects of organic and conventional farming on soil erosion. Nature 330:370–372

Reganold JP, Palmer AS, Lockhart JC, Macgregor AN et al (1993) Soil quality and financial performance of biodynamic and conventional farms in New Zealand. Science 260:344–349

Rosen CJ, Bierman PM (2005). Using manure and compost as nutrient sources for fruit and vegetable crops. http://www.extension.umn.edu/garden/fruit-vegetable/using-manure-and-compost/. Accessed 23 July 2015

Roe N, Cornforth G (1997) Yield effects and economic comparison of using fresh or composted dairy manure amendments on double-cropped vegetables. HortScience 32:462

Root RB (1973) Organization of a plant-arthropod association in simple and diverse habitats: the fauna of collards (Brassica oleracea). Ecol Monogr 43:95–124

Russelle MP, Franzluebbers AJ (2007) Introduction to symposium: Integrated crop-livestock systems for profit and sustainability. Agron J 99:323–324

Russo VM, Webber CL III et al (2007) Organic agricultural production in the United States: an old wheel being reinvented. Am J Plant Sci Biotech 1:29–35

Ryn R, Van de Kamp M, Wilson GB, Singley ME, Richard TL, Kolega JJ, Gouin FR, Laliberty L Jr, Kay D, Murphy DW, Hoitink HAJ, Brinton WF et al (1992) On-Farm composting handbook. In: Rynk R (ed) Northeast regional agricultural engineering service. Ithaca, New York, pp 6–13, 106–113

Schreiber MM (1992) Influence of tillage, crop rotation, and weed management on giant foxtail (*Setaria faberi*) population dynamics and corn yield. Weed Sci 40:645–653

Singerman A, Delate K, Chase C, Greene C, Livingston M, Lence S, Hart C et al (2011) Profitability of organic and conventional soybean production under 'green payments' in carbon offset programs. Renew Agric Food Syst 27:266–277

Schroder J (2005) Revisiting the agronomic benefits of manure: a correct assessment and exploitation of its fertilizer values spares the environment. Bioresour Technol 96:253–261

Snapp S, Swinton S, Labarta R et al (2005) Evaluating cover crops for benefits, costs and performance within cropping system niches. Agron J 97:322–332

Stoffella PJ, Kahn BA (2001) Compost utilization in horticultural cropping systems. Boca Raton, Florida, pp 95–120

Sullivan DM, Miller RO (2001) Compost quality attributes, measurements, and variability. In: Stoffella PJ, Kahn BA (eds) Compost utilization in horticultural cropping systems. Florida, Boca Raton, pp 95–120

Teasdale JR, Coffman CB, Mangum RW et al (2007) Potential long-term benefits of no-tillage and organic cropping systems for grain production and soil improvement. Agron J 99:1297–1305

Thompson W, Leege P, Milner P, Watson M et al (2002) Test methods for the examination of composts and composting. The U.S. Composting Council, Hauppauge, New York. http://www.compostingcouncil.org. Accessed 23 July 2015

Tuomela M, Vikman M, Hatakka A, Itävaara M et al (2000) Biodegradation of lignin in a compost environment: a review. Bioresour Technol 72:169–183

U.S. Department of Agriculture, Agricultural Marketing Service, National Organic Program (2010) http://www.ams.usda.gov/AMSv1.0/nop. Accessed 24 June 2015

U.S. Department of Agriculture (USDA)–National Agricultural Statistics Service (NASS). (2011). Crop Production–March 8, 2013

Worsham AD (1984) Crop residues kill weeds. Crops Soils Mag 37:18–20

Warman PR (1999) Evaluation of seed germination and growth tests for assessing compost maturity. Compost Sci Utilization 7:33–37

Woods End Research Laboratory (1999) Guide to Solvita® testing for compost maturity index. Woods End Research Laboratory Inc. Mt. Vernon, Maine. https://solvita.com/compost. Accessed 25 July 2015

Zucconi F, Forte M, Monaco A, deBertoldi M (1981) Biological evaluation of compost maturity. Biocycle 22:27–29

Chapter 12
Effective Micro-organisms (EM) as Sustainable Components in Organic Farming: Principles, Applications and Validity

Rasheed B. Balogun, Justin U. Ogbu, Ephraim C. Umeokechukwu and Risqat B. Kalejaiye-Matti

Abstract Sustainable farming systems presuppose perpetuity of yield for the present and future without compromising the environment biological and physical components where the production is taking place. One of the means of achieving this end is through the utilization of effective micro-organisms (EM) during the course of production. Thus, this paper seeks to review the rationale behind the EM concept, and X-ray recent advances in this essential aspect of modern organic agriculture systems. EM are mixed culture of beneficial micro-organisms. The concept of EM is based on the inoculation of the substrates with the intention of shifting the microbial equilibrium and thus creating an improved ecology that favors improved productivity. A couple of theories exist to justify the action of EM in agricultural production. These includes the biological suppression of pathogens theory, energy conservation theory, mineral solubilization theory, microbial ecological balance theory, photosynthetic efficiency theory, and biological nitrogen fixation theory. EM preparation was explored. Ongoing scientific experiment validating the EM technology was equally reviewed, so also its applications in different parts of the world.

Keywords Agroecosystem · Ecological balance · Effective micro-organisms · Organic farming · Sustainability

R.B. Balogun · J.U. Ogbu (✉) · E.C. Umeokechukwu
Department of Horticulture and Landscape Technology,
Federal College of Agriculture (FCA), Ishiagu 491105, Nigeria
e-mail: ogbujugo@gmail.com

R.B. Kalejaiye-Matti
Department of Animal Production Technology,
Federal College of Agriculture (FCA), Ishiagu 491105, Nigeria

D. Nandwani (ed.), *Organic Farming for Sustainable Agriculture*,
Sustainable Development and Biodiversity 9,
DOI 10.1007/978-3-319-26803-3_12

12.1 Introduction

Organic agriculture has generated increasing global interest because of the need for food safety resulting from the adverse residual effect of synthetic agrochemicals on agroproduce, produce naturalness, and wholesomeness; enhanced maturity period, sustainable yield, and the protection of the ecosystems (LEISA 2006a, b). It is presumed that organic agriculture can also provide an additional avenue for climate change mitigation through such measures as enhanced soil carbon sequestration. Besides, it is also considered ecosystem-friendly because of its emphasis on minimum tillage and reduced use of synthetic pesticides, additives, and fertilizers (De Almeiada and Fernandez 2006; Ene 1998). Organic agriculture is also expected to play a major role in fighting against desertification, preserving biodiversity, contributing to sustainable development, and promoting animal, plant, and human health. The growing interest of consumers and markets worldwide in organic products has also opened new trade opportunities for developing countries, through internationally recognized certification.

Organic agriculture as a system for crops, forestry, livestock, and fish farming emphasizes environmental protection and the use of natural farming techniques in order to enhance continuity of harvest (Gliessman 1997). It is concerned not only with the end product, but also with the entire system used to produce and deliver agricultural products. To this end, the entire farm cycle, from production and processing to handling and delivery, excludes the use of artificial products such as genetically modified organisms (GMOs) and all form of externalities or high input resources. Organic farmers rely instead on natural farming methods and modern scientific ecological knowledge in order to maximize the long-term health and productivity of the ecosystem, enhance the quality of the products, and protect the environment.

This article therefore is intended to explore the option of effective or beneficial micro-organisms' technologies among the myriads of approach to organic agriculture. This will be addressed using the following framework: Effective micro-organisms (EM), EM preparation and mode of action, principles of EM, concept of sustainable agriculture, organic agriculture, principles of organic agriculture, application of EM, and scientific validity of EM technologies.

12.2 Effective Micro-organisms

Effective micro-organisms (EM) consist of mixed cultures of beneficial and naturally occurring micro-organisms that can be applied as inoculants to increase the microbial diversity of soil, plant, livestock, and the ecosystems for sustainable performance. EM contains selected species of micro-organisms which include the following (Higa and Wididana 1991a):

- Lactic acid bacteria: *Lactobacillus plantarum*; *L. casei*; *Streptococcus Lactis*.
- Photosynthetic bacteria: *Rhodopseudomonas palustris*; *Rhodobacter sphaeroides*.
- Yeast: *Saccharomyces cerevisiae*; *Candida utilis*. (Usually known as *Torula*, *Pichia Jadinii*).
- Actinomycetes: *Streptomyces albus*; *S. griseus*.
- Fermenting fungi: *Aspergillus oryzae*; *Mucor hiemalis*.

All these micro-organisms are mutually compatible with one another and coexist in liquid or dried culture. EM is not a substitute for other management practices. It is however an added dimension for optimizing agricultural and environmental management practices (Higa and Wididana 1991b). Effective micro-organisms or EM technology is a trademarked term now commonly used to describe a proprietary blend of three or more types of predominantly anaerobic organisms that was originally marketed as EM-1 Microbial Inoculant but is now marketed by several companies under various names, each with their own proprietary blend.

12.3 Preparation of EM

Materials: sugar, water, cultured indigenous micro-organisms using local carbohydrate food, (Rice) sealed container, net mesh, paper bag plant samples—(900 g)
Procedures: 120 ball of pound rice were pack into a net mesh and then wrapped in paper bag and deposited into a trench of about 10–20 cm depth within a matured forest for micro-organism to colonize. Then, after a period of 5–7 days the cultured medium were harvested and turn into an airtight container to undergo fermentation. 250 g of sugar were added to the content. These will serve as a food to activate the micro-organisms. The cultured medium are ready for use when it gives a sweet fermented smell but if it produces a sour or rotten smell, it is a failure (Higa and James 1994). The well-fermented substance is filtered into any desirable sealed container for storage and subsequent usage. This initial filtrate is the EM 1. The concentrated filtrate is used to produce several other formulations of EM Technologies.

12.4 Theories of Effective Micro-organisms

12.4.1 Disease-suppressive Soil Theory

The term disease-suppressive soil refers to the biological means of suppressing the occurrence of plant diseases. Three examples of disease-suppressive soil are (1) the pathogen fails to become established, (2) the pathogen is present but fails to cause disease, and (3) the pathogen causes disease but declines with monoculture.

Experiments have shown that soil treated with EM 2.3.4 had a lower incidence of plant fungal diseases (*Thielaviopsis and Pseudomonas*) than the fertilized control. The suppression of plant pathogens and disease incidence is dependent on soil conditions, the plant, and which EM culture or combination of cultures is applied. This indicates that EM can induce a soil to become disease-suppressive in nature (Higa and Wididana 1991b).

12.4.2 Organic Energy Theory

In the conventional theory, organic materials added to soil undergo decomposition by micro-organisms, and minerals (nutrients) are released and become available for uptake by plants. In the organic energy theory, organic amendments are fermented by species of *lactobacillus*, and other lactic acid producing micro-organisms. This, in turn, releases amino acids and saccharides as soluble organic compounds that are absorbed intact by plants to be utilized beneficially in various metabolic pathways (Higa and Wididana 1991b). Kinjo (1990) found that the amount of amino acid produced after incubation of organic matter with EM for five days was significantly higher than the control without EM. The absorption of amino acids, sugars, and other organic compounds by plant roots has been demonstrated in plant tissue culture. Such work indicates that the plantlet, callus, or plant cell require not only macro- and micronutrients, but can also benefit from absorption of energy-yielding organic molecules such as amino acids and simple sugars. The fermentation process is often utilized in the preparation of foods, such as miso (soybean paste) and soy sauce, and in making silage for livestock.

12.4.3 Inorganic Nutrient Solubilizing Theory

Soil micro-organisms are important in decomposing organic materials and recycling their nutrients for uptake by plants. Soil productivity generally decreases as soil organic matter decreases (often through soil erosion and insufficient return of organic wastes and residues to land). When this happens, the total soil microbial population and its biodiversity also tend to decrease. Experiments (Higa and Wididana 1991b) were conducted in which a 0.1 % aqueous solution of molasses was applied to soil and to leaf surfaces of turnip (*Brassica rapa*) and green pepper (*Capsicum* spp.) as a carbon and energy source for indigenous micro-organisms. The results showed a significant increase in the number of bacteria, *actinomycetes*, and fungi in both soil and on leaf surfaces over that of the unamended control (Tables 12.1 and 12.2). The foliar-applied molasses also caused a substantial increase in the numbers of nitrogen-fixing bacteria on the surface of turnip leaves (Table 12.2). The yield of both green pepper and turnip was significantly increased by the association of the increased number of micro-organisms (Table 12.3).

Table 12.1 Effect of molasses spray applied to soil on numbers of micro-organisms

Microbial group	Dilution	Number of micro-organisms[a]	
		Control	Molasses (0.1 %)
Fungi	10^3	44.4	102
Fusarium	10^2	102	413
Bacteria	10^6	252	407
Actinomycetes	10^6	2.51	3.51

[a]Numbers per g of soil (dry weight basis). Micro-organisms were counted in soil that was planted to green pepper

Table 12.2 Effect of foliar-applied molasses spray on numbers of micro-organisms on the leaf surface of turnip

Microbial group	Dilution	Number of micro-organisms[a]	
		Control	Molasses (0.1 %)
Fungi	10^3	12.4	63.3
Fusarium	10^2	8.42	14.0
Bacteria	10^4	3.89	8.90
Actinomycetes	10^4	2.46	9.21
N fixing bacteria	10^3	1.42	10.3

[a]Numbers per g of soil (dry weight basis). Micro-organisms were counted on the leaf surface of Turnip

Table 12.3 Effect of foliar-applied molasses spray on the yield of green pepper and turnip

Treatment	Green pepper (g m^{-2})	Turnip (g m^{-2})
Control	748	3660
Molasses (0.1 %)	964[a]	4140[a]

[a]Significant difference between treatments at 5 % probability by Duncan's test

Insoluble organic phosphorus compounds that are largely unavailable to plants can often be solubilized by micro-organisms. Similar results were obtained in an experiment where various EM cultures were added to soil.

12.4.4 Balanced Population of Soil Micro-organisms Theory

The incidence and severity of plant diseases depend on soil conditions, i.e., chemical, physical, and microbiological properties; soil management (tillage, fertilizers, and pesticides), crop management (crop rotation, monoculture, and multiple cropping), and the plant cultivar (disease-susceptible and disease-resistant). These factors can greatly influence the total microbial population, its complexity, and diversity in soil. The balance in population and diversity between harmful and beneficial micro-organisms will determine the soil microbiological equilibrium; whether the soil ecosystem is favorable or unfavorable to the growth and health of plants. Generally, soils which have high populations of *actinomycetes, Trichoderma*, fluorescent pigment-producing *Psendomonas*, and other micro-organisms that are antagonistic to plant pathogens are considered to be disease-suppressive soils. Those which have large numbers of Lactobacillus and other fermentative micro-organisms (yeasts, starch digesting bacteria, and cellulose-digesting bacteria) are considered to be zymogenic soils. Soils which have large numbers of nitrogen-fixing bacteria (*Azotobacter*, *Beljerinckia*, *Derxia*, and *Spirillum*), facultative anaerobic bacteria (*Bacillus*, *Enterobacter*, *Klebsiella*, and *Clostridium*), and photosynthetic bacteria are classified as synthetic soils. When a soil has high populations of plant pathogenic fungi (*Fusarium*, *Thielaviopsis*, *Phytophthora*, *Verticillium,* and *Pythium*), it is considered to be a disease-inducing soil (Higa and Wididana 1991b).

12.4.5 Photosynthetic and Nitrogen-Fixing Theory

When EM is applied to soil or plant leaf surfaces, the populations of photosynthetic bacteria and nitrogen-fixing bacteria increase dramatically. This phenomenon is associated with the growth of more vigorous plants, higher plant yields, and improved crop quality (based on higher contents of vitamin C and sugar in fruits) compared with no EM treatment. It was thought that the high number of photosynthetic bacteria and nitrogen-fixing bacteria in soil and at leaf surfaces might enhance the plant's photosynthetic rate and efficiency, and its nitrogen-fixing capacity (Higa and Wididana 1991b). However, this has not been established experimentally, although it had been found that the net photosynthesis of *Pinus ponderosa* and *P. flexilis* tended to increase as the extent of infection by ecto-mycorrhizae increased.

Ruinen (1970) was among the first to investigate the occurrence of nitrogen-fixing bacteria on leaf surfaces. Sen Gupta et al. (1982) reported that nitrogen-fixing bacteria on leaf surfaces could markedly increase crop yields.

12.5 Organic Agriculture

Organic farming methods combine scientific knowledge of ecology and modern technology with traditional farming practices based on naturally occurring biological processes. The USDA National Organic Standards Board (NOSB) defined "Organic agriculture as an ecological production management system that promotes and enhances biodiversity, biological cycles and soil biological activity. It is based on minimal use of off-farm inputs and on management practices that restore, maintain and enhance ecological harmony" (Gold 2014). Organic agriculture is a production system that sustains the health of soils, ecosystems, and people. It relies on ecological processes, biodiversity, and cycles adapted to local conditions, rather than the use of inputs with adverse effects. Organic agriculture combines tradition, innovation, and science to benefit the shared environment and a good quality of life for all involved (International Federation of Organic Agriculture Movements (IFOAM) 2002).

Organic practices include, but are not limited to (IFOAM 2002):

- encouraging predatory beneficial insects to control pests by serving them nursery plants; and/or an alternative habitat, usually in a form of a shelterbelt, hedgerow, or beetle bank;
- encouraging beneficial micro-organisms;
- rotating crops to different locations from year to year to interrupt pest reproduction cycles;
- planting companion crops and pest-repelling plants that discourage or divert pests;
- using row covers to protect crops during pest migration periods;
- using biologic pesticides and herbicides;
- using no-till farming, and no-till farming techniques as false seedbeds;
- using sanitation to remove pest habitat;
- using insect traps to monitor and control insect populations;
- using physical barriers, such as row covers.

12.6 Organic Farmland by World Region (2000–2008)

As of 2001, the estimated market value of certified organic products was estimated to be \$20 billion. By 2002, this was \$23 billion and by 2007 more than \$46 billion (Helga and Lukas 2011). By 2012, the market had reached \$63 billion worldwide (Helga et al. 2013). Europe (2011: 10.6 million ha, which is 5.4 % of Europe's farmland and an increase of 6 % from the prior year; Europe has 29 % of the world's organic agricultural land) and North America (2011: 2.8 million ha, 7.5 % of the world's organic agricultural land) have experienced strong growth in organic farmland. In the EU, it grew by 21 % in the period 2005–2008 (Helga et al. 2013). However, this growth has occurred under different conditions. While the European Union has shifted agricultural subsidies to organic farmers due to perceived environmental benefits, the USA has not (Dimitri and Oberholtzer 2006), continuing to subsidize some but not all traditional commercial crops, such as corn and sugar. As a result of this policy difference, as of 2008, 4.1 % of European Union farmland was organically managed compared to the 0.6 % in the USA (Helga and Lukas 2011). As of 2012, the country with the most organic land was Australia (12 million ha), followed by Argentina (3.8 million ha), and the USA (1.9 million ha) (Helga et al. 2013).

12.7 Empirical Comparison Between Conventional and Organic Farming

The global food security community is shifting swiftly in support of an organic approach because it has the potential to secure a global food supply; just as conventional agriculture is today, but with reduced environmental impact and more significant quality yield. This is according to the report of Rodale Institute (2011). It was also reported that agroecological farming methods could double global food production in just 10 years. Agroecological practices, such as organic practices, attempt to mimic natural processes and rely on the biology of the soil and environment rather than synthetic sprays and other inputs. Switching to organic methods in communities where people struggle to feed themselves and their families can lead to harvest of about 180 % larger than that produced by conventional methods.

12.8 Reasons for Higher Profitability of Organic Systems

Higher market prices and premiums: even with less yields and higher production costs, organic remained more profitable due to 400 % higher market price; even with much higher costs and significantly lower yields, price premium made organic more profitable (Greene 2001); higher prices for organic accounted for 40–75 % of profits in Germany and Britain for arable farms, and 10–48 % for dairy farms (Offermann and Nieberg 2000); lower production costs: Lower production costs caused significant difference in net returns even without premiums; combination of premiums and lower production costs: Low production cost along with the 20 % premium on organic was the prime reason for higher profit margin; combination of higher yields and premiums (Gibbon and Bolwig 2007); combination of higher yields and premiums and lower production costs.

12.9 Profitability of Organic Crop Production

An Indo-Swiss research team compared agronomic data of 60 organic and 60 conventional farms over two years (Eyhorn et al. 2007; Hanson et al. 1997) and came to the conclusion that cotton-based organic farming is more profitable: Variable production costs were 13–20 % lower, inputs were 40 % lower, yet yields were 4–6 % higher in the two years, and as a consequence, gross margins for cotton were also 30–43 % higher. Although crops grown in rotation with cotton were sold without a price premium, organic farms achieved 10–20 % higher incomes from conventional agriculture. Similarly, an impact assessment study for organic cotton farmers in Kutch and Surendranagar commissioned by it had been Agrocel concluded that farmers who participated in the project enjoyed a net gain of 14–20 % resulting in higher revenues and lower costs. The updated version of the study surveying 125 organic cotton farmers concluded that 95 % of respondents saw their agricultural income rises since adopting organic agriculture, on average by 17 %, most of them attributing this largely to the reduced cost of production and increase in cost of selling. Similarly found in Andhra Pradesh that organic cotton was much more profitable, since conventional cotton did not have any profits (income was + $13 vs. −$30). In conclusion, all studies found organic cotton farming more profitable than conventional.

A long-term field study comparing organic/conventional agriculture carried out over 21 years in Switzerland concluded that crop yields of the organic systems averaged over 21 experimental years at 80 % of the conventional ones. The fertilizer input, however, was 34–51 % lower, indicating an efficient production. The organic farming systems used 20–56 % less energy to produce a crop unit; and per land area this difference was 36–53 %. The produce came off better in food preference trials and picture creating methods (Chavas et al. 2009).

A study of the sustainability of apple production systems showed that when comparing a conventional farming system to an organic method of farming, the organic system in this case is more energy-efficient (Reganold et al. 2001).

12.10 Health Costs

IFAD (2005) case studies in India showed that none of the 30 farmers interviewed in Karnataka has experienced any feelings of illness after working in the organic rice fields, whereas more than half of the conventional farmers had sometimes suffered from nausea and vomiting. In Kerala, a number of farmers were hospitalized after local groundwater was contaminated with pesticide run-off from neighboring tea estates.

12.11 Social Costs

Most studies did not evaluate the debt issue and thus did not take previous investments in agriculture into account. Some authors, however, noted that conventional farmers were significantly more indebted, especially in developing countries. Some of these authors, including Lotter et al. (2003) and Jalees (2008) noted that most conventional cotton farmers in Central India bought inputs on loan, at annual interest rates between 10 and 15 % (from cooperative societies) to over 30 % (from private money lenders). Since production costs were usually lower, the necessity in organic agriculture to take up loans was far less. As indicated, the main cause for India's extremely high farmers' suicide rate is debt servicing for start-up costs, mainly GM seeds and chemical inputs.

, According to the National Crime Records Bureau in India, between 1997 and 2005, approximately 30,000 farmers committed suicide in Maharashtra, mostly in Vidarbha region. In 2007 alone, 1211 distressed farmers took their own lives in this region, where most BT cotton is grown, due to repeated cotton failure and indebtedness. In a study done by Jalees (2008), nearly 91 % of the farmers growing BT cotton were indebted, whereas only 4 % of farmers cultivating organic cotton had debts.

12.12 Environmental Costs

The annual external costs of UK agriculture in 1996 showed £2343 million (US $3648 million), equivalent to £208/ha (US$324/ha) of arable and permanent pasture. This was 89 % of average net farm income for 1996. Significant costs arose from contamination of drinking water with pesticides, nitrate, and phosphate; from damage to wildlife, habitats, hedgerows; from emissions of gases; from soil erosion and organic carbon losses; from food poisoning; and from BSE (Rodale Institute 2011). Another study calculating the external costs of agriculture in USA (including damage to water sources, to soil and air resources, to wildlife and ecosystem biodiversity, and to human health) estimated to be at $5.7–16.9 billion annually, per cropland hectare at US$29–96 (Tegtmeier and Duffy 2004). These studies only estimated externalities that gave rise to financial costs, thus they were likely to underestimate the total negative impacts of chemical-intensive agriculture as compared to organic agriculture with less ecological impacts.

12.13 Positive Health Impacts of Organic Food

Several studies indicate that 10–60 % more healthy fatty acids (such as CLA's) and omega-3 fatty acids occur in organic dairy (Butler et al. 2008); in crops, vitamin C ranges 5–90 % more and secondary metabolites 10–50 % more in organic. Also, less residues of pesticides and antibiotics are present (Huber and van de Vijver 2009); organic food contains higher minerals and dry matter and 10–50 % higher phytonutrients (Heaton 2002); decreased cell proliferation of cancer cells was observed on extracts of organic strawberries (Olsson 2006); the Parsifal study showed 30 % less eczema and allergy complaints and less bodyweight among 14,000 children fed with organic and biodynamic food in five EU countries (Alfven 2006); in animals, organic feed leads to increased fertility (Staiger 1988) and increased immune parameters (Finamore 2004).

12.14 Positive Environmental Impacts of Organic Agriculture

Increased soil fertility: Biodynamic farms had better soil quality; greater in organic matter, content, and microbial activity; more earthworms, better soil structure, lower bulk density, easier penetrability, and thicker topsoil (Reganold 1992). Agricultural productivity doubled with the soil fertility techniques: compost application and introduction of leguminous plants into the crop sequence. More energy efficiency: Growing organic rice was four times more energy efficient than the conventional method. Organic agriculture reduces energy requirements for production systems by 25–50 % compared to conventional chemical-based agriculture. Carbon sequestration: German organic farms annually sequester 402 kg of Carbon/ha, while conventional farms had losses of 202 kg (Küstermann et al. 2008). Less water pollution: In conventional farms, 60 % more nitrate are leached into groundwater over a 5-year period (Drinkwater et al. 1998). More water capture: Enhanced organic soil structure reduces risk of floods (Lotter et al. 2003). Increased soil fauna: Organic soil fauna increases by 148 % (Dumaresq and Greene 2001). Enhanced biodiversity: Organic farms' biodiversity increases resilience to climate change and weather unpredictability. Reduced erosion: Organic agriculture reduces erosion caused by wind and water as well as by overgrazing at a rate of 10 million ha annually (Pimentel et al. 2005).

12.15 Principles of Organic Agriculture

The principles of organic agriculture are the roots from which organic agriculture grows and develops. History, culture, and community values are embedded in agriculture. The principles apply to agriculture in the broadest sense, including the way people tend soils, water, plants, and animals in order to produce, prepare, and distribute food and other goods. They concern the way people interact with living landscapes, relate to one another, and shape the legacy of future generations. The principles are integrated as a whole. They are ethical principles and it includes as enunciated by IFOAM (2002):

- The Principle of Health;
- The Principle of Ecology;
- The Principle of Fairness; and
- The Principle of Care.

12.15.1 The Principle of Health

Organic agriculture should sustain and enhance the health of soil, plant, animal, human, and planet as one and indivisible. This principle points out that the health of individuals and communities cannot be separated from the health of ecosystems—healthy soils produce healthy crops that foster the health of animals and people. Health is the wholeness and integrity of living systems. It is not simply the absence of illness, but the maintenance of physical, mental, social, and ecological well-being. Immunity, resilience, and regeneration are key characteristics of health. The role of organic agriculture, whether in farming, processing, distribution, or consumption, is to sustain and enhance the health of ecosystems and organisms from the smallest in the soil to human beings. In particular, organic agriculture is intended to produce high-quality, nutritious food that contributes to preventive health care and well-being. In view of this, it should avoid the use of fertilizers, pesticides, animal drugs, and food additives that may have adverse health effects.

12.15.2 The Principle of Ecology

Organic agriculture should be based on living ecological systems and cycles, work with them, emulate them, and help sustain them. Production is to be based on ecological processes and recycling. Nourishment and well-being are achieved through the ecology of the specific production environment. For example, in case of crops, this is the living soil; for animals, it is the farm ecosystem; for fish and marine organisms, the aquatic environment. Organic farming, pastoral, and wild harvest systems should fit the cycles and ecological balances in nature. These cycles are universal but their operation is site-specific. Organic management must be adapted to local conditions, ecology, culture, and scale. Inputs should be reduced by reuse, recycling, and efficient management of materials and energy in order to maintain and improve environmental quality and conserve resources. Organic agriculture should attain ecological balance through the design of farming systems, establishment of habitats, and maintenance of genetic and agricultural diversity. Those who produce, process, trade, or consume organic products should protect and benefit the common environment including landscapes, climate, habitats, biodiversity, air, and water.

12.15.3 The Principle of Fairness

Organic agriculture should build on relationships that ensure fairness with regard to the common environment and life opportunities. Fairness is characterized by equity, respect, justice, and stewardship of the shared world, both among people and in their relations to other living beings.

This principle emphasizes that those involved in organic agriculture should conduct human relationships in a manner that ensures fairness at all levels and to all parties—farmers, workers, processors, distributors, traders, and consumers. Organic agriculture should provide everyone involved with a good quality of life and contribute to food sovereignty and reduction of poverty. It aims to produce a sufficient supply of good-quality food and other products. This principle insists that animals should be provided with the conditions and opportunities of life that accord with their physiology, natural behavior, and well-being. Natural and environmental resources that are used for production and consumption should be managed in a way that is socially and ecologically just and should be held in trust for future generations. Fairness requires systems of production, distribution, and trade that are open and equitable and account for real environmental and social costs.

12.15.4 The Principle of Care

Organic agriculture should be managed in a precautionary and responsible manner to protect the health and well-being of current and future generations and the environment. Organic agriculture is a living and dynamic system that responds to internal and external demands and conditions. Practitioners of organic agriculture can enhance efficiency and increase productivity, but this should not be at the risk of jeopardizing health and well-being. Consequently, new technologies need to be assessed and existing methods reviewed. Given the incomplete understanding of ecosystems and agriculture, care must be taken. This principle states that precaution and responsibility are the key concerns in management, development, and technology choices in organic agriculture. Science is necessary to ensure that organic agriculture is healthy, safe, and ecologically sound. However, scientific knowledge alone is not sufficient. Practical experience, accumulated wisdom, and traditional and indigenous knowledge offer valid solutions, tested by time.

12.16 Industrial Application of Effective Micro-organisms Technologies

12.16.1 Odor Management in Livestock Industries

During the mid-1980s, livestock researchers and producers in Japan began to test EM for odor control and waste management. This research continues to the present and has found EM to be effective as a probiotic, waste treatment, and biological control agent (Kitazato Environmental Center 1994). One of the most valuable contributions of EM to the livestock industry is its deodorizing effect within confined facilities for poultry operations. EM eliminates odors by dominating the microbial ecology with organisms that exploit a fermentative pathway and therefore do not produce odorous gases (Yongzhen and Weijiong 1994). There are four different ways in which EM inoculants can be introduced into the production system in order to achieve a deodorizing effect.

1. As a probiotic additive to drinking water.
2. As a probiotic feed additive.
3. As an additive to sanitation spray water for washing the facility.
4. As a treatment added to the waste handling process.

Considering the four methods of application of EM, it approaches the problem in three of the categories described by Ritter (1981):

1. As a digestive deodorant: Probiotics is added to drinking water at a dilution ranging from 1:1000 up to 1:10,000 and can be made available to the animals continually or periodically throughout the growth cycle (Yongzhen and Weijong 1994).
2. As a feed additive: It is mixed with normal feed rations at a ratio of 1–5 % (Yongzhen and Weijong 1994). Alternatively, it can be lightly sprayed over the feed at a ratio of 1:100 (Kitazato Environmental Center 1994).
3. Non-chemical deodorant: It is used as a disinfectant to regularly spray the facility and to inoculate the litter (once a week is usually enough) with beneficial micro-organisms (Ritter 1981).

EM helps balance the microflora within the animal's digestive tract. According to Yongzhen and Weijiojng (1994), EM increases the coefficient of nitrogen utilized by the bird. At the same time, the wastes generated by the broiler or laying hen receiving EM direct-fed microbial will tend to begin fermenting while they are being produced. This represents a big advantage for the future management of the manure because it will be populated with fermentative micro-organisms rather than putrefactive and pathogenic ones. With this same purpose, it is applied to wash down the facilities and to inoculate the litter with beneficial micro-organisms. The photosynthetic bacteria in it are able to separate the hydrogen in ammonia, in hydrogen sulfide, and in hydrocarbons; it deoxidizes carbon gases and synthesizes sugars. The lactic acid bacteria in the EM produce lactic acid that kills pathogenic micro-organism. Yeast in it form alcohol and various organic acids.

Ammonia is the largest contributor to foul odors being emitted from poultry facilities. Experiments done by Yongzhen and Weijiong (1994), with groups of 400–500 broilers and laying hens, indicate that the use of EM in the drinking water reduced the ammonia concentrations within the chicken houses by 42 %. The use of EM feed reduced ammonia concentrations by 54 % and the combination of the two techniques reduced ammonia concentrations by 69 %. A case study with 30,000 adult and 20,000 young Mary and Borisbrown chickens took place. Another case study reported from the Aichi Prefecture, Japan, on a farm of 150,000 laying hens showed a significant reduction in the foul odor of the poultry houses and the dung.

Table 12.4 Hydrogen sulfide and methyl mercaptan concentrations in the sewage management plant of Naha City, Japan

	Before EM treatment (ppm)	After EM treatment (ppm)
Hydrogen sulfide	11.8	0.78
Methyl marcaptan	0.075	0.0071

(Higa and Wididana 1991a)

EM was mixed in the drinking water, it was used in 1–2 % of the feed, and it was also sprayed throughout the inside of the poultry houses once a week. The ammonia concentration in the air of chicken houses was reduced from 4.4 to 3.9 ppm after the introduction of EM in the system. It must be mentioned that the data obtained previously to the use of EM were taken with the doors open, and the data obtained after using EM were taken with the doors closed (Kitazato Environmental Center 1994). EM's deodorizing effect has also been demonstrated in the urban waste management field. In Naha City, Japan, an EM culture was introduced in the standard activated sludge management plant, adding one liter of EM to each ton of raw sewage. EM was added before the sewage entered the aeration tank. In terms of odor, hydrogen sulfide and methyl mercaptan were analyzed before and after the EM treatment (Table 12.4).

12.17 EM for Disease Prevention

EM used as a spray to wash down the facilities acts as a disinfectant of the building. In a study done in the Aichi Prefecture, Japan, after one year of the introduction of an EM culture in the production system, it became totally unnecessary to use antibiotics and disinfectants for the 150,000 laying hens. Almost all of the vaccines that had been used were no longer necessary as a result of the overall improvement of the bird health (Kitazato Environmental Science Center 1994). Disinfectants used are generally chemical products such as phenol compounds or formaldehyde, although the latter has been prohibited in some countries because it can be hazardous to the health of humans. These products are usually applied when a flock is harvested and the building is empty (Sainsbury 1992). Because of the nature of the product, EM can be used to spray down the building even when the birds are in it.

Another approach for health improvement using EM is related to its use as an additive to drinking water and feed. The gastrointestinal tract of birds may house several pathogenic micro-organisms (Larbier and Leclercq 1994). The consumption by the animals is expected to result in health improvements apparently because of competition with pathogenic microflora in the digestive tract. Anjum et al. (1996) reported greater bursa and thymus index in commercial broiler chicken supplemented with EM through drinking water and feed. According to this study, EM supported these two important lymphoid organs that make up the vital components of humoral and cellular immunity. Antibody geometric mean titer (GMT) against

Table 12.5 *E. coli* population before and after EM treatment in the sewage management plant in Naha City, Japan

	Before EM treatment	After EM treatment
E. coli	12,000 Part/ml	900 Part/ml

(Higa and Wididana 1991a)

Newcastle disease vaccine virus was 6.5 times in broilers given EM in drinking water, 3.85 times in broilers given EM feed, and 3.73 times in broilers given both EM in drinking water and feed. At the same time, the EM-treated birds which had an increase in live body weight compared to the non-treated birds presented a decrease in the following measurements: offal weight, liver index, gizzard index, intestinal weight index, intestinal length index, kidneys index, and heart index. This indicates that EM can work as a growth promoter without any associated risks. Poultry products have been blamed for the transmission of *Salmonella* spp. and other human diseases (Stern 1994). *Salmonella enteritidis* is the most reported strain causing human infection and there has been clear epidemiological association of these cases with the consumption of eggs and poultry meat (Sainsbury 1992). According to Edens et al. (1997), the colonization of lactic acid bacteria in the chicken's intestinal tract apparently controls the population of pathogenic micro-organisms such as *Salmonella* spp., *Enterococci and E. coli*. Lactic acid bacteria produce significant amounts of bacterial growth inhibitory substances such as reuterin. Reuterin has a broad-spectrum antimicrobial activity that has proven to inhibit the growth of bacteria, fungi, and protozoa.

EM contains selected species of micro-organisms that include predominant populations of lactic acid bacteria. The information about the effect of lactic acid bacteria over these pathogens suggests a possible positive response to the use of EM. Another fact that suggests the possible effectiveness of EM against these types of pathogenic micro-organisms is related to the results obtained in several studies in urban waste management. The experience in Naha City, Japan, shows significant reduction in *E. coli* populations after the introduction of EM in the system. According to (Higa and Wididana 1991a), 30 days after treating the waste water in the Gushikawa City Library, (Okinawa) with EM, *E. coli* levels were undetectable, dropping from 8500 to 0 parts/ml. EM was used in a solution of 1:1000 EM/waste water. The EM solution was flushed down in the toilets (Table 12.5).

12.18 Birds' Performance Improvement with EM

The improvement of the animal's performance after the use of EM can be related to the inoculation of the gastrointestinal tract with beneficial micro-organisms. The gastrointestinal tract of birds is host to approximately 40 species of micro-organisms with three or more different types of each one. The flora plays an important role in the digestion process. Bacterial enzymes promote the digestion of protein, lipids, and carbohydrates, and bacteria also synthesize vitamins that contribute to the nutrition of the bird (Larbier and Leclercq 1994). According to Yongzhen and Weijiong (1994), EM improves the coefficient of nitrogen absorption in the animal. After 45 days of EM treatment in day-old commercial broilers, live body weight was approximately 2004 g for broilers given EM in drinking water, approximately 1978 g for broilers given EM feed, and approximately 2022 g for broilers given EM in both ways, compared to approximately 1690 g of the control broilers.

Yongzhen and Weijiong (1994) also found that the concentration of amino acids in the feed was improved 28 % after the fermentation process with EM, indicating that EM improves the quality of the feed. A study that took place in the Aichi Prefecture in Japan with 70,000–80,000 Arbor Acre broilers using EM for two years shows an improvement in the feed conversion rate and an increase in the weight increase per day. The average broiler weight at shipment went from 2.68 to 2.9 kg. EM was given in the drinking water once a week and it was also sprayed inside and outside the chicken house before the birds were brought in (Kitazato Environmental Science Center 1994). Regarding egg quality, a study done in the Gifu Prefecture, Japan, with 30,000 adult and 20,000 young Mary and Borisbrown chickens shows the effect of working with EM for two years. EM was given to the birds as EM feed at 1 % rate. The EM-treated group had higher values than the non-treated group in the following categories: average egg weight, eggshell strength, eggshell thickness, albumen height, Haugh units, and yolk color. In this same farm, the chicken excreta is being sprayed with EM to create a fermented compost that has a good reputation as being effective in increasing crop yields (Kitazato Environmental Science Center 1994).

12.19 Poultry Litter Management with EM

Another example of the positive effects that EM has in the general management of poultry facilities is related to the quality of the organic fertilizer produced with the manure. Poultry manure is a very useful resource for the production of organic fertilizers. Hussain et al. (1994) found that the nitrogen content of poultry manure increased after composting with EM. According to Hussein et al. (1994), the amount of the time needed to obtain compost was significantly reduced after the inoculation of the piles with EM. Using EM, solid wastes from the poultry industry can be processed alone or mixed with other easily obtained organic materials. Other materials used can include legumes, rejected seed yams, fish meal, corn meal, rice husks, sawdust, carbon, and ash. The materials are chopped, mixed well, and inoculated (Table 12.6).

Table 12.6 Effect of EM on Nitrogen content of organic materials

	Nitrogen content (%)				
Organic material	Initially	After 15 days		After 45 days	
	No EM	No EM	EM	No EM	EM
Farmyard manure	0.42	0.49	0.70	0.70	0.84
Poultry manure	0.56	0.84	1.19	0.98	1.26
Wheat straw	0.35	0.42	0.49	0.49	0.56
Rice straw	0.28	0.28	0.35	0.42	0.49
City waste	0.35	0.49	0.56	0.56	0.63

Hussain et al. (1994)

12.20 Scientific Validity of EM Technologies in Agriculture

12.20.1 Livestock Trials

Acute and chronic toxicity tests, and mutagenic test of the extracts from the fermentation of plants with *EM-X* were performed in the mouse and the rat by Ke et al. (2005). In the acute toxicity test, mice were orally treated three times per day with 20-fold of concentrated EM-X for 7 days. For chronic toxicity test, the rats were orally treated with original EM-X once a day for 90 days at the dosages of 180, 120, or 60 ml/kg. At the levels tested, EM-X did not lead to significant changes in food consumption, body weight, behaviors, and stools. Hematological assays on red blood, white blood cell, hemoglobin, platelets, lymphocyte, granulocyte, middle cell, and coagulation time and the biochemical assays on aspartate aminotransferase, alanine aminotransferase, alkaline phosphatase, blood urea nitrogen, total protein, albumin, glucose, total bilirubin, creatinine, and total cholesterol did not show abnormal changes. The histological inspection of principal organs of the heart, liver, spleen, lung, and kidney did not show significant pathological changes. The delaying toxic reactions were detected 2 weeks after administration of EM-X was stopped. The mutagenic test showed that EM-X did not cause mutagenesis. Tests of micronucleus of bone marrow cell and sperm shape abnormality upon EM-X were negative. The maximal tolerance dose of EM-X was calculated to be 1800 ml/kg BW in the mouse and rat. Thus, oral administration of EM-X does not present acute and chronic toxicity and mutagenic effects in the animals.

Naqvi et al. (2000) experimented on the effect of EM4 on the health of layers. In the experiment, commercial laying (Babcock) 174 weeks old were given feed containing 1, 2, and 3 % EM4 for a period of 12 weeks. EM4 did not influence live body weight. Egg production was greater in birds given feed containing 1 and 2 % EM4 than the control ($P < 0.05$). Serum phosphorus was significantly lesser in birds given feed containing 3 % EM4 than the control birds ($P < 0.01$). Serum total protein, serum albumen, serum globulins, serum total lipids, and serum cholesterol were not influenced significantly with the EM4 treatment. The study suggests that

EM4 is a safe product for laying birds; it increases egg production when mixed in feed.

Chotisasitorn et al. (1997) in an experiment on the effect of supplementation of EM in feed on laying performance and egg quality using a 2 by 3 factorial in completely randomized design with 4 replications with a total number of 288 layers, showed remarkable results of interest. In one factor, supplementation and non-supplementation of EM were applied. In the other factor, supplementation of calcium at levels of 3, 4, and 5 % was used. Results of the study, over the 3 periods of 28 days per period, revealed that there was no significant effect of supplementation of EM on daily feed intake, body weight gain, mortality, egg mass, egg weight, albumen weight, yolk weight, egg shell weight, yolk color, and haugh unit ($P > 0.05$). But egg production ($P < 0.05$) feed per 1 dozen egg and specific gravity were highly significantly different ($P > 0.01$).

Safalaoh and Smith (1994) conducted research to evaluate the effect of using EM as an alternative to antibiotics (AB) on growth performance, feed utilization, and serum cholesterol of broilers. Dietary treatments consisted of supplementation with: AB (Zinc Bacitracin) only, EM only, AB (Zinc Bacitracin) plus EM, and control. The EM was supplemented at either 15 or 30 g/kg while the AB (Zinc Bacitracin) was added at 500 mg/kg. At six weeks of age, birds fed diets neither with the EM nor with AB had significantly ($P < 0.05$) lower weight gains (2066 g) than the rest of the treatments. Birds fed the diet containing AB and EM at 30 g/kg had significantly ($P < 0.05$) higher body weight gain (2096 g) than the rest of the treatments. The improvements in BWG were associated with slight enhancement of feed efficiency while the EM effects were more pronounced at the higher dosage (30 g/kg). The poorest feed: Gain ratio (1.82) was observed in the control. Apart from improving dressing percentage, EM supplementation also resulted in birds with low serum cholesterol levels. This study has shown that EM has growth promoting and hypocholesteremic effects and offers a potential alternative to antibiotics in broiler diets.

Osteoporosis is a disease of aging associated with bone loss that often occurs without symptoms until micro-architectural deterioration becomes so significant that bone fracture occurs. The effective micro-organism X (*EM-X*) is an antioxidant beverage derived from ferment of unpolished rice, seaweeds, and papaya with effective micro-organisms of lactic acid bacteria, yeast, and photosynthetic bacteria (containing minerals, alpha-tocopherol, lycopene, ubiquinone, saponin, and flavonoids). The levels of serum estradiol (E (2)) and the bone density of the middle and epiphysis of femurs were assessed by Ke et al. (2009) in order to determine the effect of EM-X on osteoporosis in ovariectomized rat (an animal model of postmenopausal osteoporosis). EM-X (1 ml/rat/day) was initially administrated by gavage to rats which were then allowed to consume 10 % (v/v) EM-X in water freely for 3 months. There was no statistical significance of E (2) level between sham operation group and control group, indicating that sham operation did not affect E (2) level. However, the E (2) levels in the ovariectomized rats tended to increase after treatment of EM-X for 3 months. The bone density of the middle and epiphysis of femur in both sham operation and ovariectomy group decreased with time. Rats receiving EM-X for 3 months after sham operation or ovariectomy had increased bone density of the middle of femur that was statistically significant ($P < 0.01$ and $P < 0.05$). The bone density of the epiphysis of femur in both sham operation and ovariectomy group was significantly increased, an outcome highly suggestive of the beneficial effects of EM-X on bone density of the middle and the epiphysis of femur in the rats with or without ovariectomy.

Wondmeneh et al. (2011) evaluated the effect of different administration methods of EM on the performance and serum cholesterol level of broilers at Debre Zeit Agricultural Research Center, Ethiopia. Uniform weight of mixed sex day-old-broilers of cobb-500 strain ($n = 240$) was randomly distributed to 4 treatment groups with 3 replications of 20. They were kept under a standard management condition for 49 days being subjected to treatment rations since day 10 on. Performance parameters were recorded and analyzed. Total blood cholesterol was analyzed with standard kit at the end. The result showed that there was no significant difference of EM administration methods ($P < 0.05$) on mortality of chickens during the starter (1–29 days) and finisher (30–49 days) phases. Feed consumption was found to be significantly higher for Treatment 4 (Bokashi in feed + EM in water) than the rest of the treatment groups. Weight gain was significantly higher ($P < 0.05$) for Treatment 4 (Bokashi + EM in water), during the entire period than the rest of the treatment groups. Birds fed with T4 (Bokashi + EM in water) required less feed for a unit increase in weight during the starter and finisher phases. Birds fed with T3 (Non Bokashi + EM in water) required the highest feed for a unit increase in weight. EM application in all forms resulted in significantly lower ($P < 0.05$) total blood cholesterol. EM application in both feed and water combined was the most effective in lowering the total blood cholesterol than the other application methods.

Safalaoh (1994) researched on the effects of supplementation of a microbial preparation, EM, on body weight gain, dressing percentage, abdominal fat, and serum cholesterol content of broilers. The EM was added to drinking water at a rate of 1 part EM to 1000 parts of water. The two treatments were control (0 EM) and 1 EM. Final body weights, serum cholesterol, and abdominal fat pads were determined at day 42. Dressing percentage was determined using carcass weight as a proportion of body weight. Abdominal fat was used as an indicator of the carcass' fat content and was calculated as percentage of body weight. Birds supplemented with 1 EM had significantly ($P < 0.05$) higher weight gains (2094 ± 11 g) than the control (2057 ± 15 g). Control birds had significantly ($P < 0.05$) higher feed intake (3785 ± 9 g) than the birds supplemented with 1 EM (3748 ± 13 g). However, feed efficiency, measured as feed: gain ratio, was better for the EM supplemented birds (1.79 ± 0.03) than the control (1.84 ± 0.02). Although not significantly different, serum cholesterol content was lower for 1 EM birds (3.15 ± 0.21 mmol/l) than in 0 EM birds (3.38 ± 0.17 g mmol/l). Dressing percentage was not significantly different between the two treatments, but numerically higher for the 1 EM birds than the control. Abdominal fat pad was lower for the 1-EM-treated birds than the non-EM-treated birds. The results of this study suggest that microbial preparations such as EM can be used to improve weight gain, feed utilization, and reduce abdominal fat pads, hence fat content of birds. Although not significant, the present study has shown that use of microbial preparations may have some potential to improve dressing percentage and lower serum cholesterol. However, further studies such as use of different EM concentrations are required to ascertain the results found in this study. To assess potential health benefits, research is also required to assess the effect of using EM on total cholesterol content in tissues.

Sokół et al. (2009) conducted a study on the "influence of a 14-day administration of an EM solution in drinking water to laying hens on hematological and biochemical indexes." The research was carried out on 120 hens divided into two equal groups. The birds in the experimental group were given drinking water with dissolved EM (5 % solution), and those in the control group–water without the preparation. On the 64th day of the aviculture, the hens were weighed and their blood was taken from the wing vein for hematological and biochemical examinations. Administering EM with water to hens did not influence significantly their body weight nor chosen hematological and biochemical indexes. A significant increase was found only in the number of platelets, the level of albumins, the content of total cholesterol, and the LDH activity; however, a decrease in the ALT activity was observed. Tabidi et al. (2013) investigated the effects of commercial probiotic mixtures (Liptosafe-L) via drinking water of broiler chickens in comparison with antibiotic (Newmycin) on performance of broiler chicks and growth attributes. 120 day old brioler chicks (Ross 308) withaverage weight of 41g, were subjected to a 41-day experimental period. The chicks were randomly divided into three experimental treatment groups: probiotic, antibiotic, and control group. Each group with four replicates (10 chicks per replicate). Birds in the first group were supplied with 0.5 ml of probiotic per one liter of water for the whole growth period while the antibiotic mixture is administrated at a rate of 0.20 g per liter of water along the growth period. In these two cases, the growth promoters were stopped at 7 days before slaughtering (safety period). Birds in the control group received water without growth promoter. Results of the experiment showed that there were significant differences ($P < 0.05$) in body weight gain (g), feed conversion ratio, and final body weight (g). For feed intake (g) and mortality rate (%), there is no significant difference between probiotic and antibiotic ($P < 0.05$) but both differ significantly if compared with control ($P < 0.05$). Broilers fed probiotic statistically consumed more feed over the entire experimental periods. In contrast, feed conversion efficiency did not improved in different periods in growth promoter supplemented groups compared to control birds. In conclusion, the results obtained in this study indicated that dietary inclusion of probiotic and antibiotic supported a superior performance of chicks and can be applied as antibiotic growth promoter substitutions in broilers diet.

Cross-sectional study was conducted by Dorn-In et al. (2009) to determine the prevalence of *Salmonella* and to associate management factors in fattening pigs in a production compartment of Northern Thailand. A total of 194 fecal samples and 166 environmental samples were collected from 22 fattening pig herds for isolation and identification of *Salmonella*. An additional 427 serum samples were collected from the same herds to determine *Salmonella* antibodies using ELISA.

A questionnaire was used to collect management factors likely to be associated with *Salmonella* identification. Prevalence of *Salmonella* in each sample and its confidence interval was adjusted for clustering by herds using linearization technique. A generalized estimating equation was used to determine the odds ratio and significance level for each management factor in a logistic regression model. *Salmonella* was found in all 22 study pig herds with a fecal sample prevalence of 63 % (95 % CI: 56–69 %) and a serum sample prevalence of 72 %. However, isolation results were not significantly different from ELISA results. The most isolated serotype was *Salmonella rissen* (49 %) followed by *Salmonella typhimurium* (19 %), *Salmonella stanley* (12 %), and *Salmonella weltevreden* (4 %) being significantly different in the different specimens collected ($P = 0.24$). The final logistic regression model with isolation results as outcome showed that medium herd size ($OR = 2.32$, $P = 0.003$, $P = 0.000$) was significantly associated with positive *Salmonella* isolation; with positive ELISA results; however, only the use of EM was significantly associated ($OR = 2.63$, $P = 0.011$).

12.21 Water Management and Fisheries

Water quality has received considerable attention in allocation processes for maximizing the satisfaction of various sectors. However, pollutant impurities that impede adequate supply of water have a detrimental effect on the quality and harmful for living organisms including aquatic life. For the reduction of water pollution level, various chemical and biological treatments are available but the emergence of an amazing technology of a multiculture of anaerobic and aerobic beneficial micro-organisms is presently gaining popularity due to its environmentally friendly nature. EM technology uses naturally occurring micro-organisms which are able to purify and revive nature. Applications of EM using the formula known as effective micro-organism-activated solution (EMAS) have been experimented in several rivers in Malaysia depending on the scale, location, physical, and geographical conditions with the principal objective of enhancing and improving the water quality. One of the significant contributions of EM based rehabilitation of polluted and degraded water bodies is to restore aquatic habitats and ecosystems. Existing results by Zuraini et al. (2010) of projects via EM technology in solving water quality-related problems and the nationwide campaigns in Malaysia indicated significantly the sustainability of EM technologies in aquatic ecosystem management. The role of EM-based water restoration approach for sustainability of water resources and the prospects of modeling are also discussed. Results clearly demonstrated the effectiveness of this technique for restoration of water quality of degraded/polluted river basin. Valuable lines for further research and acceptance of EM technology for the future are thus suggested as it is believed to be the key to sustained environmental improvement and offers a real opportunity for eco-innovation.

The activated sludge membrane bioreactor (MBR) has been shown to have some advantages for the processing and reclamation of domestic wastewater. Jin et al.

(2005) hypothesized that certain micro-organisms, chosen for their abilities to decompose the chemical components of raw sewage, would, when coupled with the MBR, significantly improve the stability and efficiency of this system. Environmental bacterial strains were selected which oxidize ammonia and nitrites; and produce protease, amylase, and cellulase for the development and testing of a novel biologically enhanced MBR (eMBR). We compared the eMBR with the activated sludge MBR. With the eMBR, the average values of effluent quality were chemical oxygen demand (COD), 40 mg/l (average efficiency of removal 90.0 %), and NH (4) (+)–N, 0.66 mg/l (average efficiency of removal 99.4 %). Effluent qualities met the standard and were stable during the entire 90 days of this study. For the activated sludge MBR, the COD removal rate was 91.7 %, and the NH (4) (+)–N removal (94.8 %) was less than that of the eMBR. Start-up time for the eMBR was only 24–48 h, much shorter than the 7–8 days required to initiate function of the standard MBR. The biomass concentrations of total heterotrophic bacteria and autotrophic bacteria in the eMBR did not fluctuate significantly during the course of the study. Various kinds of micro-organisms will establish an ecological balance in the reactor. Compared with the activated sludge MBR, the eMBR not only produced an excellent and stable quality of effluent but also resulted in a shorter time to start up and significantly improved the efficiency of NH(4) (+)–N removal.

Landfills are still a popular way for municipal solid waste (MSW) treatment. Leachate generated from landfills is becoming a great threat to the surroundings as it contains high concentrations of toxic substances. How to control leachate migration and to protect environmental pollution is now a concern for many environmentalists. Ding et al. (2001), in this work, isolated eight EM from wastewater, sludge, and soil samples by enrichment culturing techniques and used for leachate migration control in columns and pilot experiments. The preliminary experiments reveal that the EM could remove 25 and 40 % of chemical oxygen demand (COD) from leachate in fine sand and sabulous clay columns, respectively. An aquifer system was designed to simulate in situ control for leachate migration with EM. The EM was injected into the simulated aquifer and formed a permeable biological barrier. The experimental results showed that the barrier removed 95 % of COD and approximately 100 % inorganic nitrogen, that is, nitrate-N plus nitrite-N plus ammonia-N, from the migrating leachate. CO_2 production, redox potential, and microbial number were monitored simultaneously in the aquifer during the experiment to assess the EM activities and the effect of the bio-barrier. The data indicated that the EM isolated in this work had high activities and were effective for organic and nitrogenous contaminant removal throughout the experiment.

Han et al. (2003) in an experiment to optimize aquatic ecological structure and to regulate water quality, *Chlorella vulgaris* and effective micro-organism were added to *Exopalaemon carinicauda* pond and fishponds. The results showed that after adding *Chlorella vulgaris* to the shrimp pond and fishpond, *Chlorella vulgaris* turned into a dominant species, and its amount was 16.92 and 4.76 times of CK. The zooplankton biomass reached to 4.32 and 2.84 mg l^{-1}, increasing by 19.3 and

2.5 %, compared with CK, respectively. *Rhodospirillaceae*, photosynthetic bacteria, and yeast *Saccharomycete* in the ponds could obviously change the composition, number, ratio, and biomass of the plankton (phytoplankton and zooplankton) and adjust aquatic chemical environment. The treatment of "*Saccharomycete* + Nitrifying bacteria" decreased the concentrations of NH_4^+ obviously, which was only 44 % of CK. The BOD and COD in shrimp ponds were only 56.5 and 38.4 % of CK. The treatment could increase the dissolved oxygen and primary production in the pond.

Zhao et al. (2006) in their research utilize a special kind of carrier to immobilize effective micro-organisms B350M in a biological aerated filter (BAF) react system for treatment of oil field wastewater, which is of salinity >0.5 %, lack of N and P, and contains low organic matter. Through the biodegradation system operated for 142 days, the react system can achieve average degradation efficiency 90.5, 74.4, 85.6, 100 % for oil, TOC, COD, and H_2S, when HRT was 4 h and COD volumetric load was 1.07 kg/(m^3 × days). GC-MS results show that the organic substance in wastewater contains 27 different kind substances, a majority (23) of alkane, and a minority (4) of aromatic substances. C14H30 to C28H58 in influent could be decomposed into small molecular substance efficiently, especially the C18H38 to C28H58, and also polycyclic aromatic hydrocarbons (PAHs) such as Phenanthrene. The react system had a good diversity, because the carriers provide agreeable air and water condition for micro-organisms, to resist high salinity and toxic pollutant. Filamentous micro-organisms were observed in a great deal and will not cause foaming and bulking in BAF reactor by immobilization.

On commercial pig production farms in Southeast (SE) Asia, the liquid effluent is often discharged into rivers. The discharge is a hazard to the environment and to the health of people using water from the river either for consumption or for irrigation. Therefore, a simple percolation bio-filter for treatment of the liquid effluent was developed. Pig slurry was treated in test-bio-filters packed with different biomass for the purpose of selecting the most efficient material; thereafter, the efficiency of the bio-filter was examined at farm scale with demo bio-filters using the most efficient material. The effect of using EM added to slurry that was treated with bio-filter material mixed with Glenor KR+ was examined by Sommer et al. (2005). Slurry treatment in the test-bio-filters indicated that rice straw was better than coconut husks, wood shavings, rattan strips, and oil palm fronds in reducing BOD. Addition of EM and Glenor KR+ to slurry and bio-filter material, respectively, had no effect on the temperature of the bio-filter material or on the concentrations of organic and inorganic components of the treated slurry. The BOD of slurry treated in test-bio-filters is reduced to between 80 and 637 mg O2 l (−1) and in the demo bio-filter to between 3094 and 3376 mg O2 l (−1). The concentration of BOD in the effluent is related to the BOD in the slurry being treated and the BOD concentration in slurry treated in test-bio-filters was lower than BOD of slurry treated in demo bio-filters. The demo bio-filter can reduce BOD to 52–56 % of the original value, and TSS, COD, (chemical oxygen demand) and ammonium (NH_4^+) to 41–55 % of the original slurry. The treated effluent could not meet the standards for discharge to rivers. The composted bio-filter material has a high content of nitrogen and phosphorus; consequently, the fertilizer value of the compost is high. The investment costs were 123 US dollar per SPP which has to be reduced if this method should be a treatment option in practice.

The effect of a mixture of four indigenous bacterial genera composed of *Bacillus*, *Pseudomonas*, *Acinetobacter,* and *Flavobacterium* on egg hatchability and larval viability of *Clarias gariepinus* was investigated by Ariole and Okpokwasili (2012) at University of Port Harcourt, Nigeria. The fertilized eggs were distributed into glass Petri dishes (100 mm diameter) containing 50 ml of water at graded level of mixed indigenous probiotics ranging from 0 to 108 cells/ml. The incubation time increased from 17 h at 0 cfu/ml to 22 h at 108 cfu/ml. The mean hatching rate increased from 8.70 % at 0 cfu/ml to 53.85 % at 108 cfu/ml. The highest larval survival of 71.43 % recorded at 108 cfu/ml, where the highest hatching rate was observed, was significantly higher than the larval survival rate observed at the other concentrations. All yolk sac larvae at 0 and 101 cfu/ml died before the end of yolk sac period. These results imply that the incubation time, hatching rate, and larval survival of *Clarias gariepinus* increased with increase in bacterial load of water up to 108 cells/ml; the highest dose employed. Further investigations are needed to establish the optimal and threshold doses.

12.22 Crops Trials

Khaliq et al. (2006) conducted a field experiment to determine the effects of integrated use of organic and inorganic nutrient sources with EM on growth and yield of cotton. Treatments included: control; organic materials (OM); effective micro-organisms (EM); OM + EM; mineral NPK (170:85:60 kg); 1/2 mineral NPK + EM; 1/2 mineral NPK + OM + EM; and mineral NPK + OM + EM. OM and EM alone did not increase the yield and yield-attributing components significantly but integrated use of both resulted in a 44 % increase over control. Application of NPK in combination with OM and EM resulted in the highest seed cotton yield (2470 kg ha^{-1}). Integrated use of OM + EM with 1/2 mineral NPK yielded 2091 kg ha^{-1}, similar to the yield (2165 kg ha^{-1}) obtained from full recommended NPK, indicating that this combination can substitute for 85 kg N ha^{-1}. Combination of both N sources with EM also increased the concentrations of NPK in plants. Economic analysis suggested that the use of 1/2 mineral NPK with EM + OM saves the mineral N fertilizer by almost 50 % compared to a system with only mineral NPK application. This study indicated that application of EM increased the efficiency of both organic and mineral nutrient sources but alone was ineffective in increasing yield.

Ngele (2013) in a 2-year field experiment carried out at the research site of Federal College of Agriculture Ishiagu, Ebonyi State, Nigeria, during the 2012 and 2013 rain fed cropping season to investigate the effect of EM on the severity of nematode on sweet potato in Ishiagu southeastern Nigeria. The experimental design was randomized complete block design (RCBD) with three treatments, namely 10 l

(1 level) of EM, 20 l (2 level) of EM, and control (0 level) with three their replicates. The 87/0087 variety of sweet potato was used as the test crop. Parameters measured were number of leaves (at 4, 5, 6, 7, 8, 9 weeks after planting), vine length (at 4, 5, 6, 7, 8, and 9 WAP), nematode count before and after application of effective micro-organism and tuber yield at maturity. The results indicated a significant difference in all the parameters measured at 2 levels of application of effective micro-organisms. This was followed by 1 level of application of EM while the least was recorded in the control. It could be inferred from the results that level 2 produced the highest marketable tuber yield of sweet potato without any damage from nematodes as compared to the control.

Zhang et al. (2005) conducted a 2-year field experiment of wheat–maize rotation on a cinnamon soil of east Hebei Province, China, to study the effects of returning maize straw into field on the dynamics of soil microbial biomass, C, N, and P, and their relationships with soil nutrients and enzyme activities. The results showed that under the condition of returning maize straw combined with applying chemical fertilizer to adjust straw C/N ratio, the application of effective micro-organisms could increase soil microbial biomass, C, N, and P in each crop growth period, advance their peak time, and better regulate soil nutrient supply, compared with no application of effective micro-organisms. Soil microbial biomass had a significantly positive correlation with soil enzyme activities, but its correlation with soil hydrolysable N and available P was strongly affected by crop growth and fertilization system.

Onele (2014) in another 2-year field experiment carried out at the research site of Federal College of Agriculture Ishiagu, Ebonyi State, Nigeria, during the 2013 and 2014 rain fed cropping season to investigate the ‘influence of EM-based fermented plant extract (FPE) on the fruit yield and severity of nematodes on *Telferia occidentalis*. The experiment was randomized complete block design with four treatments and 4 replicates. The treatments are 30 g of grounded: bitter leaves (*Vermonia amagdalina*), neem leaves (*Azardirachta indica*), and Saim weed (*Chromolena odoranta*) fermented in 20 l of sugar water and activated with 10 ml of EM1 for each of the treatments. The parameters measured were number of leaves and vine length at 4, 5, 6, 7, 8, 9, and 10 WAP. Nematode counts before and after application of treatment and fruit yield at maturity. The results indicated a significant difference in all the parameters measured. Neem-based treatment (treatment 1) has the greatest yield value, followed by bitter leaf-based treatment (treatment 2) and then Siam weed-based treatment (treatment 3). The least yield value was treatment 4 (control). From the results of the experiment, neem-based treatment was recommended for adoption by farmers because of its performance and easy access to the materials for use by farmers.

Daiss et al. (2008) in their study using storage conditions recommended for conventional chard (4 °C, 90 % RH and 7 days), the chard treated with some organic preharvest treatments [effective micro-organisms, a fermented mixture of effective micro-organisms with organic matter (EM-Bokashi + EM), and an auxiliary soil product] lost considerable water (>2 %) and weight (>25 %). These results indicated that organic methods tested produce a vegetable that cannot sustain its quality when commercialized through the conventional supply chain. Nevertheless, respiration, color, pH, and titratable acidity practically remained constant during conservation. Ascorbic acid content was constant in chard treated with the different preharvest treatments and collected at 8 week after sowing (normal harvest). However, the ascorbic acid content of the control chard decreased 60 % after 7 days of storage. This vitamin diminished (35 %) in chard collected after 19 week after sowing (late harvest) during the postharvest conservation. The greatest difference in chard quality was registered between sampling dates since chard collected during the late harvest had higher levels of dry matter, sugars, acids, proteins, and ascorbic acid than chard collected during the normal harvest.

The difficult cultivation of the Saffron plant (*Crocus Sativus* L.) makes the spice of the same name made from its dried stigmas very valuable. It is estimated that some 75,000 blossoms or 225,000 hand-picked stigmas are required to make a single pound of saffron, which explains why it is the world's most expensive spice. Aytekin and Acikgoz (2008) in their study seek to identify ways of increasing the fertility and production of saffron. For this purpose, the treatment of saffron bulbs with a synthetic growth hormone—a mixture of Polystimulins A6 and K—and two different micro-organism-based materials—biohumus or vermicompost and EM—in four different ways (sole hormone, sole biohumus, sole EM and EM + biohumus) was investigated to determine whether these treatments have any statistically significant effects on corms and stigmas. The results indicated that EM + biohumus were the most effective choice for improved saffron cultivation.

12.23 Bioremediation Trials

Ekpeghere et al. (2012) experimented on the application of loess balls containing EM to the remediation of contaminated harbor sediments, and to thereby elucidate the functions of EM in remediation. Changes in physicochemical, biochemical, and microbiological parameters were measured to monitor the remediation process at a laboratory scale. Treatment with high concentrations of EM stock culture and EM loess balls (4 %), and a low concentration of EM loess balls (0.1 %) that contained molasses (0.05 %) contributed to more rapid removal of malodor. Acetic acid, propionic acid, valeric acid, carbonic acid, and lactic acid were rapidly removed in the presence of molasses (0.05 % w/w) as a carbon nutrient source, indicating enhanced EM activity by amendment with molasses. Fermentation of molasses by EM showed that more acetic acid was produced compared with other organic acids, and that the majority of organic acids were eventually converted to acetate via intermediate metabolites. Sediment bioremediation tests showed that there was no significant difference in eubacterial density with the control and the treatments. However, the density of a *Lactobacillus* spp. in sediments treated with 0.1 and 4.0 % EM loess balls was significantly higher than the control, which indicated the bioaugmentation effect of EM loess balls in the polluted sediments. Treatment with EM loess balls and an appropriate amount of molasses, or other nutrients, will facilitate the remediation of polluted marine sediments by mal odor removal, via EM degradation or utilization of offensive organic acids. The study showcased EM functions during the bioaugmentation process, both in terms of organic acid metabolism and the dynamics of the engineered microbial community.

Chen et al. (2009) conducted an experiment in which heterotrophic nitrification-aerobic denitrifier bacteria CPZ24 was isolated from the livestock wastewater by the way of the limiting dilution combined with the chromogenic medium screening methods. This bacterium was gram positive rod. The colonies of the strain were orange-red. It was identified as *Rhodococuus pyridinivorans* according to its morphological and physiological properties and the analysis of its 16S rDNA gene. Studies on its function of heterotrophic nitrification and aerobic denitrification results showed that all NH_4^+–N is removed and the removal rate of TN is 98.70 % in heterotrophic nitrification. This high effective micro-organisms with nitrogen removed is able to realize simultaneous nitrification and denitrification. The removal rate of NO_3^-–N by this strain is 66.74 % and the removal rate of TN is 64.27 %. It can perform the whole process of bacteria denitrification independently.

Zhou et al. (2008) conducted research to test the damage to DNA of EM by heavy metal ions $As_3^+, Cd_2^+, Cr_3^+, Cu_2^+, Hg_2^+, Pb_2^+$, and Zn_2^+, as well as the effects of EM bacteria on wastewater treatment capability when their DNA is damaged. The approach applied in this study was to test with COMET assay the damage of EM DNA in wastewater with different concentrations of heavy metal ions $As_3^+, Cd_2^+, Cr_3^+, Cu_2^+, Hg_2^+, Pb_2^+, Zn_2^+$, as well as the effects of EM treated with $As_3^+, Cd_2^+, Cr_3^+, Cu_2^+, Hg_2^+, Pb_2^+$, and Zn_2^+, on COD degrading capability in wastewater. The results showed that the damage of the DNA of EM was negatively correlated with their treatment capability and that EM bacteria maximum tolerant concentrations of these heavy metal ions was at 0.05 mg/l for As_3^+, 0.2 mg/l for Hg2+, 0.5 mg/l for Cd_2^+, Cr_3^+, and Cu_2^+, and 1 mg/l for Pb_2^+, and Zn_2^+.

Mukred et al. (2008) investigate on the bioremediation of polluted groundwater, wastewater aeration pond, and biopond sites using bacteria isolated from these sites located at the oil refinery Terengganu, Malaysia. Out of 62 isolates, only 16 isolates: from groundwater (8) and wastewater aeration pond (3) and biopond (5) were chosen based on growth medium containing 1 % (v/v) Tapis crude oil. Only four isolates: *Acinetobacter faecalis*, *Staphylococcus* spp., *Pseudomonas putida, and Neisseria elongata* showed percentage biodegradation of crude oil more than 50 % after 5 days using a mineral salts medium. The effect of physical parameters (temperature, pH and agitation) on growth by all four strains showed a maximum growth in MSM medium with 1 % Tapis crude oil at 37 °C with pH 7 and agitation of 130 rpm.

Izallalen et al. (2008) in their study on *Geobacter sulfurreducens* strain engineered for increased rates of respiration inferred that *Geobacter* species are among the most effective micro-organisms known for the bioremediation of radioactive and toxic metals in contaminated subsurface environments and for converting organic compounds to electricity in microbial fuel cells. They hypothesized that faster rates of electron transfer could aid in optimizing these processes. Therefore, the Optknock strain design methodology was applied in an iterative manner to the constraint-based, in silico model of *Geobacter sulfurreducens* to identify gene deletions predicted to increase respiration rates. The common factor in the Optknock predictions was that each resulted in a predicted increase in the cellular ATP demand, either by creating ATP-consuming futile cycles or by decreasing the availability of reducing equivalents and inorganic phosphate for ATP biosynthesis. The in silico model predicted that increasing the ATP demand would result in higher fluxes of acetate through the TCA cycle and higher rates of NADPH oxidation coupled with decreases in flux in reactions that funnel acetate toward biosynthetic pathways. A strain of *G. sulfurreducens* was constructed in which the hydrolytic, F (1) portion of the membrane-bound F (0) F (1) (H (+))-ATP synthase complex was expressed when IPTG was added to the medium. Induction of the ATP drain decreased the ATP content of the cell by more than half. The cells with the ATP drain had higher rates of respiration, slower growth rates, and a lower cell yield.

Genome-wide analysis of gene transcript levels indicated that when the higher rate of respiration was induced, transcript levels were higher for genes involved in energy metabolism, especially in those encoding TCA cycle enzymes, subunits of the NADH dehydrogenase, and proteins involved in electron acceptor reduction. This was accompanied by lower transcript levels for genes encoding proteins involved in amino acid biosynthesis, cell growth, and motility. Several changes in gene expression that involve processes not included in the in silico model were also detected, including increased expression of a number of redox-active proteins, such as ctype cytochromes and a putative multicopper outer-surface protein. The results demonstrate that it is possible to genetically engineer increased respiration rates in *G. sulfurreducens* in accordance with predictions from in silico metabolic modeling. This is the first report of metabolic engineering to increase the respiratory rate of a micro-organism.

12.24 Conclusion

Principles, applications, and validity of EM as a sustainable component in organic farming systems were reviewed given the global consumer shift to organic produce. This resulted from scientific evidences that justifies its naturalness, safety, and wholesomeness; ecological sustainability, profitability, longer shelf live in storage, higher food value in terms of nutrient, stability of soil fertility, improved maturity period, continuity of yield et cetera. The article further explored the concept of organic agriculture and the underlying principles, EM production, and mode of action; justifying theories, scientific validity, and applications.

References

Alfvén T (2006) Allergic diseases and atopic sensitization in children related to farming and anthroposophic lifestyle—The PARSIFAL study. Allergy. 2006 April, 61(4):414–21. Institute of Environmental Medicine, Karolinska Institute, Stockholm, Sweden

Anjum A, Hussaln T, Rizvi F, Gilani G, Javaid T (1996) Influence of effective microorganisms on health and immune system of broilers under experimental conditions. In: 5th conference on the technology of effective microorganisms at Sara Buri, Thailand on 10–11 Dec. Organized by APNAN and INFRC. P-5

Aytekin A, Acikgoz AO (2008) Hormone and microorganism treatments in the cultivation of saffron (*Crocus sativus* L.) plants. Molecules 13(5):1135–1147

Ariolel CN, Okpokwasili GC (2012) The effect of indigenous probiotics on egg hatchability and larval viability of *Clarias gariepinus*. www.ambi-agna.net

Butler et al (2008) Fatty acid and fat-soluble antioxidant concentrations in milk from high- and low-input conventional and organic systems: seasonal variation. J Sci Food Agric 88:1431–1441

Chavas JP, Posner JL, Hedtcke JL (2009) Organic and conventional production systems in the wisconsin integrated cropping systems trial: II. Economic and risk analysis 1993–2006. Agron J 101(2):288

Chen PZ, Wang LG, Wang YC, Li J, Ding W, Ren TZ, Li SP (2009) Screening and denitrification characteristics of a heterotrophic nitrification-aerobic denitrifier bacteria. Huan Jing Ke Xue 30 (12):3614–3618

Chotisasitorn S, Somchai C, Seksom A, Attawoot P (1997) Using an effective micro-organisms supplementation in layers. Kasetsart J Nat Sci 31:363–367

Daiss N, Lobo MG, Gonzalez M (2008) Changes in postharvest quality of Swiss chard grown using 3 organic pre harvest treatments. J Food Sci 73(6):S314–S320

De Almeiada SG, Fernandez GB (2006) Economic benefit of a transition to ecological agriculture. LEISA Mag 22(2):28–29

Dimitri C, Oberholtzer L (2006) EU and US organic markets face strong demand under different policies. Amber Waves

Dorn-In S, Fries R, Padungtod P, Kyule MN, Baumann MP, Srikitjakarn L, Chantong W, Sanguangiat A, Zessin KH (2009) A cross-sectional study of Salmonella in pre-slaughter pigs in a production compartment of northern Thailand. Presv Vet Med 88(1):15–23

Ding A, Zhang Z, Fu J, Cheng L (2001) Biological control of leachate from municipal landfills. Chemosphere 44(1):1–8

Drinkwater LE, Wagoner P, Sarrantonio M (1998) Legume-based cropping systems have reduced carbon and nitrogen losses. Nature 396(6708):262–265

Edens FW, Parkhurst CR, Casas IA, Dobrogosz WJ (1997) Principles of ex ovo competitive exclusion and in ovo administration of Lactobacillus reuteri. Poultry Science. Poultry Science Association, Inc., vol 76, pp 179–196, Savoy, IL

Ekpeghere KI, Kim BH, Son HS, Whang KS, Kim HS, Koh SC (2012) Functions of effective microorganisms in bioremediation of the contaminated harbor sediments. J Environ Sci Health Tox Hazard Subst Environ Eng 47(1):44–53

Ene LSO (1998) Sustainable agricultural development and a changing environment. In: Proceeding of the 32nd animal conference of the agricultural society of Nigeria, Held at Federal College of Agriculture, Ishiagu, Ebonyi State. 15–21

Eyhorn F, Ramakrishnan M, Mader P (2007) The viability of cotton-based ¨ organic agriculture systems in India. Int J Agric Sustain 5:25–38

Finamore A (2004) Novel approach for food safety evaluation. Results of a pilot experiment to evaluate organic and conventional foods. J Agric Food Chem 52:7425–7431

Greene K (2001) U.S. organic farming emerges in 1990s: adoption of certified systems. U.S. Department of Agriculture, Economic Research Service. Bulletin 770:24

Gibbon P, Bolwig S (2007) The economics of certified organic farming in tropical Africa: a preliminary assessment. DIIS Working Paper 2007:3, January 2007

Gliessman SR (1997) Agroecology: ecological processes in sustainable agriculture. CRC Press [Formerly Ann Arbor Press] Florida, USA 15–16

Gold M (2014) What is organic production? National Agricultural Library, USDA

Han S, Zhang Z, Yu L (2003) Using chlorella and effective microorganisms to optimize aquatic ecological structure and to regulate water quality. Ying Yong Sheng Tai Xue Bao 14(1):101–104

Hanson JC, Lichtenberg E, Peters SE (1997) Organic versus conventional grain production in the mid-atlantic: an economic and farming system overview. Am J Altern Agric 12(1):2–9

Heaton S (2002) Assessing organic food quality: Is it better for you? In: in Powell Jet al (eds) Proceedings of the UK Organic Research 2002 Conference, 55–60. Organic Centre Wales, Institute of Rural Studies, University of Wales Aberystwyth

Helga W, Lernoud J, Robert, H (2013) The world of organic agriculture: statistics and emerging trends 2013. Research Institute of Organic Agriculture (FiBL) and the International Federation of Organic Agriculture Movements (IFOAM)

Helga W, Lukas K (2011) The world of organic agriculture. Statistics and emerging trends 2011. FiBL, Bonn; IFOAM, Frick

Higa TD, James P (1994) Beneficial and effective micro-organisms for a sustainable agriculture and environment. Jpn Int Nat Farmi Res Cent 7

Higa T, Wididana GN (1991a) Changes in the soil microflora induced by effective micro-organisms. In: Parr JF, Horniock SB, Whiteman CE (eds) Proceedings of the first international conference on Kyusei nature farming, pp 153–162. US Department of agriculture Washington, D.C., USA

Higa T, Wididana, GN (1991b) The concept and theories of effective microorganisms. In: Parr JF, Hornick SB, Whiteman CE (eds) Proceeding of the first international conference on Kysuesi nature farming, pp 118–124. US Department of Agriculture, Washingtion, D.C., USA

Huber M, van de Vijver L (2009) Overview of research linking organic production methods and health effects in the lab, in animals and in humans. Department of Health Care and Nutrition, Louis Bolk Institute, Netherlands

Hussain T, Jilani G, Yaseen M, Abbas MA (1994) Effect of organic amendments and EM on crop production in Pakistan. In: Parr JF, Hornick SB, Simpson ME (eds) Proceedings of 2nd international conference on Kyusei nature farming, pp 132–139, Oct 7–11, 1991, Piracicaba, SP, Brazil. Pub. USDA. Washington, D.C

IFOAM (2002) Principles of organic agriculture. Basic Stand Organic Agric Process 1–4

Izallalen M, Mahadevan R, Burgard A, Postier B, Didonato R Jr, Sun J, Schilling CH, Lovley DR (2008) Geobacter sulfurreducens strain engineered for increased rates of respiration. Metab Eng 10(5):267–275

Jin M, Wang XW, Gong TS, Gu CQ, Zhang B, Shen ZQ, Li JW (2005) A novel membrane bioreactor enhanced by effective microorganisms for the treatment of domestic wastewater. Appl Microbiol Biotechnol 69(2):229–235

Ke B, Liang YF, Zhong ZX, Higa T, Aruoma OI (2005) Evaluation of the toxicity and safety of the antioxidant beverage effective microorganisms-X (EM-X) in animal models. Environ Toxicol Pharmacol 20(2):313–320

Khaliq A, Abbasi MK, Hussain T (2006) Effects of integrated use of organic and inorganic nutrient sources with effective micro-organisms (EM) on seed cotton yield in Pakistan. Bioresour Technol 97(8):967–972

Kinjo S (1990) Studies on effective utilization of organic matter by lactic acid fermentation. M.S. Thesis, Department of Agriculture, University of the Ryukyus, Okinawa, Japan

Kitazato Environmental Science Center (1994) What is EM? EM is Non Toxic. 4

Küstermann B, Kainz M, Hülsbergen KJ (2008) Modeling carbon cycles and estimation of greenhouse gas emissions from organic and conventional farming systems. Renew Agric Food Syst 23:38–52. 30 Lampkin N.H

Larbier M, Leclercq B (1994) Nutrition and feeding of poultry. Nottingham University Press, U.K. 305

LEISA (2006a) Farming matters: understanding sustainable agriculture. LEISA Mag Spec Issues 4–10

LEISA (2006b) Changing farming practices. LEISA Mag 22(2):4

Lotter D, Seidel R, Liebhardt W (2003) The performance of organic and conventional cropping systems in an extreme climate year. Am J Altern Agric 18:146–154

Mukred AM, Abd-Hamid A, Hamzah A, Yusoff WM (2008) Growth enhancement of effective microorganisms for bioremediation of crude oil contaminated waters. Pak J Biol Sci 11 (13):1708–1712

Naqvi ZH, Mustaq ul Hassan M, Chaudry Z, Akram M, Ahmad R (2000) Effect of EM4 on health of layers. Pak J Biol Sci 3(9):1516–1518

Ngele CP (2013) Effect of effective microorganisms (EM) on the severity of nematode on Sweet Potato (*Ipomea batata*) in Ishiagu South Eastern Nigeria. Federal College of Agriculture Ishiagu, Ebonyi State, Nigeria. Ongoing Research Progress Report

Niggli U (2009) Low greenhouse gas agriculture: mitigation and adaptation potential of sustainable farming systems. FAO Rev. 1

Offermann F, Nieberg H (2000) Economic performance of organic farms in Europe. Organic farming in Europe. Econ Policy, vol 5. Universität Hohenheim

Olsson ME (2006) Antioxidant levels and inhibition of cancer cell proliferation in vitro by extracts from organically and conventionally cultivated strawberries. J Agric Food Chem 54:1248

Onele II (2014) Influence of effective microorganisms (EM) based fermented plant extract (FPE) on the fruit yield and severity of nematodes on *Telferia occidentalis*. Federal College of Agriculture Ishiagu, Ebonyi State, Nigeria. Ongoing Research Progress Report

Pimentel DP, Hepperly P, Hanson J, Douds D, Seidel R (2005) Environmental, energetic and economic comparison of organic and conventional farming systems. Bioscience 55(7):573–582

Reganold Glover JD, Andrews PK, Hinman HR (2001) Sustainability of three apple production systems. Nature 410(6831):926–930

Reganold JP (1992) Effects of alternative and conventional farming systems on agricultural sustainability. Washington State University, Department of Crop and Soil Sciences

Ritter WF (1981) Chemical and biochemical odor control of livestock wastes: a review. Can Agric Eng 23:1–4, Canadian Society of Agricultural Engineering, Ottawa, Summer

Rodale Institute (2011) The farming systems trial: celebrating 30 years. Rodale Institute 13:19

Ruinen J (1970) The phyllosphere. V. The grass sheet, a habitat for nitrogen fixing microorganisms. Plant Soil 33:661–671

Safalaoh ACL (1994) Body weight gain, dressing percentage, abdominal fat and serum cholesterol of broilers supplemented with a microbial preparation. EM Technologies Inc

Safalaoh ACL, Smith GA (1994) Effective microorganisms (EM) as an alternative to antibiotics in broiler diets: effect on broiler growth performance, feed utilisation and serum cholesterol. In: Proceedings of conference of greater productivity and cleaner environment through Kyusei nature farming, EM Technologies Inc.

Salsbury D (1992) Poultry health and management, 3rd edn. Blackwell Scientific, London 214p

Sen Gupta B, Nandi AS, Sen P (1982) Utility of phyllosphere nitrogen fixing microorganisms in the improvement of crop growth II. Wheat. Plant and Soil 68:69–74

Sokół R, Michalczyk M, Spodniewska A, Barski D (2009) The influence of administering "effective microorganisms" to pullets on chosen haematological and biochemical blood indexes. Pol J Vet Sci 12(4):519–522

Sommer SG, Mathanpaal G, Dass GT (2005) A simple biofilter for treatment of pig slurry in Malaysia. Environ Technol 26(3):303–312

Staiger D (1988) The nutritional value of foods from conventional and biodynamic agriculture. IFOAM Bull 4:9–12

Stern NJ (1994) Mucosal competitive exclusion to diminish colonization of chickens by *Campylobacter jejuni*. Poultr Sci. Poultry Science Association, Champaign, IL

Tabidi MH, Mukhtar AM, Hassan IM (2013) Effects of probiotic and antibiotic on performance and growth attributes of broiler chicks. Global J Med Plant Res 1(1):136–142

Tegtmeier EM, Duffy M (2004) External costs of agricultural production in the United States. Int J Agric Sustain 2(1)

Wondmeneh E, Adey M, Tadelle D (2011) Effect of effective microorganisms on growth parameters and serum cholesterol levels in broilers. Livestock Research Directorate, National Poultry Research Case Team, Debre Zeit Agricultural research Center, Ethiopian Institute of Agricultural Research, Ethiopia

Yongzhen N, Waijiong L (1994) Report on the deodorizing effect of effective microorganisms (EM) in poultry production, 402–440. Beijing, China

Zhang D, Han Z, Li D, Liu W, Gao S, Hou D, Chang L (2005) Effects of returning maize straw into field on dynamic change of soil microbial biomass C, N and P under different promoted decay condition. Ying Yong Sheng Tai Xue Bao 16(10):1903–1908

Zhao X, Wang YM, Ye ZF, Ni JR (2006) Biodegradation of oil field wastewater in biological aerated filter (BAF) by immobilization. Huan Jing Ke Xue 27(6):1155–1161

Zhou S, Wei C, Liao C, Wu H (2008) Damage to DNA of effective microorganisms by heavy metals: impact on wastewater treatment. J Environ Sci 20(12):1514–1518

Zuraini Z, Sanjay G, Noresah MS (2010) Effective microorganisms (EM) technology for water quality restoration and potential for sustainable water resources and management. In: 2010 international congress on environmental modelling and software modelling for environment's sake, fifth Biennial meeting, Ottawa, Canada

Chapter 13
Season Extension in Organic Systems

W.B. Evans

Abstract Season extension technology is a great tool for farmers looking to increase market presence and control environmental variables. In this chapter, we address season extension technologies in the context of organic and near-organic production systems. There are regulatory, monetary, and ecological questions that go with any farming endeavor. Season extension technologies and systems also have these questions. It will be left to the reader to draw their own conclusions and make their own choices about the appropriateness of the technology in a given system. It can be noted, though, that the more intense and effective the season extension technology employed, the more likely it is to be among the more resource-intensive tools available. This is a somewhat logical conclusion when one considers a technology such as choosing a heat-tolerant cultivar of tomato for summer production versus using artificial shading systems to produce a less heat-tolerant cultivar during the heat of summer. However, even the most resource-intensive technologies have significant positive aspects about them. Vertical farming using hydroponics is perhaps the most resource-intensive system discussed in the following pages. These systems allow production of huge amounts of food in very small areas, millions of pounds per acre in some cases. The systems can be very resource and capital intense. They present huge regulatory questions as well, questions that regulators have weighed in on and will continue to weigh in on. Farmers, advocates, and communities will have to discuss these and other systems to find their place in the local, regional, and global food systems of the future. Organic farming will continue to be a part of that system; in what form remains to be worked out. One more thing, before we begin looking at season extension in organic systems directly. Although the ranks of organic gardeners and hobbyists outnumber organic farmers by a huge margin, this chapter is slanted toward commercial, for profit farms. These include the very smallest herb farm selling to five local restaurants, to the very largest farms growing four crops on dozens of acres. Most of what follows has applicability to all sizes of organic and near-organic

W.B. Evans (✉)
Mississippi Agricultural and Forestry Experiment Station, Truck Crops Branch, Crystal Springs, MS, USA
e-mail: wbe1@msstate.edu

D. Nandwani (ed.), *Organic Farming for Sustainable Agriculture*, Sustainable Development and Biodiversity 9,
DOI 10.1007/978-3-319-26803-3_13

farms, with a large amount being relevant to all those farming edible horticulture crops (fruits, vegetables, herbs, and edible flowers). Now, on to season extension in organic systems.

Keywords Succession crops · Temperature management · Mulches · Shade cloth · Weed

13.1 Components of Season Extension Systems

Season extension systems usually contain more than one tool or component. These are often combined to increase the effectiveness of the system for one purpose or another. In general, it is valuable to consider several systems and components when deciding how best to use season extension technology. The first main components are the natural ecosystem of the farm and surrounding areas. By this, we mean the soil, topography, weather, climate, pests, water, and light. The second is the inputs brought to bear by the farmers. These include mulches to cover the soil and row covers to cover plants. They also include at least three types of structures. The first is greenhouses, protected growing structures with natural or natural and supplemental lighting combined with imposed heating and cooling systems. In greenhouses, crops are usually grown in artificial soil-like mixes or in liquid solutions in a system called hydroponics or, if misted, aeroponics. The second is high tunnels or hoop houses (Fig. 13.1). These are built like greenhouses, with light-transmitting

Fig. 13.1 Polyethylene mulch-covered raised beds in a high tunnel just prior to planting. *Note* sides and ends up for ventilation. Drip irrigation tape, not visible in the photograph, was placed under the mulch when the beds were made

plastic or other glazing material on a supporting superstructure. In contrast with greenhouses, most crops in high tunnels are raised in soil on the floor of the high tunnel, although containerized production is gaining popularity in these systems. High tunnels do not have elaborate bench and container systems, nor do they have significant artificial heating and cooling systems. The tunnels are cooled by opening the sides and/or the roof covering. Growers do sometimes install electric openers for the walls, and many will use a small amount of supplemental heat if a freeze event threatens a crop. The third structure is the newest for commercial farming, the indoor and vertical farm system. These systems use artificial lighting and heating, combined with hydroponics to raise crops in buildings without significant natural light, heating, or ventilation.

13.2 Why Season Extension

Season extension includes all practices that allow production of crops during periods of the year in which they would otherwise not be produced. This broad definition of season extension is important because it allows us to consider more than just frost protection before and after the frost-free growing season. It allows us also to consider shading and crop cooling to improve cool season crop production in warm weather, management practices that hasten or slow crop maturation, and several other aspects of altering the cropping system to produce crops over a longer period than the local conditions would normally allow.

What are some of the benefits of season extension? First and foremost for many growers is temperature management and control. Usually, this is thought of as increasing temperatures on the shoulders of the local frost-free period, the few weeks before the last spring frost and the first fall frost of each season. This rolls into the second benefit, getting crops in the ground or into the market early or late in the season. Early season vegetables can command much higher prices than main-season harvests. The same is true of late-season harvests. Being in the market first, last, and/or longest has strong competitive advantages in wholesale and retail markets. If customers are more likely to stay with a seller, they start with at the beginning of the season than switch, this would be another argument for getting to market early in the season. With the early market and higher prices, gross and net returns to the farm also increase. Market prices for retail and wholesale produce are higher when supply is scarce. For local produce and for wholesale produce with limited supplies out of local season, prices can be substantially higher than when an item is abundant in the market. For these and other factors, Conner et al. (2009) concluded that extending the season with high tunnel production may help increase the viability of Michigan farms selling at farmers' markets.

On the other hand, season extension technology does have costs, sometimes significant ones. These must be accounted for to make sure the costs are outweighed by the potential profits. Otherwise, season extension will not be a wise use of resources. It would be a bad business decision to design a system to come to market early if the traditional local farmers market outlet was not scheduled to open until later in the season, for instance. Season extension also has what economists call opportunity costs. The farmer and workers could be using the capital, time, and other inputs used in extending the growing season to do something other than manage a season extension system. They could be repairing equipment, attending trainings, or any number of other things in preparation for the regular portion of the growing season. The capital used to buy row covers or build protective structures would also not be available for other purchases, at least not until crops in the season extension system were sold. These capital, time, labor, and supply costs must all be accounted for when considering the use of season extension technologies.

Back on the supporting side, there are other reasons to consider season extension beyond the simple profit question. Season extension technology also increases a farmer's control of yield and quality. Greenhouses and high tunnels exclude rainfall, reducing weed pressure, and plant disease incidence, and often excluding many insects. Low tunnels, cloches, mulches and shading can also do this. Season extension can also reduce risk in a farm operation. This is important for many fruit, vegetable, and flower growers whose dependence on the weather is profound. Season extension systems can reduce weather risks by altering temperature, water, nutrients, and pests in advantageous ways. They can reduce single-crop risk by increasing the diversity and number of plantings. They can even reduce environmental risk by improving resource-use efficiency. They can also allow a grower to enter a market that was closed to them. For instance, crops not normally suited to a particular location can sometimes be grown successfully using the season extension technologies available.

13.3 Costs and Returns

Season extension technology runs from a few hundred dollars (US) per hectare to tens and even hundreds of thousands of dollars per hectare. Products like spun-bound propylene row covers can cost under $\$1.00/m^2$. High tunnels currently run anywhere from just few dollars a square meter for homemade styles, to installed tunnels with automatic roll up sides and doors that can cost \$25 to $\$75/m^2$ in the USA. With these kinds of costs, can this be profitable for growers? Yes, provided the system is well designed and managed, and the grower takes care to select proper crops and markets.

A low-to-mid-range high tunnel in the USA runs about $\$2.50/ft^2$. or about $\$23.00/m^2$. That is $\$230,000\ ha^{-1}$! To make that work, a grower has to manage the tunnel as a very high value piece of farmland. Using average field yields from Knott's Handbook for Vegetable Crop Producers (Maynard and Hochmuth 1997) and Atlanta (GA) and New York City wholesale prices from the USDA Agricultural Marketing Service in Spring 2015 (USDA AMS 2015a, b), one can calculate the potential wholesale value of many vegetable crops. For season extension to pay for itself, it must increase the dollars generated by the cropping system more than it costs the growers to install and manage the technology.

Can season extension systems do this? Yes, season extension can increase returns per unit of area, time, capital, and labor. Season extension costs money, sometime a lot. But the net cost is often quite profitable. Is it always profitable? No, there are many instances where it does not make sense to use season extension. For things like high tunnels and greenhouses, it often makes poor economic sense to raise long-term crops of relatively low value per square foot. Consider pumpkins (Cucurbita pepo L. or C. maxima (Duch.) and C. moschata (Duch. ex Pior)). This crop can take over 100 days to grow and the vines can be over four meters long. The fruit is too heavy to grow on a trellis, so it usually runs on the ground. A very good yield of 30 T ha^{-1} would result in a yield of 3 kg m^{-2}. If the retail market value is US\$0.50 kg^{-1}, this equals \$1.50 m^{-2}. If the crop is in the ground for 100 days, the daily gross return is 1 % of this, or only \$0.015 m^{-2} d^{-1}. Contrast this with leaf lettuce (*Latuca sativa* L.). Green leaf lettuce takes 45 days to mature from transplant and takes about 0.1 m^{-2} $plant^{-1}$ to grow. If a grower can get a retail value of \$3.00 for a single lettuce, this equals \$30.00 m^{-2} over 45 days, or \$0.66 m^{-2} d^{-1}, forty-four times the gross return of the pumpkin crop. Even though the lettuce crop costs more to grow due to transplant costs and other factors, the net returns after expenses still make it a much better candidate for using season extension techniques than pumpkins are in most cases. The same sorts of numbers can be generated for any crop using local prices.

In another example, we can try to decide if polyethylene mulch is a good choice for strawberry and pumpkin production. First, let us look at the cost of mulch/running meter. If we assume it is \$0.10/m^{-1}, and we have strawberry rows on 1.2 m centers, that gives us 8333 row ft ha^{-1}. The mulch will cost \$833.34 ha^{-1}. For pumpkins on with 3 m between rows, we need 3334 m ha^{-1}, at a cost of \$333.34 ha^{-1}. If the strawberry crop yields 5000 kg ha^{-1} at \$5.00 kg^{-1}, that is a gross return of \$25,000 ha^{-1}, less the mulch for a gross of \$24,666. No two crops generate the same return. It, therefore, is part of the season extension system for the grower to develop a crop mix that uses the technology appropriately and creates the desired overall income potential while remaining compatible with the overall management systems, regulations, and values of the farm.

So, what are some other ways that season extension can benefit the farm and farmer? What are some of the components available for season extension and how are they used. How do these fit with organic agriculture and when might they be in conflict? We will now explore more of the benefits of season extension, present an overview of many of the current and developing technologies, and address some of the cultural and regulatory issues that touch these and the choices a farmer must make related to their use.

13.4 Season Extension Can Make Good Use of Limited Space

Because season extension technologies can be used to improve or optimize growing conditions, they allow one to make the better use of limited space. For growers with limited land areas, or limited areas suitable for crops, extending the season can allow them to grow more products in the space they have. This is achieved by increasing the number of crop cycles in a year, reducing the time it takes to mature a crop, and/or producing transplants that allow faster crop cycles in the field. Crops are often grown at closer spacing under protected culture than they are in open fields, increasing yields per square meter.

13.5 Multiple and Succession Crops

Season extension can allow for multiple crops and more succession cropping on a farm. Retailers will often say that you cannot sell something off an empty shelf. The same is true for growers of fruits and vegetables. Money is made when crops are growing. However, organic production has certain characteristics that can modify this idea and that require special management consideration. First, looking at the impact of season extension, a farm in a temperate area can often increase the production season by four to six weeks in both the spring and the fall, for a total of eight to twelve more weeks of frost-free growing season, fifty-six to eighty-four days. This can equal one more crop cycle in the season, or even two if crops like radishes (*Raphanus sativus* L. Radicula group) or mustard (*Brassica juncia* L. Czernj. and Coss, var. *crispifolia* Bailey) are grown.

When more crops are planted over a season, this changes the ecology of the farm. One may need to adjust overall farm management when including season extension techniques. Planting succession crops may mean more tillage or tractor passes on a farm. The grower may need to pay closer attention to tillage timing to avoid damaging wet soils in the spring and fall. As more material is removed over the course of multiple crop harvests, more nutrients are removed as well. This may require adjustments in fertilizer and nutrient management. Soil moisture, evaporation rates, crop growth rates, and rainfall are almost always different in the extended season than they are in the main season; adjustments in irrigation and other management practices may be needed to optimize water status in the crop system.

Using season extension technology can significantly alter pest and weed pressure, often but not always, to the benefit of the farmer. Understanding these changes is very important in organic systems. Many insects and weeds are seasonal in their biology, growing, and reproducing based on day length and temperature. When we use season extension techniques, we are often growing a crop at a time of year when its main pests are not at their peak of growth or infestation. Often insect pressure is less in the cooler seasons than in the warmer seasons. Insects are less active and grow more slowly at the lower temperatures and day lengths seen in the spring and fall. In areas with significant freezing temperatures, some pest outbreaks can be further reduced by a frost event outside the controlling structure. This results in fewer pests moving in.

13.6 Soil Moisture

The effects of season extension on soil moisture also influence organic matter and its mineralization, in a somewhat complicated manner. Organic agriculture relies heavily on nutrients cycled through the organic matter fraction of soil. With the exception of histosols which are derived from decaying plant residues, soils are mostly made up of decomposed rocks, with an organic matter fraction of deposited and decaying plant and animal matter. In organic systems, the organic matter is managed to serve as an important nutrient supplier and buffer. Carbon in soil organic matter is lost from the soil through decomposition and oxidation, processes that are optimized by moderate moisture and warm temperatures. The decomposition and oxidation of the organic matter leads to minerals becoming available through mineralization, conversion of elements to mineral forms from organic molecules such as proteins, fats, and sugars. Because plants can only take up nutrients in the mineral forms, this process is very important for optimizing plant growth. On the other hand, mineralized nutrients can also be subject to loss though leaching, uptake by weeds, and other avenues that elements held in the organic form are generally not subject to. This is part of what gives organic matter such power in soils; it can serve as a powerful storage pool for plant nutrients, buffering the system against deficiencies and providing nutrients to the plants relatively slowly as the organic matter decomposes. Depending on the type of season extension technology used, organic matter can be either increased, preserved, or decreased over time. Season extension systems that lead to cooler summer soils can retain soil organic matter better than ones that warm the soil without adding or protecting organic matter, such as using plastic mulch, tunnels, or greenhouses. An irrigated high tunnel may lead to increased average soil temperature and more optimal soil moisture status over open field conditions, especially in temperate zones. This may actually lead to increased organic matter oxidation over open field conditions. In tropical zones, a high tunnel can block excessive rainfall during crop production, potentially reducing oxidation rates over that in the open field. In cooler areas or with shading, a reduction in excess soil moisture under a high tunnel could lead to preservation of organic matter. So, the suitability of high tunnels (and other season extension technologies) is somewhat site specific. A grower needs to consider the goals of the system and capabilities of the natural and imposed system they are working with in order to properly plan for and use season extension technologies.

In high tunnels, irrigation and nutrient applications are often confined to the cropped areas, irrigation by the use of drip irrigation systems and nutrients by banding or other methods of application and incorporation that minimize the amount of nutrients and organic matter applied to pathways and ends of the rows. The lack of moisture in the non-cropped areas reduces weed germination and growth. This would not be the case in an open field where rainfall may allow for ample moisture in the non-cropped portions of the field for some or all of the season.

13.7 Temperature Management and Control

One of the greatest influences on crop growth and quality is temperature, of both the air and the soil. Even in a somewhat naturally managed system like an organic one, temperature can be modified quite easily, adding flexibility and quality to production. Temperature conditions during seedling production can have a huge impact on final yields after transplanting, and sometimes not as one might expect. Kalisz and Siwek (2006) showed that unheated greenhouse conditions (Tmin 4–6 °C over two experiments) during production of Chinese cabbage [Brassica pekinensis (Lour.) Rupr.] seedlings resulted in higher yields at harvest compared to yields from seedlings grown in a heated greenhouse (Tmin 12–14 °C over two experiments) and increased the percent marketable heads in the field. However, the seedlings grown at the lower temperature regime also produced more bolted heads once transplanted to the field than transplants raised in the heated control conditions. The heads harvested from seedlings grown at lower temperatures had lower thiocyanate concentrations than those from the heated control, as well.

It must be noted that temperature management can be very exacting. For instance, consider the question of what is the critical temperature above which plants should be kept to optimize growth. This differs among crop species, of course. In one series of studies looking at how much carbon was actually incorporated into strawberry leaves over time, Maughan et al. (2015) showed that net assimilation rates (NAR) in strawberry (*Fragaria* × *ananassa*) leaves exposed to temperatures at or below −5 °C could not achieve the same NAR as leaves maintained at 10 °C, even 28 days after a single day of low temperature exposure. Studies like this remind us that it a critical component of managing season extension technologies is the need to be consistent. A single occasion of not covering a sensitive crop on a cold night might undue all the work done to that point by a grower.

13.8 Frost Protection

Perhaps the greatest role of season extension technologies in temperate areas is for frost protection. Frost is the formation of ice on surfaces and occurs when the air temperature is at or below 0 °C (32 F) when it reaches the dew point, the point when the relative humidity is 100 %. Plant tissues do not freeze at 0 °C. They contain salts, sugars, and other solutes that act to lower the freezing point of the tissues below 0 °C. When frost-sensitive plant tissue reaches its freezing point, cell damage, and crop loss can occur. In most cases, frost protection technology reduces a crop's susceptibility to cold injury. They do this mainly by raising the temperature around the crop through reduced radiational cooling. This prevents ice formation on and especially in the tissues by keeping the tissues above their freezing point. Of course, this protection is not infinite; frost protection systems will only provide a few to several degrees of protection. Combining strategies can increase the amount of frost protection provided by the system, but again, the protection is not infinite. It is also not identical for each event or location. Wind, soil moisture, and other factors influence how much frost protection a system can provide. It should also be noted that, in some cases, frost mitigation strategies can backfire and actually increase damage by increasing tender growth early or late in the season.

13.9 Bacterial Ice Nucleation

This is how most precipitation is formed. Water condenses on a microscopic, solid particle in the air or on a surface. More water is then attracted to the newly formed droplet and the droplet grows. In a cloud, when the weight of the droplet exceeds the ability of the air to hold it up, it falls to earth as rain. We can spray crops with ice-nucleating bacteria and induce the formation of a frost protective layer of water or ice.

13.10 Water Application

The freezing point of water is 32 °F/0 °C. When it freezes, large amounts of heat are given off. In addition, as long as there is liquid water around the ice as it forms, the ice will remain at a constant temperature of 32 °F/0 °C. We can use these facts of physics and chemistry to our advantage, especially when we combine them with a little bit of plant physiology. Plants are made up mostly of water. This water contains both solids and solutes, such as dissolved sugars and salts. These solutes act like salt does on an icy sidewalk. They lower the freezing point of the water they are held in. Indeed, the water in most plant tissues does not freeze until it reaches

28 °F/−2 °C or even less, and it is when water freezes in plant cells that the most damage occurs.

Weed ecology is often very different in an extended season situation. In systems extending cool weather production, winter annuals such as chickweed (*Stellaria media*), henbit (*Lamium amplexicaule*), brassica weeds (mustards, etc.), and some grasses can become a concern where they are not a concern in the traditional main season. For those using season extension in the field, cultivation of weeds can be challenging in winter and early spring when soils tend to have more moisture and tillage can significantly damage the soil structure. Living or non-living mulches placed between rows can help with this. However, it is very important to know the weeds and to use sufficiently dense mulching material to adequately suppress weeds.

13.11 Season Extension and Labor

Season extension can allow a farmer to start some crops before others, extending planting season and taking advantage of planting resources over an extended period of time. Labor that might be used to plant over four weeks in a traditional spring planting season might be used for eight and sometimes more weeks with season extension technologies in place. This can make a farm more attractive to student, part-time, and migrant labor pools where these are available because a worker can be confident that they will have many weeks of work in one location, rather than a few. These same workers may then be retained for maintenance and harvest of spring crops, moving right into fall planting and a possibly extended fall season. For a farm's best workers, this can be very attractive. This can also be very attractive to any skilled labor and management employees who often benefit greatly from having more steady work through much of the year.

On the other hand, it is also true that season extension often means labor is indeed needed for a longer part of the year. Where labor is in short supply or is difficult to train or retain, using season extension technologies can increase labor challenges. Adding workers for short periods of harvest, planting, or other labor-intensive periods can be difficult in some farming regions. Where this is the case, having more of these periods during the year adds to this difficulty, although, as mentioned above, having more consecutive weeks of work can sometimes be a competitive advantage in hiring good workers.

It is important to remember that most season extension technologies involve a bit of user skill in order to get the maximum benefit out of the technology. High tunnels and greenhouses need to be operated at the right temperatures and irrigated properly. Mulches have to be selected and applied properly for maximum benefit. Row covers may need to be placed (with care) and removed (also with care) daily for certain crops. This can use a lot of labor early and late in the day, and the labor for all these technologies needs to be appropriately trained and supervised to maximize crop quality, crop yield, and farm efficiency.

13.12 Crop Selection and Timing for Season Extension

Generally, systems using season extension technologies lend are best suited to higher value crops or crops to which value-added post-harvest to justify the capital and labor costs associated with season extension. We find season extension systems in horticulture mostly: vegetables, fruits, flowers, and nursery crops. Grain crops, turf, and large tree crops like walnuts (*Juglans nigra* L.) that grow large and slowly, are not usually considered for most season extension technologies. Suitable crops have high value per day of production in most cases. This is an important concept, the concept that the most valuable crop still may not be suited to season extension if the return per square meter per day does not compensate the grower adequately. Then, there are other crops, like wine grapes (*Vitis vinifera* L.) that can have some of the highest values per hectare of any crop, but because it is so dependent on microclimate, soils and "terroir," the totality of the ecosystem influencing the grapes, it is not usually considered for anything other than mulching the rows or possibly some frost management technologies.

13.13 Appropriateness of Season Extension Technologies

There are also times when a technology or system is appropriate in one situation but not in another. This can be illustrated with the recent work of Sideman (2015). Their team grew sweet potatoes (*Ipomea batatas* L.) in New Hampshire, well north of the traditional sweet potato production areas from Washington, DC and south. They used biodegradable polyethylene mulch to warm the soil and produce a marketable crop of roots. Yields from their plots were nationally competitive and more than sufficient to allow a grower to harvest and sell profitably into a local retail market. However, their system in their location would not likely be appropriate for large production for the national wholesale or processing markets. The system they tested allowed them to plant earlier than they would have in bare New Hampshire soil, and to harvest before the first frosts in September. The expense of the mulch was acceptable and even profitable because the system was designed to serve a local market at a reasonable retail price. That same system might not have been an appropriate choice for a large grower in an established sweet potato growing area because sweet potatoes grown in the more traditional southern states would benefit little from warming the soil early in the season. There are plenty of growing days for the crop to finish without season extension technology in these areas and because the crop is stored after harvest, there is less need to be first or last in the wholesale sweet potato marketplace than there would a market for, say, leaf lettuce.

13.14 Natural Season Extension

Growers can use farm topography and microclimates as natural way to extend their season. When designing a farm layout or considering buying farm land, one can consider elevation, soil type, air and water drainage off sloping ground, understory plantings, and tillage as parts of the season extension system.

13.15 Positioning Crops in Full Sun

In terms of natural season extension, this seems at first blush that it does not need to be mentioned. However, many growers, especially small ones, have plots with light, moderate, or even heavy natural shade. As a general rule, fruiting crops of all types grow and yield best when raised in full Sun. Leafy crops and many root crops can perform well under some shade. In terms of season extension, a crop grown in full Sun will access warmer soils than those grown in shade under otherwise similar conditions. Generally, the soil in full Sun will dry out to a tillable state faster than shaded soils. On the other side, shaded soils may cool more slowly at night than those in the wide open areas without shade. Crops under partial shade are also subject to less damage from light frosts due to reduced radiational cooling under the shade than in open areas.

13.16 Shade Cloth

For most applications, standard black shade cloth is used. Photosynthetic light saturation occurs at a level below that of full sunlight, meaning that no more photosynthesis occurs at higher light levels than at the saturation point. Shading that reduces light levels to below saturation can reduce carbon fixation and assimilation rates, and thus crop growth. This can delay maturity, reduce yield, and/or influence crop quality. However, some plants can greatly benefit from shading, even to levels somewhat below the light saturation point. Shade can reduce soil and plant temperatures during times of high light and temperature. For example, lettuces grown under shade can have reduced leaf temperatures without reduced yield in times of high air temperatures. By cooling the plant and soil, shading can also reduce water loss from plants and soils.

Shading can improve soil organic matter retention by reducing oxidation. Soil respiration rates are generally lower at lower soil temperatures. This can slow oxidation of organic matter since this is governed by biological activity in the soil. This means that applied composts and manures and incorporated cover crops will often breakdown more slowly under shade than under full Sun. This preserves the physical benefits of organic matter additions, but also slows the mineralization of nutrients in the organic matter. Nutrients in organic matter, either existing in the soil or added by the farmer, must be converted from organic forms found in the tissues and partially decomposed tissues of plants and animals into what are called non-organic forms. These are generally ions and salts that can be taken up by plant roots. For example, proteins contain significant amounts of nitrogen, but the nitrogen is in the amino acids that make up the protein. This nitrogen needs to be liberated and converted into nitrate or ammonium molecules to be taken up by plants. Similarly, phosphorus in cell membranes and other compounds needs to be converted to phosphate ions dissolved in the soil solution for the plant roots to take it up.

Of course, not all shading is equal. Mulching a soil is different than shading over the crop canopy. An organic mulch, such as chopped bark, can provide 100 % shade and reduced soil temperatures in a warm season while at the same time itself decomposing at the soil/mulch interface, resulting in more organic matter and higher mineralization and organic matter oxidation levels than in adjacent un-mulched soils. On the other hand, a black polyethylene mulch placed tightly over the soil under similar conditions and can provide nearly 100 % shade while warming the soil and holding in soil moisture. All things being equal, this would likely increase organic matter decomposition and nutrient mineralization compared to that under un-mulched soil nearby.

13.17 Mulches

Plastic mulches can increase spring and fall soil temperatures, and white plastic mulch can lower summer soil temperatures. Roll-type paper mulch can raise early soil temperatures but often not as much as well-laid black polyethylene mulch does. Bark, sawdust, and leaf mulches tend to lower soil temperatures for most of the season and help retain some heat in the fall. For small growers using cardboard, newsprint, or shredded office paper, these tend to lower soil temperatures in all seasons, making them valuable in situations where lowering soil temperatures can be valuable, such as for summer root crops. There are also red and other colors of plastic mulch that have crop-specific value. Translucent types that allow some light through them tend to warm the soil more than black mulch. Greer and Dole (2006) provide a very good overview of findings related to plastic mulches used in vegetable crop production. Papers reviewed included findings on yield effects, disease and insect infestations, and crop quality, as well as the many reasons that authors of the papers used to explain the effects they reported.

Mulches influence soil temperature and moisture. Teasdale and Mohler (1993) reported that natural mulch of hairy vetch or rye reduced soil moisture loss and soil temperatures. Plastic sheet mulches usually warm soil; organic mulches often lead to lower soil temperatures than are found in adjacent un-mulched, bare soils. Warmer soils mean faster reactions and more mineralization of OM if the mulch helps retain soil moisture. In open high tunnels, moisture loss may exceed losses in outside soils in cooler areas, resulting in greater losses of OM and more rapid mineralization, but mineralization rates can decline greatly as soil moisture declines, even at substantially high temperatures. However, the latter will not be common because most agricultural situations will include irrigation, resulting in elevated mineralization as soils in a tunnel structure warm. Plastic mulches, combined with ridged beds were shown to significantly increase soil moisture and temperature in corn and increase crop water-use efficiency. This means the mulched treatments allowed more corn to be produced with less water use, something the authors said would be especially valuable in semiarid areas (Zhou et al. 2009). Ramakrishna et al. (2006) found that plastic mulch increased soil temperatures in a Vietnamese groundnut (*Arachis hypogea* L.) trial by up to 6 °C. In their work, both the polythene and rice straw mulches helped hold in soil moisture. They suggested that the abundantly available rice straw could be of great value to groundnut production in Vietnam.

13.18 Mulch Color

Plastic mulch color influences crop yield and quality. Strawberries grown on black plastic mulch have been shown to have higher starch and soluble carbohydrate concentrations than fruit raised on red mulch (Wang et al. 1998). Insects can be confused by reflected light. Plastic mulches treated to be silver and reflect light have been shown to reduce pest incidence and sometimes disease transmission. For instance, beetle populations in cucumber and squash have been reduced by using polyethylene mulch with a reflective surface (Caldwell and Clark 1999). The reflections reduce the amount of cucumber beetles infesting the crop. Since they are the vector for bacterial wilt, the incidence of the wilt is reduced. However, once the vines cover most of the mulched surface, light is no longer reflected in a way that confuses the insects and the potential for higher numbers of insects and subsequent infection increases. Similar results have been found on many crop and insect combinations, including whitefly in watermelons (Simmons et al. 2010). In their work, Caldwell and Clark (1999) found that the reflective mulch treatments did not need to be sprayed with pesticides, as did the black plastic mulch treatments. This means a grower can use this system to market pesticide-free products, claiming a premium in many markets and making this system actually more profitable than the black plastic mulch system.

13.19 Mulches and Weeds

A ground mulch used for season extension will greatly reduce weed growth in the covered area. A few weeds, such as purple and yellow nutsedge (*Cyperus rotundus* and *C. esculentus*, respectively) can penetrate and grow through a plastic mulch or a thin layer of organic mulch, but for the most part, weed growth is greatly limited by mulch. Remember, too, that most weed seeds need light to germinate (Wesson and Wareing 1969). After the initial cultivation and mulch placement, the soil is not disturbed any further in the mulched area. This results in few weeds emerging in the mulched area. The farmer then only has to control weeds between the mulched areas. Natural mulches made of different plant materials have different decomposition rates. This can affect how they influence soil temperatures and weed growth, as was found in a comparison of hairy vetch and rye (Teasdale and Mohler 1993), who found that the vetch decomposed more quickly, leading to more light penetration to the soil surface and increased soil moisture loss and weed growth compared to that under the rye mulch.

13.20 Row Covers

Row covers are any number of flexible sheeting materials that cover a crop and often the surrounding soil surface to modify the temperature around the crop and sometimes to provide a barrier from insects. They can also provide a bit of protection from wind and desiccation. They may cover one row or plant at a time, or they may cover thousands of square meters/feet at a time. The original plant covers were likely glass cloches, French bell-jar-like plant covers that were set over individual plants for frost protection. In the mid-twentieth century, many growers and farmers in the USA relied on HotKaps and similar paper cloches. These are translucent paper domes placed over individual plants for early spring frost protection. Today, HotKaps are still sold in many garden catalogs and retail outlets but do not find much use on commercial farms. They and other individual cloches are used mainly by hobby gardeners because commercial sheeting designed for row covers are cheaper and easier to use in most commercial plantings. The individual paper cloches do still find favor with some organic growers that prefer the paper product over plastic, cheesecloth, the loosely woven cotton fabric.

Today, most row covers are intended to cover hundreds or thousands of plants with one piece. Most of these row covers are made of synthetic materials that are not biodegradable. The most common products today are made from spun-bound polypropylene. This comes in different weights per square meter. The heavier the fabric, the greater its ability to protect from cold. Some are rated to keep plants safe down to 20–24 °F (−5 to −7 °C).

When row covers were placed on primocane-fruiting blackberries in late winter to warm the air around the crop, early season yields were increased (Strik et al. 2008). However, seasonal changes in crop and air temperature responses to row covers have been reported. Hamasaki (2013) confirmed that air temperatures under row covers were higher than the ambient air in the summer time, and lower than the ambient air temperature beginning in early autumn. The author postulated that the lower sun angle of autumn reduced heat loads and thus allowed air temperatures to become lower under cover than in the surrounding air as the reduced temperature of the covering material acted to cool the air underneath it. This resulted in soil temperatures greater than air temperatures under the covers in winter, too. This, the authors wrote, has additional implications for the management of row covers in high tunnels. Using nonwoven fabrics as row covering or as a tunnel cover has also been shown to reduce ozone injury in susceptible vegetables (Matsumaru 1994).

The combination of row cover and mulch may or may not create synergistic benefits. In many cases, it has been shown that plastic a row mulch is enough to improve yield of vegetable crops and cover over the plants does not add additional yield or other value. Examples of this include watermelons on black plastic covered raised beds in Mexico (Ibarra-Jimenez et al. 2005). However, row covers can have value in situations where insect exclusion is desired or where a bit of frost protection will improve early crop survival and growth. Slitted row covers made of clear polyethylene have been shown to increase early growth and yield of vegetables on black plastic mulched beds. Straw and living mulches have been shown to reduce movement of Colorado potato beetle (*Leptinotarsa decemlineata*) larvae into eggplants, but they also reduced eggplant growth compared to that achieved by using black plastic mulch (Stoner 1997). The authors speculated that a combination of black plastic mulch with natural mulch between the rows may provide the benefits of both: good crop growth and reduced pest movement into the crop.

Row cover and mulch treatments often increase early season crop growth. The advantages will often disappear when the crop covers most or all of the mulched portion of the field. This is sometimes seen in vine crops such as watermelons. Early crop growth increases, but this may or may not translate into increased yield at the end of the season. Spun-bound row covers can reduce insect pressure when used alone or with plastic mulch. When used with clear plastic mulch in a Mexican cantaloupe field with high insect and insect-vectored disease pressure, yields were up to four times those achieved with one component or under bare soil conditions (Orozco-Santos et al. 1995).

13.21 Plastics Management and Regulation

Much of season extension technology involves one or more aspects of the plasticulture system described by Lamont (2005). In this system, enhancements to growth, yield, and sustainability are made through the appropriate blending of modern technologies and sound-integrated farm management. For organic systems, this means the selective use of plastics when appropriate for enhancing overall environmental stewardship, food safety and quality, worker safety, and community well-being. In both regulated and unregulated organic management systems, the use and disposal of agricultural plastics becomes an important consideration. Permitting systems that allow growers to use plastic mulches can have requirements related to disposal. The International Federation of Organic Movements requires that mulches and covers be removed from the soil and cannot be burned [IFOAM Norms 4.6.4 (IFOAM 2015)]. Plastic mulches and row covers get dirty with use and may not be accepted by standard recycling outlets available for other consumer and industrial recyclable items. In many cases, plastic mulches are landfilled. However, there are others that have agricultural recycling efforts that take plastic mulch. There are also some areas where mulch and other used plastics can be burned for fuel directly or after some processing to make them suitable for the incineration system being employed. In 2011, Sullivan brought to focus several still ongoing issues related to the use of plastic mulches in organic agriculture. Sullivan brings to light the conundrum of adjacent jurisdictions allowing and not allowing the use of compostable plastic mulch in organic production. He points out that Canada and Europe allow it, but the USDA does not, because the rules do not allow the use of synthetically modified plant and animal derivatives. Starch-based compostable plastic is thought to violate this premise.

There are great debates among organic production advocates and farmers as to the appropriate role of plastics in organic cropping systems. In the USA, many biodegradable plastics are not permitted in USDA Certified Organic production because they contain starch made from Genetically Modified Organism (GMO) corn that has herbicide resistance or other genes inserted from other organisms into its DNA. Because GMOs are not legal, the introduction of the starch to an organic system conflicts with the organic regulations. However, the use of polyethylene mulch, polypropylene row covers, drip irrigation tape, and other plastics is legal, provided the products are removed completely from the field at the end of the season or there useful period. It can be very difficult to remove plastic mulch from fields due to tearing while it is being pulled up, but growers are required to remove all traces of plastic to be in compliance with USDA National Organic Standards (USDA NOS) (USDA 2015a, b). Disposal after the plastic is removed is another issue. Some areas of the country have agricultural plastic recycling, others do not. Mulch and irrigation tape are particularly hard to recycle because they are quite dirty after use and removing the dirt is difficult for both the farmer and the processor.

13.22 Containers as Season Extension Vehicles

Growers can use containers for season extension. Containers have interesting effects in season extension applications. Being above the ground, the media in a container can be more subject to low and high temperatures than soil in a field. When freezing temperatures occur, the substrate in an above ground pot can freeze solid. For some cool season crops at moderately cold temperatures, this may not be a problem. However, for most warm season crops, freezing of the soil could result in complete crop death at temperatures that might only kill the tops on a field-grown plant.

Containers can allow one to have crops growing in the spring before wet soils dry out. They can be used in areas of seasonal water shortages, where only small amounts of water are available in certain seasons. In this instance, a water storage system that can supply small amounts of water, enough for containers but not for field production, may be installed. In this case, the general rules that soils and mixes are approximately 25 % water by volume in most agricultural systems and that crops loose can provide a general guide for how much crop can be raised in containers using a given water storage system.

13.23 Hydroponics for Season Extension in Organic Systems

True hydroponics, growing plants in a liquid nutrient solution rather than a soil or soilless substrate, were originally not permitted in USDA Certified Organic production systems because there is no way to maintain or improve soil health in such a system. Hydroponics are now permitted, although some certifiers will not certify hydroponic systems for organic production. It can be very difficult to produce quality plants with liquid organic materials. These materials often have solid fractions or precipitates that clog emitters. It can also be very hard to find a single material or blend of materials that provide the right amount and balance of all the essential elements. This can lead to nutrient imbalances and less than optimal crop performance. Unlike many fertilizer salts, almost all organic nutrient sources provide more than one element. The few exceptions that exist, like chicken (*Galus galus*) feathermeal, are considered only a nitrogen source. But it and other organic nitrogen sources, like alfalfa (*Medicago sativa*) meal, also contain small (feather meal) or large (alfalfa) amounts of several other essential elements. With organic hydroponic systems, one has to find blends of soluble materials and these, too, usually contain many, if not all, of the essential elements, making it very difficult to blend them in a fertilizer solution that delivers what the plants need without over or under delivering one or more element. Common ingredients for liquid feeds such as seaweed extracts, fish emulsions and hydrosylates, and compost or manure teas and extracts all have this issue, making their management in hydroponic systems very difficult.

13.24 Indoor Farming

Indoor farming is perhaps the ultimate season extension vehicle. In these systems, crops are raised under artificial lights, often stacked in racks of trays, using hydroponic or near-hydroponic systems. The temperature, lighting, irrigation, and nutrient delivery are carefully controlled to optimize production. These systems can be designed and managed to meet the US Department of Agriculture's National Organic Standards, although that was not always the case. Even today, some USDA-approved certifying agents will not certify hydroponic and vertical farms as organic (Coleman 2015), believing that soil-based farming and improving soil health are critical parts of running an "organic" farm.

The advantages of these indoor farms are in their huge production potential per square meter. Imagine the best field-based yields, multiplied by the number of stacked trays and you can easily envision yields of lettuce exceeding 1000 $MT.ha^{-1}$ in just one cropping cycle. Another advantage is that they can be placed close to large populations and markets, where field-based farming would be nearly impossible. One of the biggest negatives of these farms for growers their establishment costs. Simple vertical farming systems can cost hundreds of dollars a square meter to establish. This does not include cost of the building, plumbing, electrical systems, seeds, fertilizer, and labor, just the physical rack system. These systems also require a very intense and detailed management approach. Like a greenhouse, and perhaps even more so, these systems require monitoring every day, unfettered commitment to sanitation and quality control, scientific management of resources and systems, and execution of a sound business and marketing plan.

Indoor and vertical farming systems lend themselves to the production of high value crops with quick cycles, such as lettuce, greens, and herbs. Even crops with very high revenue potential like tomatoes may not be suitable for these systems because of the number of days the take to come into production and the vertical space they command. Efforts are underway to design cultivars and systems that will grow a broader range of crops. There is also some controversy about their "organic" nature, even when they are managed using practices approved by a certifying agency. A large certified vertical farm opened near Chicago in 2013 (Anonymous 2013). The system is certified by the US Department of Agriculture, but the article quotes at least one expert as having reservations about the project due to its intense lighting and relatively large carbon footprint, among other things. The zero-waste system, as it is described, uses aquaculture to overcome some of the questions about nutrient availability and balance in hydroponic organic systems.

13.25 Pest Management in High Tunnels, Greenhouses, and Indoor Farms

Pottorff and Panter (2009) present a wide range of pest control strategies for high tunnel systems. They argue that high tunnels are excellent systems to deploy biological controls and to practice the entire suite of integrated pest management strategies, some of which are all but impossible in field and orchard situations. This is because a high tunnel system is a relatively closed system and modest in size. Greenhouses and indoor vertical farms share some of these characteristics and IPM should be the pest management system of choice in these situations, too. As Pottorff and Panter state for high tunnels, greenhouse, and vertical farms should use exclusion, sanitation, and other non-chemical methods to minimize the introduction of pests, diseases, and weeds to the system. This should greatly reduce, and in some cases, eliminate the need for sprays, organic or not. Using insect screening, planting resistant cultivars, directing irrigation only to crop rows to reduce weed germination and growth are among many non-chemical strategies growers can use for prevention and control of pests, diseases, and weeds in high tunnels.

13.26 Sanitation

Cleanliness and disinfestation can be a challenge in any production system. In an organically managed system, many common greenhouse sanitizing agents are not allowed. In US Certified Organic greenhouse production, sodium hypochlorite, or bleach, is allowed for sanitation, though only in the context of good sanitation practices. Several other synthetic sanitizing agents are not allowed. So, this sends the producer back to the general practices of organic agriculture of using prevention, exclusion, scouting, and rapid management.

13.27 Season Extension in Tropical and Equatorial Regions

In these regions, the value of season extension technologies may be in moderating both temperature and moisture. Mulching is used to retain soil moisture in dry weather. High tunnels and row covers can keep heavy rains from physically flooding a crop or creating a high-moisture environment in which fungal pathogens grow. In high-altitude areas such as the Andes, small holders used high tunnels with concrete or rock pony walls that support the glazed structure. The mass of rock or concrete takes in daytime heat from the Sun and radiates it at night, providing a warmer environment for growth than the outside does. Shading can also be an important addition to a cropping system in the high light environments near the equator. Often, farms will grow crops as an understory to a thinned forest or a cultivated tree crop of some sort. The shade extends the season for these growers without too much cost. In such instances, there are trade-offs that only the grower can decide to make. Many smaller growers make the conscious decision to grow several crops in the same area and are quite pleased.

13.28 European Regulations of Season Extension Technologies

Organic regulations in Europe fall under Council Regulation (EC) No. 834/2007. Currently, EU rules specifically ban hydroponic production in organic production but do not otherwise specifically address organic greenhouse or other protected culture. The EU rules emphasize local sourcing and soil health maintenance. There appear to be plans in the works to draft language related to organic greenhouse in Europe but these are not complete as of this writing. Readers are advised to check locally with their certifying agent for advice and recommendations before adding or significantly modifying any season extension practices in a certified organic system.

The International Federation of Organic Movements (IFOAM) Norms 4.4.10 (IFOAM 2015) specifically ban hydroponic and other systems that are not soil based, "a plant must spend its life in the soil." The IFOAM Standard for Organic Production and Processing restricts the use of artificial light to propagation and some extension of daylight. They also require the use of renewable energy sources for heating and other utilities in production when such utilities are employed. This has significant implications for greenhouse production, vertical farming, and even some high tunnel growers. For example, one could not be in compliance with the Norms if one had non-renewably sourced electricity controlling thermostatically controlled high tunnel wall openers [IFOAM, International Federation of Organic Movements].

Acknowledgment This material is based upon work that is supported by the National Institute of Food and Agriculture, U.S. Department of Agriculture, Hatch project under Accession Number 232043. MIS-149141. Any opinions, findings, conclusions, or recommendations expressed in this publication are those of the author and do not necessarily reflect the view of the U.S. Department of Agriculture.

References

Anonymous (2013) World's largest vertical farm is certified organic. GreenFutures. Oct 2013, No. 90

Caldwell JS, Clark P (1999) Repulsion of cucumber beetles in cucumber and squash using aluminum-coated plastic mulch. HortTech 9(2):247–250

Coleman E (2015) Why you can't have organic food without soil. Civil Eats, 13 April 2015. http://civileats.com/2015/04/13/why-you-cant-have-organic-food-without-soil/. 24 July 2015

Conner DS, Montri AD, Montri DN, Hamm MW (2009) Consumer demand for local produce at extended season farmers' markets: guiding farmer marketing strategies. Renew Agric Food Syst 24(4):251–259

European Union (2008) Commission Regulation (EC) No 889/2008. http://eur-lex.europa.eu/legal-content/EN/TXT/HTML/?uri=CELEX:32008R0889&from=EN. 29 July 2015

Greer L, Dole JM (2006) Aluminum foil, aluminium-painted, plastic, and degradable mulches increase yields and decrease insect-vectored viral diseases of vegetables. HortTechnology 13 (2):276–284

Hamasaki T (2013) Effect of floating row covers on soil and air temperatures in different seasons. J Agric Meteorol 69(2):65–79
Ibarra-Jiménez L, Munguía-López J, Lozano-del Río AJ, Zermeño-González A (2005) Effect of plastic mulch and row covers on photosynthesis and yield of watermelon. Aust J Exp Agric 45 (12):1653–1657
IFOAM (2015) The IFOAM norms for organic production and processing, Version 2014. International Federation of Organic Movements. http://infohub.ifoam.bio/sites/default/files/ifoam_norms_version_july_2014.pdf. 28 July 2015
Kalisz A, Siwek P (2006) Yield and quality of spring Chinese cabbage as affected by different temperature conditions during seedling production. Folia Horticulturae 18(2):3–15
Lamont WJ Jr (2005) Plastics: modifying the microclimate for the production of vegetable crops. HortTechnology 15(3):477–481
Matsumaru (1994) Effects of photochemical oxidants on crops. Bull Chiba Prefectorate Agric Exp Station 27:1–77
Maughan TL, Black BL, Drost D (2015) Critical temperature for sub-lethal cold injury of strawberry leaves. Sci Hortic 183:8–12
Maynard DN, Hochmuth GJ (1997) Knott's handbook for vegetable growers, 4th edn. Wiley, New York
Orozco-Santos M, Perez-Zamora O, Lopez-Arriaga O (1995) Floating row cover and transparent mulch to reduce insect populations, virus diseases and increase yield in cantaloupe. Fla Entomol 78(3):494–501
Pottorff LP and KL Panter (2009) Integrated pest management and biological control in high tunnel production. HortTech 19(1):61–65
Ramakrishna A, Tam HM, Wani SP (2006) Effect of mulch on soil temperature, moisture, weed infestation and yield of groundnut in northern Vietnam. Field Cr Res 95(2–3):115–125
Sideman R (2015) Performance of sweetpoato cultivars grown using biodegradable black plastic much in New Hapshire. HortTech 25(3):412–416
Simmons AM, Kousik CS, Levi A (2010) Combining reflective mulch and host plant resistance for sweetpotato whitefly (Hemiptera: Aleyrodidae) management in watermelon. Crop Prot 29 (8):898–902
Stoner KA (1997) Influence of mulches on the colonization by adults and survival of larvae of the Colorado potato beetle (Coleoptera: Chrysomelidae) in eggplant. J Ent Sci 32(1):7–16
Strik BC, Clark JR, Finn CE, Buller G (2008) Management of primocane-fruiting blackberry to maximize yield and extend the fruiting season. Proc. IX Inter. Symp Rubus Ribes Acta Hort 777:423–428
Sullivan D (2011) Mixed messages from regional certifiers bring to the fore the issue of bioplastics in National Organic Program. BioCycle 52(3):36
Teasdale JR, Mohler CL (1993) Light transmittance, soil temperature, and soil moisture under residue of hairy vetch and rye. Agron J 85(3):673–680
USDA AMS (2015a) National Organic Program. Code of Federal Regulations part 205. http://www.ecfr.gov/cgi-bin/text-idx?c=ecfr&sid=3f34f4c22f9aa8e6d9864cc2683cea02&tpl=/ecfrbrowse/Title07/7cfr205_main_02.tpl. 25 July 2015
USDA AMS (2015b) Wholesale market price reports. http://www.ams.usda.gov/AMSv1.0/ams.fetchTemplateData.do?template=TemplateO&rightNav1=ViewU.S.TerminalMarketPriceReports&topNav&leftNav&page=FVMarketNewsTerminalMarketReportsMore. 30 July 2015
Wang SY, Galletta GJ, Camp MJ, Kasperbauer MJ (1998) Mulch types affect fruit quality and composition of two strawberry genotypes. HortSci 33(4):636–640
Wesson G, Wareing PF (1969) The role of light in the germination of naturally occurring populations of buried weed seeds. J Exp Bot 20(2):402–413
Zhou L-M, Li F, Jin S, Song Y (2009) How two ridges and the furrow mulched with plastic film affect soil water, soil temperature and yield of maize on the semiarid Loess Plateau of China. Field Cr Res 113(1):47–51

Chapter 14
Consumer Perception of Organic Food and Product Marketing

Prabodh Illukpitiya and Pramatma Khanal

Abstract Consumer demand for organically produced goods has grown continuously in the USA since United States Department of Agriculture (USDA) established the national standards for organic production and processing in 2002. Although USDA does not maintain official statistics on US organic food sales, industry data suggest that the market share of organic sales held by various food categories has been remarkably stable over the last decade. The main reason for purchasing organic food products is an expectation of a healthier and environmentally friendly means of production. In general, organic buyers tend to be older and highly educated than those who do not buy them. In addition, consumers' trust in the authenticity of the goods and price are also issues. However, the main barrier to increase the market share of organic food products is consumer information. Increased consumer awareness of organic labeling and their trust in organic labels as well as increasing the availability and range of organic food products may be the most effective way of increasing their market share. Organic food sales in the USA have increased from approximately \$11 billion in 2004 to an estimated \$27 billion in 2012. Organic food products are still gaining ground in conventional supermarkets as well as natural foods markets, and organic sales accounted for more than 3.5 % of total US food sales in 2012. Significant price premiums exist for fresh organic produce and organic milk, the two top organic food sales categories, compared with conventional products, reflecting short supply and higher organic production costs. Public investment in organic agriculture facilitates wider access to organic food for consumers and helps farmers capture high-value markets and boost farm income.

Keywords Organic produce · Consumers · Willingness-to-pay · Price premium · Market

---

P. Illukpitiya (✉) · P. Khanal
Department of Agricultural and Environmental Sciences, Farrell Westbrook Agricultural Complex, 3500 John A. Merritt Blvd, Nashville, TN 37209, USA
e-mail: pillukpi@tnstate.edu

D. Nandwani (ed.), *Organic Farming for Sustainable Agriculture*, Sustainable Development and Biodiversity 9,
DOI 10.1007/978-3-319-26803-3_14

14.1 Introduction

Consumer demand for organically produced goods has grown continuously in the USA since USDA established national standards for organic production and processing in 2002. While Americans economized on their food purchases during the 2007–2009 recession, including purchases of organic products, growth in demand for organic products rebounded quickly following the recession. Industry analysts estimate that US organic food sales were $28 billion in 2012 (over 4 % of total at-home food sales), up 11 % from 2011. USDA has begun organic regulation of nonfood agricultural products—for example, laundry detergent with organic coconut oil, aloe vera, and other ingredients—which accounted for another $2.2 billion in organic sales in 2011 (USDA 2013).

Although USDA does not maintain official statistics on US organic food sales, industry data suggest that the market share of organic sales held by various food categories has been remarkably stable over the last decade. Produce (fruits and vegetables) and dairy are still the top two organic food categories, accounting for 43 and 15 % of total organic sales in 2012; their standing has been relatively unchanged in recent years. Packaged foods, beverages (including soymilk), and breads/grains have 9–11 % market shares, down slightly from a decade ago. The meat, fish, and poultry category, which is dominated by poultry sales, gained the most over the last decade but still represents just 3 % of total organic sales (USDA 2013).

As a result of the increasing consumer concerns about food safety and environmental quality, organic food has rapidly emerged as an important food industry in the USA and many other countries since the early 1980s (Chang and Zepeda 2005; Lohr 1998; Thompson 1998). For example, the total retail sales of organic food and beverages in the USA rose from $178 million in 1980 to $1 billion in 1990, $7.8 billion in 2000, and $23 billion in 2008. In relative terms, the share of organic food and beverages in total food and beverage retail sales in the USA increased from 1.9 % in 2003 to 2.5 % in 2005 and reached 3.5 % in 2008 (Wang et al. 2010). The growth in organic food is also reflected in the increasing availability of organic food products in mainstream supermarkets as well as in local food stores and farmers' markets (Timmons et al. 2008).

Food consumption in most developed countries has attained a saturation point in quantity terms, and consumer food choices are broader than in the past. The result is a more diversified consumption. In this saturated market environment, distribution channels, marketing activities, diversification strategies, and food quality are increasingly important. In addition, consumers have become more concerned about nutrition, health, and the quality of food they eat. As a consequence, organic product's production and consumption have grown in recent years.

Rapid growth in the organic sector has highlighted issues that need to be addressed: shortages of organic raw materials such as organic grain and organic sugar and competition from food marketed as "locally grown" or "natural."

A shortage of affordable organic ingredients or products, such as corn and soybeans for livestock feed, left organic producers unable to meet market demand (Green et al. 2009).

14.2 Consumer Perception of Organic Products

The main reasons for purchasing organic food products are an expectation of a healthier and environmentally friendly means of production. Morasso et al. (2000) discussed as simple associated risks. Accordingly, emotional and cognitive factors are influencing evaluation, risk perception, and changing consumer attitudes (Fig. 14.1). In general, organic buyers tend to be older and higher educated than those who do not buy them. In addition, consumers' trust in the authenticity of the goods and price are also issues. However, the main barrier to increase the market share of organic food products is consumer information. The main motives to purchase organic food products are health and environmental benefits, plus support for local or small farmers. In addition, an important factor that was revealed as a barrier to the development of organic foods is consumer information. Increased consumer awareness of organic labeling and their trust in organic labels as well as increasing the availability and range of organic food products may be the most effective way of increasing their market share. According to Sangkumchaliang et al. (2012) "the organic buyers in Chiang Mai Province in Thailand tend to be older,

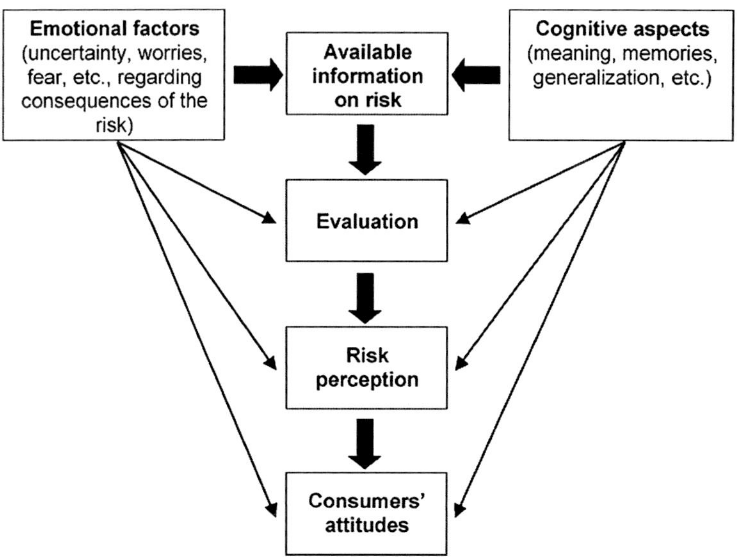

Fig. 14.1 Simple and interpretative model for consumer attitude toward food-associated risks (*Source* Magkos et al. 2006)

highly educated, and more likely to have children in their household than those not purchasing organic food products." The study found that the groups of buyers and non-buyers have significant differences in demographic characteristics. However, age, household size, children in household, and education level seemed to have an effect on the perceptions of consumers. The main barrier of organic foods market share is the information available and consumer awareness.

Many surveys of consumer attitudes and characteristics have been conducted to identify the reasons for the increased trend of consumption of organic food (Thompson 1998). The preference for organic food has been associated with multiple factors that in general (Fig. 14.2) reflect an increased interest toward personal health, animal welfare, and environmental protection (Makatouni 2002). Health-related issues seem to assume greater importance than other concerns and notions about food safety are fundamental for purchasing organics (Magnusson et al. 2003). This perception is mainly due to the principles associated with the organic production. Organic farming system is a production system that avoids or largely excludes the use of synthetic fertilizers, pesticides, growth regulators, and livestock feed additives (Soil Association 1997).

To the maximum extent feasible, organic farming systems rely on crop rotations, crop residues, animal manures, legumes, green manures, off-farm organic wastes, and aspects of biological pest control to maintain soil productivity, supply plant nutrients, control insects, weeds, and other pests (Soil Association 1997). The non-use of synthetic chemicals and a number of other environmentally sound

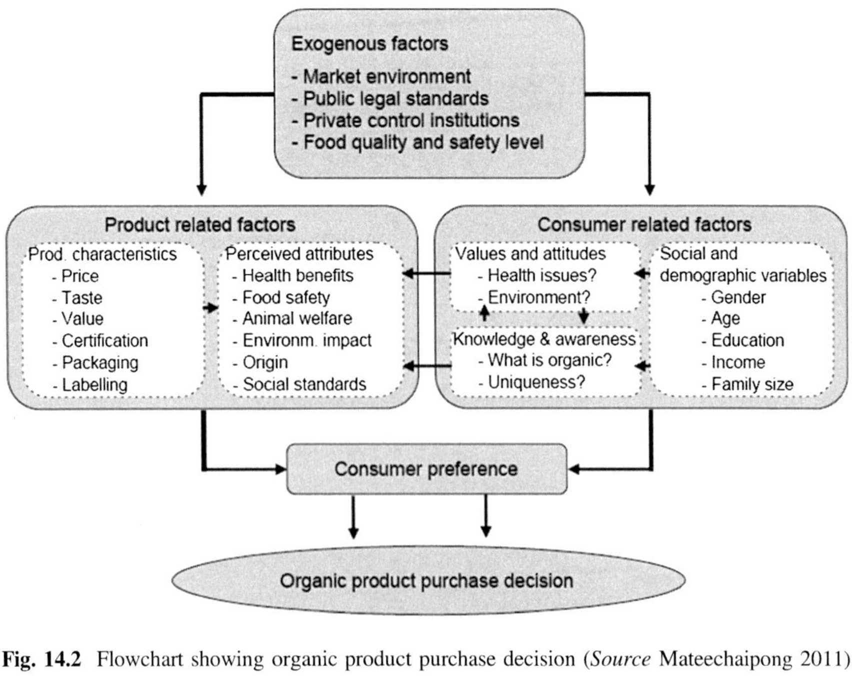

Fig. 14.2 Flowchart showing organic product purchase decision (*Source* Mateechaipong 2011)

techniques practiced by organic farmers remain part of the allure of the organic movement, and underlie consumer belief that organic food is virtually free of the hazards found in conventional produce (Marcus 2001). According to Shukla (2001), it is the consumer perception of the quality and safety that primarily drives the continuously growing demand of organic food products. In fact, in some cases (e.g., organic baby food), the organic label is by far the most important characteristic that consumers value in food; its nutrient content being far less appreciated (Harris 1997). Producers have different views regarding organic products. According to a survey results (Carter et al. 2014), some of the respondents believe that organic production reduces worker exposure to chemicals, decreasing the environmental impact of agricultural production, and personal values contribute to the use of organic practices, with 86 % or more of respondents agreeing that these factors motivate them to choose organic practices over conventional practices. A majority of producers pursue USDA organic certification to differentiate their products from other products claiming to be "natural" or "sustainable," with 75 % of respondents listing this factor as "important" or "very important." Eighty-five percent of producers agree or strongly agree that National Organic Program regulations increase consumer confidence in products marketed as "organic," while a smaller percentage (59 %) indicates that the regulations directly increase consumer's understanding of the difference between conventional and organic products. Almost half of the producers (49 %) express that National Organic Program regulations should more precisely specify allowed and prohibited substances. Seventy percent disagree with allowing more synthetic substances under the regulations (Carter et al. 2014).

14.3 Consumers' Willingness-to-Pay for Organic Products

The production of organic food raises the cost of production relative to conventional products, and the production of organic products with higher percentages of organic ingredients also raise the cost of production. Hence, a key issue is whether consumers are willing to pay more for these products. The notion of willingness-to-pay could be defined as the amount of money represented by the difference between consumers' surplus before and after adding or improving a given food product attribute.

There exist a number of studies which show that consumers exhibit a greater willingness-to-pay (WTP) for organic products over conventional products with identical appearance (Govindasamy and Italia 1999; Williams and Hammitt 2000; Piyasiri and Ariyawardana 2002; Nouhoheflin et al. 2004; Darby et al. 2008; Liu et al. 2009; Jesse and Huffman 2012; Owusu and Anfari 2013). For example, according to a research, the estimated mean WTP price premiums for 1 kg of organic lettuce and watermelon compared to conventional watermelon and lettuce were US$1.04 and US$0.46, respectively. The median WTP premium for 1 kg of organic lettuce is US$1.26 and that of organic watermelon is US$0.48 (Owusu and Anfari 2013).

According to Jesse and Huffman (2012), participants were willing to pay higher prices for an organic product with high levels of organic purity. Also, individuals with more education were willing to pay more for organic relative to conventional products and addition household income (per capita basis) increases willingness-to-pay for organic products up to $76,100. However, their willingness-to-pay decreased as per capita household income increased above $76,100. The consumers who purchased organic food had their household average monthly expenditure on organic food in 2002 was $69.30, or 19.90 % of their average monthly food expenditure of $357.90. A comparison of the sample statistics with the available demographic information of the Vermont population suggests that this group of respondents included slightly more individuals with higher education levels and higher household income and fewer individuals with children. The leading reason for purchasing organic food was that organic food is healthier followed by organic food can help small farmers, is better for the environment, is safer, and tastes better.

Nouhoheflin et al. (2004) employed the hedonic pricing approach, which is an indirect method of valuation, to assess consumer perceptions and willingness-to-pay premiums for organic vegetables compared to conventional vegetables in Benin and Ghana. Their empirical findings revealed a consumer willingness-to-pay of more than 50 % price premium for chemical-free vegetables. Empirical literature on consumer surveys reveals that consumers' socioeconomic characteristics such as age, gender, level of education, income level, household size as well as the level of consumers' awareness and perceptions, product price, taste, size, freshness, and cleanness tend to influence consumers' willingness-to-pay (WTP) for organic food products. Govindasamy and Italia (1999) showed that younger consumers, regardless of gender, paid higher premiums for organic products. Consistent with this finding, Liu et al. (2009) found an inverted-U-shape relationship between age and consumer WTP, indicating that WTP for additive-free foods increases with age but decreases as age increases beyond a threshold age. However, Darby et al. (2008) found no significant impact of age on consumer WTP. Some consumer studies have shown females in particular to be more willing to pay higher premiums for safe foods (Williams and Hammitt 2000).

Higher educated consumers are expected to pay higher price premiums for organic foods since they tend to appreciate issues of preventive health care through the consumption of chemically free food products better than consumers with no education (Piyasiri and Ariyawardana 2002). Darby et al. (2008) also found education to be positively correlated with WTP statistically.

14.4 Organic Product Marketing

Organic farming is a growing sector in the world, which is encouraged by the government and many private initiatives. Therefore, production is expected to rise to meet the growing demand in the domestic market for organic foods (see Fig. 14.3 for general marketing strategies). The increased range of healthy foods and the establishment of certificates for pesticide controlled vegetables indicate that there is a potential market. Since the late 1990s, US organic production has more than doubled, but the consumer market has grown even faster. Organic food sales have more than quintupled, increasing from $3.6 billion in 1997 to $21.1 billion in 2008. More than two-thirds of US consumers buy organic products at least occasionally, and 28 % buy organic products weekly, according to the Organic Trade Association (USDA 2009). This fast-paced growth has led to input and product shortages in organic supply chains, and several new issues—concern about premium-priced product sales in a tight US economy, as well as competition from new environmental labels—are emerging in the organic industry. While new producers have emerged to help meet demand, market participants report that a supply squeeze is constraining growth for both the individual firms and the organic sector overall.

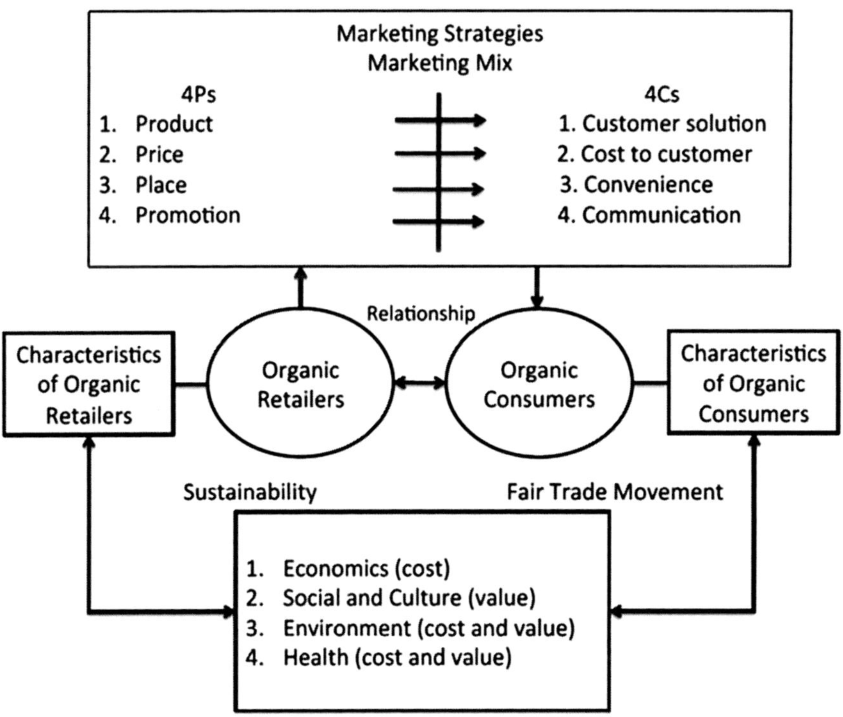

Fig. 14.3 General marketing strategies for organic product (*Source* Mateechaipong 2011)

Partly in response to shortages in organic supply, Congress in 2008 included provisions in the Food, Conservation, and Energy Act that, for the first time, provide financial support to farmers to convert to organic production. On the production side, high costs, especially labor costs, and the difficulty in shifting from conventional to organic farming are also limiting factors (Vetter and Christensen 1996; Hamiti et al. 1996). Furthermore, food availability and seasonality influence marketing activities and make it difficult to establish appropriate retailing outlets. Higher costs of production and retailer margins jointly may result in higher prices than consumers are willing to pay for organic food attributes.

Organic food sales in the USA have increased from approximately $11 billion in 2004 to an estimated $27 billion in 2012 (Fig. 14.4) according to Nutrition Business Journal (USDA-ERS 2013). Organic food products are still gaining ground in conventional supermarkets as well as natural foods markets, and organic sales accounted for more than 3.5 % of total US food sales in 2012. Markets for organic vegetables, fruits, and herbs have been developing for decades in the USA, and fresh produce is still the top-selling organic category in retail sales. Although the annual growth rate for organic food sales fell from the double-digit range in 2008 as the US economy slowed, its 7.4 % growth rate in 2012 was more than double the annual growth rate forecast for all food sales in 2012.

Significant price premiums exist for fresh organic produce and organic milk, the two top organic food sales categories, compared with conventional products, reflecting short supply and higher organic production costs. Even if price premiums for organic products can be maintained, the public-goods nature of environmental services, such as biodiversity and water quality, implies that prices do not reflect the true social value of these services. Public investment in organic agriculture

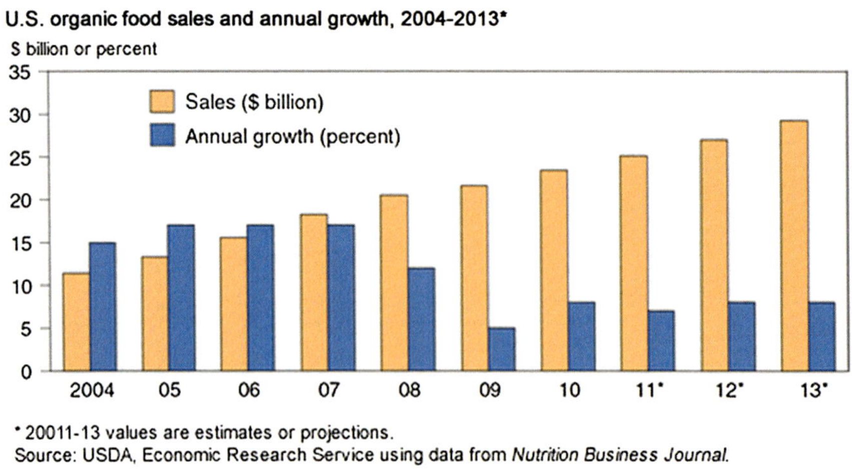

Fig. 14.4 Growth trends in US organic food sales (*Source* USDA 2013)

facilitates wider access to organic food for consumers and helps farmers capture high-value markets and boost farm income, as well as conserve nonrenewable natural resources and protect US soil and water (Green et al. 2009). For example, price coupons may encourage some consumers who have not purchased organic food to try organic apples. Also, information on the reasons for purchasing organic food and reasons for not purchasing organic food, reported in this study, can help organic farmers and organizations develop effective marketing strategies and educational materials for promoting organic food. For example, information and educational materials on the benefits of organic food and on certification regulations and procedures may change the negative attitudes of some consumers who have not purchased organic food.

References

Carter DP, Weible CM, Siddiki SN et al (2014) Certified organic producer perceptions of U.S. organic regulations and organic certifying agents. School of Public Affairs, University of Colorado Denver

Chang H, Zepeda L (2005) Demand for organic food in Australia: results from a focus group study. J Food Distrib Res 36:223–224

Darby K, Batte MT, Ernst S et al (2008) Decomposing local: a conjoint analysis of locally produced foods. Am J Agric Econ 90 (USDA, 2013):476–486

Govindasamy R, Italia J (1999) Predicting willingness-to-pay a premium for organically grown fresh produce. J Food Distrib Res 30:44–53

Greene C, Dimitri C, Lin B et al (2009) Emerging issues in the U.S. organic industry. EIB-55. U.S. Dept. of Agriculture, Economic Research Service

Harris JM (1997) Consumers pay a premium for organic baby foods. Food Rev 20:13–16

Hamiti T, Laajimi A, Albisu, LM (1996) Adopción de Tecnología en la Agricultura Sostenible. Documento de Trabajo, 96/5. Unidad de Economía Agraria (SIA-DGA) Zaragoza

Jesse LS, Huffman WE (2012) Willingness-to-pay for organic food products and organic purity: experimental evidence. Iowa State University. Department of Economics Working paper no. 120017

Liu Y, Zeng Y, Yu X (2009) Consumer willingness to pay for food safety in Beijing: a case study of food additives. In: Contributed paper prepared for presentation at the international association of agricultural economists conference, Beijing, China, 16–22 Aug 2009, pp 6

Lohr L (1998) Implications of organic certification for market structure and trade. Am J Agric Econ 80:1125–1129

Magkos F, Arvaniti F, Zampelas A (2006) Organic food: buying more safety or just peace of mind? a critical review of the literature. Crit Rev Food Sci Nutr 46:23–56

Magnusson MK, Arvola AA, Hursti UK et al (2003) Choice of organic foods is related to perceived consequences for human health and to environmentally friendly behavior. Appetite 40:109–117

Makatouni A (2002) What motivates consumers to buy organic food in the UK? Results from a qualitative study. Br Food J 104:345–352

Marcus MB (2001) Organic foods offer peace of mind–at a price. US News World Rep 130:48–50

Mateechaipong C (2011) Developing marketing strategies for organic products employed by small and medium retailers in Bangkok (Masters dissertation)

Morasso G, Bolognesi C, Duglio E, Musso M (2000) Pesticides as food contaminants: a pilot project for correct public information. Trends Food Sci Technol 11:379–382

Nouhoheflin T, Coulibaly O, Cherry AJ et al (2004) Consumers' perception and willingness to pay for organic vegetables in Benin and Ghana. Paper presented at the inaugural symposium of the African association of agricultural economists, Nairobi, Kenya

Owusu V, Anifori MO (2013) Consumer willingness to pay a premium for organic fruit and vegetables in Ghana. Int Food Agribusiness Manage Rev 16(1):67–86

Piyasiri AGSA, Ariyawardana A (2002) Market potentials and willingness to pay for selected organic vegetables in Kandy. Sri Lankan J Agric Econ 4(1):107–119

Sangkumchaliang P, Huang WC (2012) Consumers' perceptions and attitudes of organic food products in Northern Thailand. Int Food Agribusiness Manage Rev 15(1):90–91

Shukla VKS (2001) Organic foods: present and future developments. INFORM 12:495–499

Soil Association (1997) Standards for organic food and farming: including standards for registration and certification, general crop production, livestock husbandry, conservation, processing, packing and distribution, importing. Revision 11 Bristol: Soil Association, Organic Marketing Co

Thompson GD (1998) Consumer demand for organic foods: what we know and what we need to know. Am J Agric Econ 80:1113–1118

Timmons D, Wang Q, Lass D (2008) Local foods: estimating capacity. J Extension 46(5), Article 5FEA7. Available at: http://www.joe.org/joe/2008october/a7.php

USDA (2013) Growth patterns in the U.S. organic industry. Available http://www.ers.usda.gov/amber-waves/2013-october/growth-patterns-in-the-us-organic-industry.aspx#.VNORo53F_n8. Retrieved 01 Feb 2015

USDA-ERS (2009) Emerging issues in the U.S. organic industry. Available http://www.ers.usda.gov/publications/eib-economic-information-bulletin/eib55/report-summary.aspx. Retrieved 01 Feb 2015

USDA-ERS (2013) Consumer demand drives growth in the organic food sector. Available http://www.ers.usda.gov/data-products/chart-gallery/detail.aspx?chartId=35003. Retrieved 01 Feb 2015

Vetter H, Christensen A (1996) Evil ecologists. IX EAAE Paper presented at the IX EAAE Congress, Edinburgh

Wang Q, Sun J, Parsons J (2010) Consumer preferences and willingness to pay for locally grown organic apples: evidence from a conjoint study. HortScience 45(3):376–381

Williams PRD, Hammitt JK (2000) A comparison of organic and conventional fresh produce buyers in the Boston Area. Risk Anal 20(5):735–746

Chapter 15
Sustainable Urban Agriculture: A Growing Solution to Urban Food Deserts

Sarada Krishnan, Dilip Nandwani, George Smith and Vanaja Kankarta

Abstract Urban Agriculture is here to stay and is defined as *the growing, processing, and distribution of food and other products through intensive plant cultivation and animal husbandry in and around cities*. Growing food and non-food crops in and near cities contributes to healthy communities by engaging residents in work and recreation that improves individual and public well-being. Urban agriculture integrates multiple functions in densely populated areas offering an alternative land use. In addition to food production, urban agriculture also offers a wide range of other functions such as energy conservation, waste management, biodiversity, nutrient cycling, microclimate control, urban greening, economic revitalization, community socialization, human health, preservation of cultural heritage, and education.

Keywords Urban agriculture · Sustainability · Community-supported agriculture · Community gardens · Vertical farms

15.1 Introduction

With the global population anticipated to reach over nine billion by the year 2050, the role of urban agriculture in global food security has become an important discussion topic. When dealing with the topic of urban and peri-urban agriculture, there are many definitional challenges with the terms referring to a diverse range of activities to include crops, livestock, poultry, and aquaculture production and

---

S. Krishnan (✉)
Department of Horticulture and Center for Global Initiatives, Denver Botanic Gardens, Denver, CO 80206, USA
e-mail: KrishnaS@botanicgardens.org

D. Nandwani · G. Smith · V. Kankarta
Department of Agricultural and Environmental Sciences, College of Agriculture, Human and Natural Sciences, Tennessee State University, Nashville, TN 37209, USA

D. Nandwani (ed.), *Organic Farming for Sustainable Agriculture*, Sustainable Development and Biodiversity 9,
DOI 10.1007/978-3-319-26803-3_15

ranging in scale from rooftop gardens to larger cultivated open spaces (Thebo et al. 2014). Defined in simple terms, urban agriculture is defined as the growing, processing, and distribution of food and other products through intensive plant cultivation and animal husbandry in and around cities (Bailkey and Nasr 2000). The main feature that distinguishes urban agriculture from rural agriculture is its integration into the urban economic and ecological system (Mougeot 2000). Throughout history, integration into the urban system has been critical for the persistence of urban agriculture. Even though the nature of cities and of urban food supply systems has changed over time, the interaction of urban agriculture with the rest of the city and rural production and imports remain as true today as it was thousands of years ago (Mougeot 2000). Van Leeuwen et al. (2010) trace the history of urban green spaces to 600 BC with the functional use of the green spaces changing over time.

Urban agriculture is considered an alternative agriculture movement advocating major shifts toward a more ecologically sustainable agriculture compared to the conventional paradigm of large-scale, highly industrialized agriculture (Sumner et al. 2010). Core beliefs and values underlying the two approaches of conventional versus alternative urban agriculture systems, respectively, are as follows: (1) centralization versus decentralization; (2) dependence versus independence; (3) competition versus community; (4) domination of nature versus harmony with nature; (5) specialization versus diversity; and (6) exploitation versus restraint (Sumner et al. 2010). Urban agriculture takes into account the vital role of culture and values in agriculture and their connection to sustainability (Sumner et al. 2010). Urban agriculture also improves access to fresh, nutritious food, helping combat childhood obesity, diabetes, and poor nutrition prevalent in many urban communities, access to rare foods that support cultural heritages of immigrant communities, and provides social benefits through improving interracial relationships and decreasing crime (as cited in Lovell 2010).

In densely populated areas, urban agriculture offers an alternative land use integrating multiple uses (Lovell 2010). Even though urban agriculture has historically been an important element of cities in many developing countries, only recently has this become a growing movement in cities of developed countries including the USA (Lovell 2010). Functionally, urban agriculture serves different purposes in developed versus developing countries, with an emphasis on recreation in developed countries and food security in developing countries (Pearson et al. 2010). In developed countries, urban agriculture takes place often on small pieces of land tucked away in corners of cities either rented or on own parcel of land with community gardens maintained by a group or community such as school gardens (Van Leeuwen et al. 2010). In addition to recreational benefits and home consumption, urban agriculture in these regions also serves to create social interaction and production of organic or otherwise healthy foods (Nugent 2000). In contrast, in developing countries, urban agriculture is practiced to generate self-employment and direct revenue or savings leading to greater social stability (Van Leeuwen et al. 2010). By 2015, the population of 26 major cities is expected to exceed 10 million requiring about 6000 tons of food to feed the city. Taking into account the complex space economy of urban cities, multitasking to promote enhanced use of farmland within urban areas can become a win-win situation (Van Leeuwen et al. 2010).

15.2 Sustainability Factors

Urban agriculture integrates three main principles of sustainability:

1. Environmental health: Sustainable urban agriculture is supportive of environmental health in that it requires low input of water and low to no use of fertilizers and pesticides.
2. Economic profitability: Sustainable urban agriculture reduces transportation costs of shipping between local producers to local markets.
3. Social wellness: Sustainable urban agriculture provides opportunities for social interaction and individual recreational opportunities.

Table 15.1 Impacts of sustainable urban agriculture

	Sustainability	Benefits/impacts
1	Environmental	*Pollution* • Urban planting helps clean up the air and water and builds resilience of aquatic environment • Reduces heat and noise in urban areas • Recycles urban waste and uses as nutrients for the plants *Biodiversity* • Protects and improves biodiversity of urban areas • Increases ecosystem resilience *Climate change* • Reduces global heat and improves microclimate • Carbon sequestration
2	Economic	*Creating avenues* • Creates employment opportunities • Increases business and expands urban economy *On-farm benefits* • Higher yields and returns from the land • Reduced food miles
3	Social	*Community engagement* • Community development/building social capital • Increases awareness, education and youth development, and recreational opportunities • Food security and access • Increases access to land
4	Health	• Good health and fitness • Good food and health literacy • Improved overall well-being (Mental Health and Physical Activity)

Deelstra and Girardet (2000), Nugent (2000), Pearson et al. (2010), Van Leeuwen et al. (2010)

Sustainable urban agriculture plays an essential part in addressing a city's problems in innovative ways (Brown and Carter 2003). Through urban agriculture's efforts to green cities, environmental stewardship is enhanced. When inner city residents have the ability to grow and market their own food through farmer's markets, providing opportunities for entrepreneurs and commercial farmers, this leads to economic development and community revitalization. Through access to and greater control over food systems, social welfare of urban dwellers is improved through improvement in individual health and an enhanced sense of empowerment (Brown and Carter 2003).

The various impacts of urban agriculture leading to the attainment of the three sustainability factors are listed in Table 15.1.

15.3 Types of Sustainable Urban Agricultural Systems

Pearson et al. (2010) break down urban agricultural production into three scales: (1) microcommercial scale (green roofs, green walls, courtyards, backyards, and street verges); (2) mesocommercial scale (community gardens and allotments, urban parks); and (3) macrocommercial scale. Table 15.2: Characteristics and benefits of sustainable urban agriculture, listed below, compares differing types of urban agricultural systems (Fig. 15.1):

Notes: Definitions of Physical Structure:

1. *Whole Building* refers to practicing sustainable urban agriculture indoors in buildings. The buildings can be unused buildings that are retrofitted to grow sustainable urban agriculture produce under artificial and/or natural lighting conditions, primarily for commercial sale.
2. *Rooftop* refers to growing sustainable urban agriculture produce on structurally suitable and accessible rooftops in urban areas.
3. *Wall Structure* refers to a structurally sound exterior or interior wall that is suitable and available to grow types of agricultural climbing and potted plants, such as tomatoes, grapes, peppers, and vining pea plants.
4. *Wall Hangers* refer to a system of plants in pots fastened to exterior or interior walls and connected to a constructed irrigation/feeding pipe system.
5. *Commercial or Communal Greenhouses* refer to greenhouse production of sustainable agricultural products on a small communal to large commercial scale. These structures can be homemade with recycled materials for communal use, or large scale using prefab construction methods for commercial use.
6. *Free-standing Frame Structure* refers to vertical structures at a range of scales, including some that are constructed out of repurposed materials including PVC pipes, empty barrows, fence posts, and scrap metal frames welded together.
7. *Patio Pots* refer to growing vegetable and fruits in small pots in limited spaces such as patios, for recreation and private consumption

Table 15.2 Characteristics and benefits of sustainable urban agriculture

Form and function		Impacts		
Physical structure form	Characteristic	Social	Economic	Environment
Whole building (1)	Commercial, high-tech solar power interior lighting	Disconnected from urban context and people	High potential commercial success	Neutral to the urban environment
Rooftop (2)	Community oriented with commercial potential	Community access where safety not a concern	Can be for profit and provides access to fresh and affordable urban food	Functions as a filter for air pollutants and adds greenery to urban areas
Wall structure as support element (interior and/or exterior) (3)	Community oriented with commercial potential	Community access where safety not a concern	Can be for profit or provides access to fresh and affordable urban food source	Functions as a filter for air pollutants and adds greenery to urban areas
Wall hangers for pots (4)	Community oriented with commercial potential	Community access where safety not a concern	Can be for profit or provides access to fresh and affordable urban food source	Functions as a filter for air pollutants and adds greenery to urban areas
Commercial or communal greenhouse (5)	Community oriented with commercial potential	Community access where safety not a concern	Can be for profit or provides access to fresh and affordable urban food source	Functions as a filter for air pollutants and adds greenery to urban areas
Free-standing frame structure (6)	Small-scale communal or commercial	Communal produce gives access to fresh affordable food	Extra produce can be sold for a small profit	Minimal impact and adds to reducing pollution

15.3.1 Commercial Farms

Brown and Carter (2003) report three categories of metropolitan farms: (1) recreational farms which sell less than \$10,000 annually in less than 100 acres; (2) adaptive farms which sell \$10,000 or more annually of high-value products and are 100 to 200 acres in size; and (3) traditional farms which sell greater than \$10,000 annually of high-value products and are greater than 200 acres. In reality, typical urban agricultural operations operate on fewer than 25 acres. Urban farmers create direct access to food through farm stands and farmers' markets, increasing the amount of food dollar going into their own pocket and helping the local

Fig. 15.1 Sustainable urban agriculture system model. *Credit* Sustainablelifestyles.wordpress.com

economy (Brown and Carter 2003). Many urban commercial farms tend to be located in the suburbs and help create strong community connections through pick-your-own operations, corn mazes, petting zoos, school tours, and farm stands (Brown and Carter 2003).

15.3.2 Community-Supported Agriculture (CSA)

According to the US Department of Agriculture (USDA 2015), in basic terms, community-supported agriculture (CSA) consists of a community of individuals who pledge support to a farm operation so that the farmland becomes, either legally or spiritually, the community's farm, with the growers and consumers providing mutual support and sharing the risks and benefits of food production. Typically, members or "shareholders" of the farm or garden pledge in advance to cover the anticipated costs of the farm operation and farmer's salary. In return, they receive shares in the farm's bounty throughout the growing season, as well as satisfaction gained from reconnecting to the land and participating directly in food production. Members also share in the risks of farming, including poor harvests due to unfavorable weather or pests. By direct sales to community members, who have provided the farmer with working capital in advance, growers receive better prices for their crops, gain some financial security, and are relieved of much of the burden of marketing.

The concept of CSA was introduced to the USA in 1984 by Jan VanderTuin from Switzerland with two CSA projects in operation by 1986 (Adam 2006).

Currently, there are over 1300 registered farms across North America (as cited in Sumner et al. 2010). By providing a wide range of healthy, locally grown food, reconnecting consumer with the producer, and fostering a sense of local environment and human stewardship, CSA farms aim to address the broad goals of sustainable agriculture (McIlvaine-Newsad et al. 2004). In a case study of Fourfold Farm, CSA in southwestern Ontario, Canada, Sumner et al. (2010) identified three main motivating themes central to the CSA: civic engagement, community, and the celebration of local food.

Case Study Example of Community-Supported Agriculture (CSA)

The most popular CSA projects known as "Local harvest" with various locations are widely spread out in the USA. CSA is one of the popular ways for consumers to buy local, seasonal food directly from a farmer. The basic principle on which it operates is that a farmer offers a certain number of "shares" to the public which may consist of a box of vegetables. Interested consumers purchase a share through a membership or subscription and in return receive a box of seasonal produce each week throughout the farming season. This arrangement is rewarding to both the farmer and the consumers.

Advantages for farmers:

- Provides farmers the opportunity to market his/her produce early in the season
- Early payments for the sale and increases in his/her cash flow
- Provides opportunity to meet people and know who eats the food they grow

Advantages for consumers:

- Provides fresh, on-farm produce with all nutritional benefits
- Opportunity for public to visit the farm and know how the food is grown
- Helps develop relationship with the grower

Local harvest has had a tremendous impact on thousands of families who love to eat fresh and healthy food that is grown locally and easily accessible.

15.3.3 Community Gardens

Community gardens (also called allotment gardens in Europe), serving those who do not have access to private garden plots, usually are large lots of land divided into smaller plots for individual/household use for production of edibles such as vegetables, fruits, and herbs, but also with ornamental plants dispersed among the gardens. Ownership of these lots varies from a municipality, an institution, a community group, a land trust, or private ownership (Brown and Carter 2003). The objectives of community gardens today go beyond food production (Adam 2011). They build a sense of community by providing gathering spaces for local residents, enhancing social interactions. Additionally, most community gardens mandate sustainable or organic growing methods, contributing to environmental sustainability (Adam 2011). Some challenges associated with starting and maintaining a community garden include ensuring security, being accepted by the wider community, coexisting with wildlife, ensuring land tenure, securing labor, addressing self-sufficiency, addressing zoning issues, and securing gardening inputs such as water, tools, garden supplies, and compost (Adam 2011).

15.3.4 Community Orchards

In simple terms, a community orchard is an orchard that is cared for by some community of people and not being managed for private profit (Ames 2013). The community orchard movement began in England in the early 1990s when abandoned privately owned orchards in danger of being lost due to neglect or land development were saved by local citizens concerned about green space, survival of old varieties, local history, and healthy eating. Ownership of such an orchard varies with the land either deeded or leased to a local municipality, a charitable trust, a "friends of the orchard" group, a food co-op, or the residents of a group housing project (Ames 2013). In the USA, community orchards have been created by repurposing public land, usually in a public park, with open public access (Ames 2013). Even though community orchards in the USA were not started with the goal of existing orchard preservation, the main reasons are preservation of heirloom varieties, education, access to food, aesthetics, and creating a sanctuary (Ames 2013). Due to the relative permanence of orchards, planning a community orchard requires forethought. In order to reduce long-lasting problems, varieties need to be chosen with care to eliminate chronic pest and disease issues, soil amended properly before planting and develop a pruning, fertilizer and training regimen (Ames 2013).

Case Study Example of Community Orchards

Bloomington Community Orchard is a nonprofit organization operating in partnership with the Bloomington Parks and Recreation Department. It is located in Bloomington, Indiana. The orchard is devoted to growing fruit for the community, and growing community "orcharding" skills through educational opportunities. It is a publicly owned orchard, maintained by volunteers, and the harvest is available to everyone in the community. It is one of the only projects of its kind in the nation, always in the forefront of sustainable living and community building. Volunteers have worked hard to carry out this orchard to grow the fruits and harvest them. This orchard contributes to the Bloomington's food security, inspires joyful community engagement, and educates the citizens while making sustainability delicious. The orchard design illustrated in Fig. 15.2 shows the arrangement of different fruit species in the orchard.

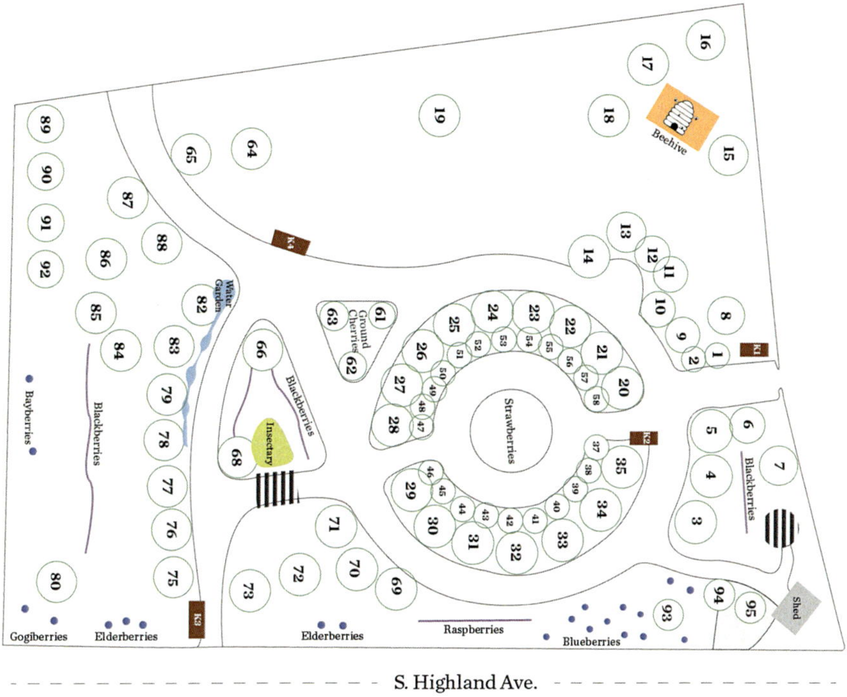

Fig. 15.2 Map showing arrangement of different fruit species at Bloomington Community Orchard design. *Credit* http://www.bloomingtoncommunityorchard.org/site/about/

15.3.5 Backyard Gardens

Plots around homes, including balconies, decks, and rooftops, constitute urban backyard gardens (Brown and Carter 2003). In the USA, as many as a quarter of the households have gardens with most backyards raising their own food seasonally to supplement their diets (Brown and Carter 2003). Even though these gardens in much of North America do not equate to subsistence farming in developing countries, in many cases backyard harvests have stretched food budgets of low-income families and their network of family and friends (Brown and Carter 2003).

15.4 Future of Sustainable Urban Agriculture

In some cases, it is true that the past is the future. Vertical farming is rooted in the agricultural practices of the nineteenth century. During World War I and later during World War II, people in communities throughout the USA were urged and supported by government officials to grow food in whatever spare land was available in the cities and towns of the nation, as a method of augmenting the scarce provisions available at that time. The future of sustainable urban agriculture has its roots in the past as the following examples of emerging sustainable urban agriculture will show.

15.4.1 Mulberry Dyke Fish Pond Model System of China

The Mulberry Dyke Fish Pond Model System of China provides a clear demonstration of a symbiotic environmental relationship between water and landscape ecosystems. This fishpond complex is a system developed by farmers in the Pearl River Delta region of China to make full use of available land and water resources present in and close to crowded rural communities. This model is an integrated ecosystem that brings into full play the productive potential of humans and their environment and promotes the different branches of agriculture.

The Mulberry Dyke Fish Pond Model was first development in sixteenth century in China during the late Ming Dynasty in the Pearl River Delta. The mulberry dyke fishpond complex contains two interrelated systems of dyke and pond; the dyke is the land ecosystem for the growth of mulberry trees, whereas the pond is the water ecosystem, consisting of fish and aquatic plants. Mulberry leaves are fed to the silkworms, whose excreta are used as fish food, and the fertile pond mud consisting of fish excreta, organic matter, and chemical elements is brought up from the bottom and used as manure for the mulberry trees. The pond mulch is rich in organic matter and nutrients. When placed on the dyke where the mulberry trees are planted, the pond mulch decomposes and fertilizes the mulberry trees (Lee 2004, pp. 2–3).

15.4.2 Vertical Farms

For every 1° rise in atmospheric temperature, it is estimated that 10 % of the land currently under crop cultivation will be lost. One answer to reversing this situation is controlled environment agriculture (Despommier 2011). Vertical farms have been proposed as a solution for future cities to grow most of the food inside city limits in ultra-efficient greenhouses (Vogel 2008) and, regardless of location, can be applied to every urban center (Despommier 2011). Despommier (2011) identifies the following rationale for creating vertical farms in urban centers over conventional agriculture: year-round produce; lack of crop loss due to weather events; no use of fossil fuels to harvest, transport, and refrigerate; no use of pesticides and herbicides; job creation in urban centers; lesser water use; and limited spoilage from excessive handling. This approach to food production is largely environment independent and hence immune to climate change (Germer et al. 2011). Well-designed greenhouses

Fig. 15.3 Vertical farming models aeroponic and hydroponic growing systems. *Credit* Dr. Dilip Nandwani (2015)

should use as little as 10 % of the water and 5 % of the area required by conventional farms (Vogel 2008). In addition to vertical farming in greenhouse structures, another logical step is retrofitting 2–6 story tall buildings with aeroponic and hydroponic growing systems with provisions for adequate light to create free-standing vertical farms (Despommier 2011) (see Fig. 15.3). The development of this technology is still in its infancy. A vertical farm called Vertical Harvest now under construction in Jackson, Wyoming, will be one of the world's first vertical farms (accessed on August 10, 2015, at http://www.fastcoexist.com/3042610/a-vacant-lot-in-wyoming-will-become-one-of-the-worlds-first-vertical-farms#9).

15.4.3 Food Forests

The urban food forest is a gardening or land management systems that mimics natural ecosystems. This system is a designed community of mutually beneficial plants and animals intended to produce food for human consumption (Jacke and Toensmeier 2005). There are multiple levels/layers (7–9) of trees and shrubs in different stages which may include fruit and nut trees in the canopy down to ground cover crops and even vines (see Fig. 15.4).

The food forest approach to sustainable urban agriculture adapts modifications of the permaculture techniques. This process is a tested and proven sustainable system that requires low maintenance and minimal watering. It is a largely self-maintaining, inclusive, and highly productive system of multi-storied trees, shrubs, grasses, flowers, pollinators, soil, water, and community. Food forests help improve the environment we live in; build stronger, more resilient communities; and provide a host of economic benefits as well. They also help create more sustainable communities that are healthy and enjoyable to live in that helps increase productivity, yields, resource use efficiency, resilience, and resistance to pest and diseases (Jacke and Toensmeier 2005).

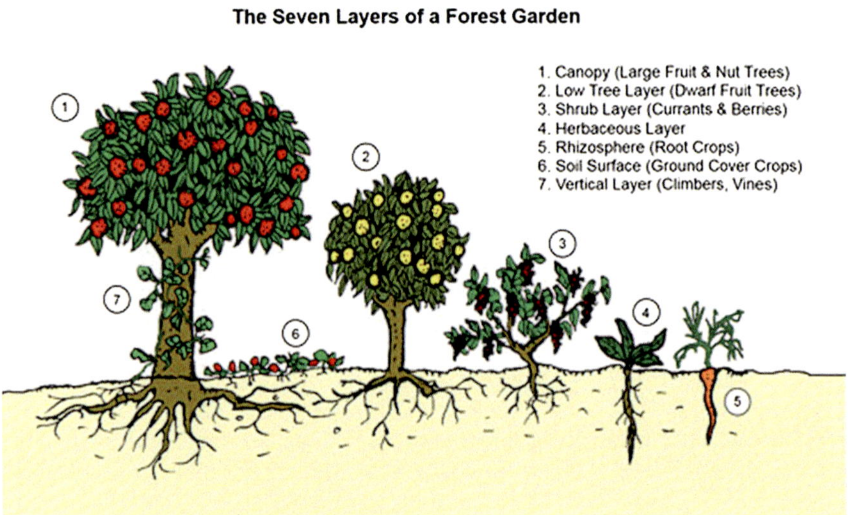

Fig. 15.4 Seven layers of a food forest (accessed on August 10, 2015, at http://deepgreenpermaculture.com/permaculture/permaculture-design-principles/7-small-scale-intensive-systems/)

Benefits of an Urban Food Forests

Increased urbanization replaced trees and other vegetation with impervious land cover (pavement, concrete, etc.). This unsustainable practice impacts negatively on humans and ecosystems. Alternatively, integrating sustainable urban landscape planning and management improves urban landscape sustainability (Clark and Nicholas 2013). Following is a brief synopsis of the benefits of urban food forests.

1. **Economic**

   (a) **Ecotourism**: The presence of urban green spaces has been shown to increase the economic value of an area, which among other benefits can promote ecotourism.

   (b) **Storm water management**: Urban areas are extensively covered with a mix of impervious surfaces causing impaired water quality and reduced groundwater recharge, flooding, and soil erosion. Urban food forests are very effective in handling storm water.

   (c) **Reduces urban heat effect**: Urban areas with impervious surfaces including concrete pavement, asphalt, and compacted soils cause urban areas to become at least 3 °C warmer than their surroundings. This heat island effect is caused by the impervious layers of pavement and concrete mentioned above. Urban street trees and food forests can help reduce the urban heat island effect through shading and cooling the air through a process called evapotranspiration (Millward and Sabir 2011).

(d) **Forest products**: Urban food forests also provide food, timber, and medicinal plants for the community.
(e) **Aesthetic value**: Food forests are valuable landscapes that enhance the community and add to the value of surrounding homes. In addition, food forests create recreational opportunities such as bird-watching and hiking.

2. **Environmental**

(a) **Pollution**: Food forests are composed of important vegetation that can assist in cleaning up environmental pollution in urban areas, thereby protecting human health (Currie and Bass 2008).
(b) **Biodiversity**: Food forests help protect and improve biodiversity of urban areas, contributing to the elimination of vegetation and tree pest and disease infestations. This increases ecosystem resilience. Urban trees and plants improve water quality and resilience of aquatic environment while providing many recreational opportunities.

3. **Social**

(a) **Urban Forests**: Urban forests aid in creating strong resilient communities through the benefits that have been identified herein and by strengthening the community fabric.
(b) **Food Security**: Urban food forests assist in ensuring food security by providing renewable sources of food including nuts, fruits, berries, mushrooms, herbs, and other edible plants to individuals and food banks. Food banks have recently been filling a crucial role in sustaining and improving food security in urban areas.
(c) **Educational Opportunities**: Food forests are potentially a great resource providing unique educational opportunities for school children and community gardeners. Improving our valuable green spaces, including food forests, will help build an even stronger community and foster a sense of belonging. This assists in connecting peoples with the food system, and how our food produces come to our tables.
(d) **Human Health**: Urban green spaces improve human health with clean and fresh air and the surroundings.

Future Scope of Food Forests

Urban food forests are more than just green spaces in cities. Food forests not only provide recreational services but also provide valuable products and services to improve human health. They can also provide valuable economic, environmental, and social benefits. In urban areas, green infrastructure programs have recently been thriving, planting trees, restoring habitat, and developing trails. Some of these efforts such as storm water management, as well as food security projects including community gardens, rooftop vegetable gardens, and public orchard, are being accepted and appreciated by the community.

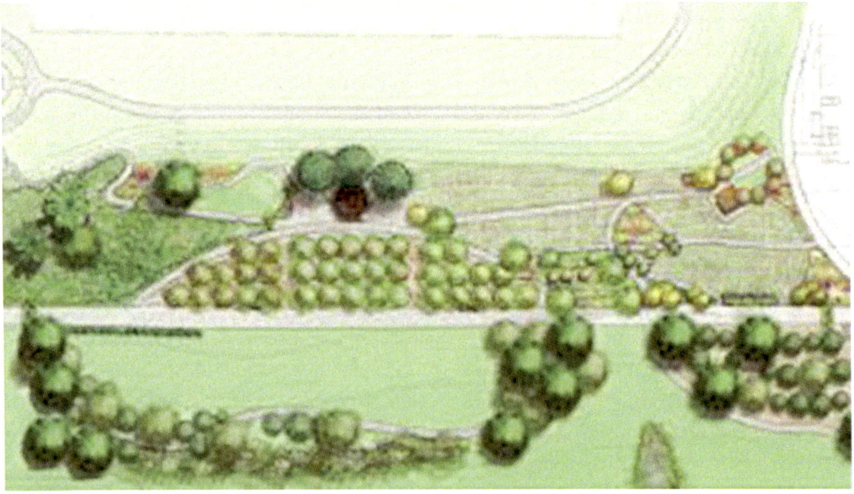

Fig. 15.5 Schematic design of seven-acre Beacon Food Forest Permaculture Project (http://www.beaconfoodforest.org/project.html)

Case Study

Beacon Food Forest Permaculture Project is located in Seattle, Washington. The goal of the project is to design an urban food forest that allows the community to gather together, grow their own food, and rehabilitate their local ecosystem. They hope to offset some of the environmental impacts of agriculture, improve local food security, and provide educational opportunities.

The project is located on a 7-acre site in the Beacon Hill neighborhood, near downtown Seattle. The plan includes an edible arboretum, a berry patch, a nut grove, a community garden, a gathering plaza, and a kid's area. The full 7-acre design is shown in Fig. 15.5.

The site development includes water supply and construction of planting beds with community involvement. The plantings also include dwarf trees, shrubs, and perennials. The main challenge faced by them was selection of types of plants to include. The Beacon Food Forest site plan includes a food tree arboretum, which has "unusual" fruit trees that can grow in Seattle (Bingle 2013). Overall, the project has been successful in addressing and including a collection of needs voiced by this diverse community.

15.5 Policies

Food charters have become popular tools in pursuing the urban agriculture agenda (Hardman and Larkham 2014). Hardman and Larkham (2014) define food charters as a set of principles which bring together the local authority, community, private sector, and other actors pursuing an agenda to increase food security within their locale. Food charters, though designed as aspirational agreements for increasing urban agriculture, often end up creating a legacy ensuring that dialogue is maintained and enhanced into the future (Hardman and Larkham 2014).

Urban agriculture, though has many successes and offers many opportunities, is not without issues. Many activities have been established through grass roots efforts, inspiring change through bottom-up approach. One of the greatest constraints is limited access to land for those interested in growing food and the lack of secure tenure on land (Lovell 2010). There needs to be better regularization on conditions for acquiring land for agriculture in cities, which are currently often informal (Nugent 2000). These issues need to be addressed through incorporation into planning and policies at all levels. Top-down efforts might be needed to help improve coordination of urban agriculture activities (Lovell 2010). Policies also need to address issues related to markets and labor. These would address the uncertainty faced by urban producers wishing to market their output by identifying viable downstream microindustrial activities as well as create semiskilled labor opportunities (Nugent 2000). Urban farmers are disadvantaged in getting credits due to lack of collateral, which also needs to be addressed through local governments (Nugent 2000). Establishing critical policies will legitimize urban farming as a viable economic activity in many cities (Nugent 2000).

15.6 Conclusion

In an effort to take control of food security, social issues, and environmental degradation in their communities, residents in many major cities have undertaken urban agricultural activities that create opportunities to provide food, jobs, environmental enhancement, beautification, education, inspiration, and hope (as cited in Sumner et al. 2010). It is estimated that 800 million city dwellers around the world, including developed countries, engage in agriculture to feed themselves and their families with reports that in some Latin American and African cities, up to a third of the vegetable demand is met by urban production (as cited in Sumner et al. 2010). Urban agriculture is here to stay. Urban agriculture, in the form of the alternative paradigm as defined by Sumner et al. (2010), carries enormous cultural potential creating sustainable systems of farming. Urban agriculture is reshaping the landscape of cities contributing to worldwide strategies to create sustainable cities (Van Leeuwen et al. 2010).

References

Adam KL (2006) Community supported agriculture. National Sustainable Agriculture Information Service Publication. Available at www.attra.ncat.org/attra-pub/PDF/csa.pdf

Adam KL (2011) Community gardening. National Sustainable Agriculture Information Service Publication. Available at www.attra.ncat.org/attra-pub/PDF/communitygarden.pdf

Ames GK (2013) Community orchards. National Sustainable Agriculture Information Service Publication. Available at www.attra.ncat.org

Bailkey M, Nasr J (2000) From brownfields to greenfields: producing food in North American cities. Commun Food Secur News. Fall 1999/Winter 2000:6

Bingle L (2013) Beacon food forest—a new future for community gardens. Retrieved on April 25, 2014 from: http://www.pacifichorticulture.org/articles/beacon-food-forest/

Brown KH, Carter A (2003) Urban agriculture and community food security in the United States: farming from the City Center to the urban fringe. A primer prepared by the Community Food Security Coalition's North American Urban Agriculture Committee (29 pp)

Clark KH, Nicholas KA (2013) Introducing urban food forestry: a multifunctional approach to increase food security and provide ecosystem services. Landscape Ecol 28:1649–1669

Currie BA, Bass B (2008) Estimates of air pollution mitigation with green plants and green roofs using the UFORE model. Urban Ecosystems 11:409–422

Deelstra T, Girardet H (2000) Urban agriculture and sustainable cities, pp 43–65. Available at http://citeseerx.ist.psu.edu/viewdoc/download?doi=10.1.1.168.4991&rep=rep1&type=pdf

Despommier D (2011) The vertical farm: controlled environment agriculture carried out in tall buildings would create greater food safety and security for urban populations. J Consum Prot Food Saf 6:233–236

Germer J, Sauerborn J, Asch F, de Boer J, Schreiber J, Weber G, Muller J (2011) Skyfarming an ecological innovation to enhance global food security. J Consum Prot Food Saf 6:237–251

Hardman M, Larkham PJ (2014) The rise of the 'food charter': a mechanism to increase urban agriculture. Land Use Policy 39:400–402

Jacke D, Toensmeier E (2005) Edible forest gardens: ecological vision and theory for temperate climate permaculture. Chelsea Green Publishing, White River Junction, VT

Lee YK (2004) Mulberry dyke pond model, China: a sustainable traditional method of land-water ecosystem. Department of Geography, The University of Hong Konk, Pokfulam Road, Hong Kong

Lovell ST (2010) Multifunctional urban agriculture for sustainable land use planning in the United States. Sustainability 2:2499–2522. doi:10.3390/su2082499

McIlvaine-Newsad H, Merrett CD, McLaughlin P (2004) Direct from farm to table: community supported agriculture in Western Illinois. Cult Agric 26(1&2):149–163

Millward AA, Sabir S (2011) Benefits of a forested urban park: what is the value of Allan Gardens to the city of Toronto, Canada? Landscape Urban Plann 100:177–188

Mougeot LJA (2000) Urban agriculture: definition, presence, potentials and risks, pp 1–42. Available at http://www.ruaf.org/sites/default/files/Theme1_1_1.PDF

Nugent R (2000) The impact of urban agriculture on the household and local economies, pp 67–97. Available at http://wentfishing.net/farmlit/Theme3.pdf

Pearson LJ, Pearson L, Pearson CJ (2010) Sustainable urban agriculture: stock take and opportunities. Int J Agric Sustain 8(1&2):7–19

Sumner J, Mair H, Nelson E (2010) Putting the culture back into agriculture: civic engagement, community and the celebration of local food. Int J Agric Sustain 8(1&2):54–61

Thebo AL, Drechsel P, Lambin EF (2014) Global assessment of urban and peri-urban agriculture: irrigated and rainfed croplands. Environ Res Lett 9:114002 (9 pp). doi:10.1088/1748-9326/9/11/114002

USDA (2015) Defining community supported agriculture. Available at http://www.nal.usda.gov/afsic/pubs/csa/csadef.shtml. Accessed 28 Feb 2015

Van Leeuwen E, Nijkamp P, de Noronha Vaz T (2010) The multifunctional use of urban greenspace. Int J Agric Sustain 8(1&2):20–25

Vogel G (2008) Upending the traditional farm. Science 319:752–753

Index

D. Nandwani (ed.), *Organic Farming for Sustainable Agriculture*,
Sustainable Development and Biodiversity 9,
DOI 10.1007/978-3-319-26803-3

Printed in the United States
By Bookmasters